MW01259495

SHIP KILLER

THOMAS WILDENBERG and NORMAN POLMAR

SHIP KILLER

A HISTORY OF THE AMERICAN TORPEDO

NAVAL INSTITUTE PRESS
Annapolis, Maryland

This book has been brought to publication with the generous assistance of Edward S. and Joyce I. Miller.

Naval Institute Press
291 Wood Road
Annapolis, MD 21402

World War II torpedoman drawing by Fred Freeman from *United States Destroyer Operations in World War II* (1953).

Library of Congress Cataloging-in-Publication Data
Wildenberg, Thomas
Ship killer : a history of the American torpedo / Thomas Wildenberg and Norman Polmar.
p. cm.
Includes bibliographical references and index.
ISBN 978-1-59114-688-9 (hardcover : alk. paper) 1. Torpedoes—United States—History. I. Polmar, Norman. II. Title.
V850.W55 2010
359.8'2517—dc22

2010036621

Printed in the United States of America on acid-free paper

14 13 12 11 10 9 8 7 6 5 4 3 2
First printing

*Sink 'em All**

Sink 'em all, Sink 'em all
Tojo and Hitler and all
Sink all their cruisers and carriers too
Sink all their tin cans and their stinking crews . . .

* Raucous ditty composed by U.S. submariners in World War II and sung to the tune of "Bless 'em All," by Irving Berlin. The tune was played by dockside Navy bands as submarines departed Pearl Harbor at the beginning of their war patrols.

Contents

Perspective

Torpedoes have sunk more ships in the past 100 years than any other weapon except for mines. Indeed, torpedoes have probably sunk more oceangoing ships than have naval gunfire and aerial bombs combined. While most historical accounts center on submarine attacks with torpedoes, the weapon has also been used with great effect by surface ships and aircraft. The success of torpedoes has won battles and campaigns—they were the principal weapon used in the sinking of three Japanese battleships in World War II, including the "super battleships" *Yamato* and *Musashi*, the largest warships of that conflict. Aerial torpedoes also sank three American battleships at Pearl Harbor, and aerial and submarine torpedoes were responsible for sinking four British and three Italian battleships and battle cruisers during the war. Scores of aircraft carriers, cruisers, destroyers, and naval auxiliaries also fell victim to torpedoes. But torpedoes are complex weapons that can be prone to failure if not designed or maintained properly as exemplified by the very high failure rate of U.S. submarine torpedoes during the early months of World War II from December 1941 through March 1943. The problem of water entry also made exacting demands on launching aircraft, forcing them to fly low and slow—a formula that exacted heavy losses such as those experienced by U.S. airman during the Battle of Midway on 4 June 1942.

Since the start of the 20th Century there have been a thousand books published about submarines and on the order of another thousand discussing aircraft attacks on ships. The principal weapon of most of those submarine attacks and many of the aerial attacks—both by land- and carrier-based aircraft—was the torpedo. Yet, only a handful of books have been published about torpedoes. The only major works about torpedoes in the past 50 years have been Robert Gannon's *Hellions of the Deep* (1996), which concentrates on U.S. torpedo problems and solutions during World War II, Edwin Gray's *The Devil's Device* (1975, 1991), which is primarily the story of torpedo pioneer Robert Whitehead, who did his work in the late 19th Century in Austria, and E. N. Poland's *The Torpedomen* (1993), which chronicles the development of the torpedo in the Royal Navy. The limitation of these books can be seen in Gray's work, which omits any discussion of the decades-long negotiations between the U.S. Navy and Whitehead, the eventual licensing agreement with the U.S. Navy, and the subsequent development of the Bliss-Leavitt torpedo in America.

A few other books have provided significant information about torpedoes. For example, Willem Hackmann's *Seek and Strike* (1984), is a useful albeit official British history of anti-submarine warfare that has some discussion of U.S. post–World War II torpedo development.

None of these works, with the exception of Gannon's *Hellions of the Deep*, emphasizes the development of the torpedo in the U.S. Navy, and that book only covers a very limited period during World War II; nor does it chronicle tactical implications of the torpedo failures. Thus, no comprehensive treatment of the torpedo in the U.S. Navy exists. It is for this reason that we choose to produce this book, which documents the development and employment of the U.S. torpedo from its inception to the present.

We begin with a brief description of the weapons developed for "submarines" prior to the beginning of the 20th Century—the efforts of Americans David Bushnell and Robert Fulton, the spar torpedo of the Civil War era, and the U.S. Navy's attempts to imitate the Whitehead torpedo. Then, from the beginning of the 20th Century, we discuss American torpedo development in peace and in war and their use—from submarines, surface warships and small combatants (PT boats), and aircraft (fixed-wing and helicopters). Also addressed are the technologies and politics involved in torpedo development and various unusual efforts to deliver torpedoes via such innovative schemes as RAT, ASROC, DASH.

In addition, we have provided brief "sidebars" to indicate contemporary foreign torpedo development efforts and use in combat.

Today torpedoes are the principal weapon of the world's submarines for attacking surface ships and other submarines; similarly, torpedoes are anti-submarine weapons of the world's air and surface forces. Despite the development of rockets and missiles, there is no naval weapon in existence or known to be in development that will replace the torpedo.

Thomas Wildenberg
Norman Polmar

Acknowledgments

The authors greatly appreciate the assistance provided to this project by Rear Admiral Jay Cohen, USN (Ret.), former attack submarine commander; Alan Ellinthorpe, acoustics expert; Glenn E. Helm, Director of the Navy Department Library; Edwin Finney Jr., photographic section, Naval Heritage and History Command; Janis Jorgensen and Sandy Schlosser of the Naval Institute's photo library; K. J. Moore, submarine technologist; and Richard Russell, Director of the Naval Institute Press, who asked us to write this book.

We would also like to thank Kate Epstein for reviewing several chapters of the manuscript and for sharing her research on Bliss-Leavitt torpedoes.

At the Naval Institute Press we were assisted by Jan Jorgensen, Chris Onrubia, and Susan Corrado, and Maryam Rostamian who laid out the book.

And, of course, we are in debt to Mr. and Mrs. Edward Miller for their support of this project.

Torpedo Nomenclature

In the early years torpedo types were identified by manufacturer, size, and mark number. The latter was used to differentiate different models of or modification of the same torpedo. The mark number became the standard model designation in 1913 when all U.S. torpedoes were reclassified and the modification number, or "mod," was introduced to designate modifications to the basic torpedo. Thus since 1913 all torpedo designs were identified by sequentially applied mark numbers. By convention these numbers were shown as Roman numerals until 1943, when they began being replaced by Arabic numbers. For consistency, and to avoid confusion, the authors use Arabic numbers for all mark numbers assigned after 1913, beginning with the Bliss-Leavitt Mark 1 torpedo. All previous models are designated using the Roman numeral system then extant in the U.S. Navy.

Abbreviations and Designations

ACR	Armored cruiser*
ADCAP	Advanced capability torpedo
AGP	Motor torpedo boat tender*
AKV	Aircraft cargo ship*
ALWT	Advanced Lightweight Torpedo
AM	minesweeper*
AO	Oiler*
AS	Submarine tender*
ASROC	Anti-Submarine Rocket
ASTOR	Anti-Submarine Torpedo
ASW	Anti-Submarine Warfare
AV	Seaplane tender*
BB	Battleship*
BuAer	Bureau of Aeronautics
BuOrd	Bureau of Ordnance
CA	Heavy cruiser*
CAPTOR	Encapsulated Torpedo
CBASS	Common Broadband Advanced Sonar System
CG	Guided missile cruiser*
CGN	Guided missile cruiser (nuclear-propelled)*
CL	Light cruiser*
CLAA	Anti-aircraft cruiser*
CM	Minelayer*
CSS	Confederate States Ship
CV	Aircraft carrier*
CVE	Escort aircraft carrier*
CVL	Light aircraft carrier*
CZ	Convergence Zone (acoustic)
DASH	Drone Anti-Submarine Helicopter
DD	Destroyer*
DDE	Escort destroyer*
DDG	Guided missile destroyer*
DDK	Hunter-killer (ASW) destroyer*
DDR	Radar picket destroyer*
DE	Destroyer escort (ocean escort post–World War II)*
DLG	Guided missile frigate*
DLGN	Guided missile frigate (nuclear-propelled)*
DTMB	David Taylor Model Basin
FRAM	Fleet Rehabilitation and Modernization
GUPPY	Greater Underwater Propulsion Project
HBX	High Brissance Explosive
HUSL	Harvard Underwater Sound Laboratory
LAMPS	Light Airborne Multipurpose System
LHT	Lightweight Hybrid Torpedo
Mod	Modification
MTB	Motor torpedo boat
NDRC	National Defense Research Committee
NEARTIP	Near Term Torpedo Improvement Program
NRAC	Naval Research Advisory Committee
NOTS	Naval Ordnance Test Station
PBXN	Plastic-bonded explosive
Pk	Probability of kill
PT	Motor torpedo boat*

PTC	Motor submarine chaser[*]
PTF	Fast patrol boat[*]
RAF	Royal Air Force (British)
RAT	Rocket-Assisted Torpedo
RDX	Cyclotrimethylenetrinitramine, main ingredient in plastic explosive
RETORC	Research Torpedo Configuration
RUM	Ship-to-Underwater Missile
SALT	Strategic Arms Limitation Talks
SCEPS	Stored Chemical Energy Propulsion System
SKINC	Sub-Kiloton Insertable Nuclear Component
SS	Submarine[*]
SSBN	Ballistic missile submarine (nuclear-propelled)[*]
SSK	Hunter-killer submarine[*]
SSN	Submarine (nuclear-propelled)[*]
SUBROC	Submarine Rocket
TB	Torpedo boat[*]
TBD	Torpedo boat destroyer
TBS	Telephone-Between-Ships (radio)
TDC	Torpedo Data Computer
TF	Task Force
TNT	Trinitrotoluene
Torpex	An explosive based on trinitrotoluene (TNT) that is 50% more powerful than TNT
USS	United States Ship
UUM	Underwater-to-Underwater Missile
VT	Torpedo squadron (aviation)

*Official U.S. Navy ship classification.

SHIP KILLER

CHAPTER ONE

The First Torpedoes

Bushnell, Fulton, and the Civil War (1775–1865)

The word "torpedo" first entered into the lexicon of naval warfare around 1805, when it was coined by Robert Fulton. He used it to describe the gunpowder-filled kegs he used to sink the brig *Dorothea* in an experiment conducted near Deal, England. The 200-ton brig was blown up on 15 October 1805 as a demonstration of the explosive power of 180 pounds of powder placed under the bottom of a vessel.

Unlike the self-propelled weapon now associated with this term, Fulton's torpedoes were nothing more than buoyant, moored mines triggered by a clockwork mechanism that activated a firing pistol to set off the 180-pound charge of gunpowder. Fulton adopted the term torpedo from the Cramp fish (*Torpedo electricus*), which kills its prey by an electric shock. The name stuck and was soon in use by the naval community to describe any weapon intended to be exploded beneath a hostile ship.

Although Fulton gave the name to these new weapons, the idea for using underwater explosives in warfare was first conceived by David Bushnell during his college years at Yale sometime between 1771 and 1775. Bushnell's first experiment involved two ounces of gunpowder, which he exploded underwater to demonstrate that gunpowder would burn when submerged. He successively detonated larger quantities under rafts to demonstrate the destructive effects of placing a charge under a vessel. Having demonstrated the effectiveness of such a device, he turned his attention to designing a submersible that could transport and affix the explosive charge to the bottom of an enemy warship. The result was the *Turtle*, the first American submersible.

Bushnell's submersible was built in Saybrook, Connecticut, in 1776 with the aid of his brother, Ezra. It was an egg-shaped craft constructed of oak timbers that were carefully joined and caulked at the joints. These were held together with iron bands and covered with pitch to make the craft watertight. It was operated by one man, who operated the hand and foot valves that controlled descent and ascent by alternately admitting water or pumping it out of the ballast tank installed at the craft's bottom. It was propelled by a hand-operated screw and had a large rudder for directional control. Once submerged, the craft contained enough air for 30 minutes of operation.

Secured to the stern of the submersible was a wooden case containing at least 100 pounds of gunpowder that was to be secured to the underside of the intended target by means of a long screw, also worked by the operator from the inside the submersible. Once attached, the explosive charge was to be fired by a clockwork mechanism activated by a lanyard pulled by the operator as he paddled away from the intended victim.

After undergoing extensive test trials in the safe waters of the Connecticut River off Saybrook, the *Turtle* was secretly transported to New York Harbor in the

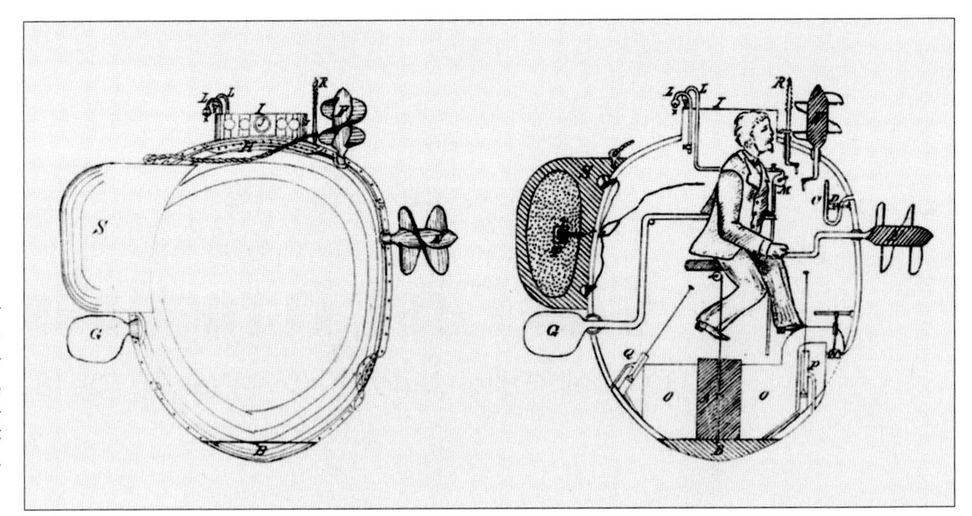

David Bushnell's submarine torpedo boat *Turtle* of 1776 is shown in this drawing by Lieutenant Commander F. M. Barber, made in 1885 from a description left by Bushnell. Note the variety of controls and pedals the operator had to work. (U.S. Navy)

spring of 1776 and stationed at the Battery, which was still under the control of the Colonial forces. There, Ezra trained to operate the craft until he and David were satisfied that he had successfully mastered the skills needed to maneuver the craft. Bit Ezra became ill with a fever and died before he had an opportunity to use the *Turtle*. As a replacement, Bushnell recruited Ezra Lee, then a sergeant in the Continental Army, who had been among those men who had come forward to volunteer after Ezra Bushnell's death.

Lee trained for two months before setting out on the night of 7 September 1776 to attack the 64-gun ship-of-the-line *Eagle*, flagship of the British naval forces in New York, then anchored off Staten Island. Bushnell waited until near midnight, when the moon and tide were favorable, before commencing the operation, which began by having a rowboat tow the *Turtle* toward the *Eagle*.

Half way to the target, the rowboat stopped, Lee climbed in, fastened the hatch, and began the trek across the remaining distance to the *Eagle*. Lee brought the *Turtle* alongside the warship, submerged, and continued until he was under the vessel. He pumped out a small quantity of water from the ballast tank until a jarring bump indicated he had positive buoyancy and was up against her hull. Lee then began turning the screw that was intended to attach the explosive charge to the *Eagle*. For the next few minutes, Lee tried to work the screw into the ship's hull. But the *Eagle*'s timbers were too hard to penetrate or the boring device had struck a bolt or iron brace. When Lee attempted to shift the *Turtle* to another position beneath the hull, he lost contact with the target. By now Lee was exhausted and the outgoing tide threatened to take the small craft out to sea. Desperately he ejected all of the remaining ballast water and headed for the Battery. With the ballast gone, one-third of the *Turtle*'s hull stuck out of the water, making it clearly visible in the growing light of dawn.

At that time two British soldiers set out from Governor's Island in a patrol skiff to investigate the object sighted in the water. To divert the patrol and to lighten his craft, Lee released the explosive charge and, picking up speed, reached the Battery safely. A short time later the early-morning silence was shattered by a spectacular explosion of the gun powder charge that aroused the British fleet.

Subsequent efforts to use the *Turtle* against the *Eagle* and a British frigate were thwarted by the tides and tricky currents. Bushnell accordingly decided to move to a more favorable location. He loaded the little craft in a sloop headed for Connecticut. They were intercepted by a British frigate, however, which, according to the British, sank the sloop and its precious cargo. The loss of the *Turtle* ended Bushnell's efforts at underwater navigation, although he also attempted to sink a British ship with floating mines. But these attacks were also unsuccessful.

Bushnell's *Turtle* is depicted during its unsuccessful attempt to attack HMS *Eagle*—Admiral Howe's flagship—anchored off Governor's Island, New York, on 7 September 1776. This was history's first submarine attack. (U.S. Navy)

The concept of a submarine torpedo boat was resurrected in 1797 by Robert Fulton as a means of overcoming the unchallenged power of the Royal Navy. Fulton's interest in such a vessel was probably inspired by Bushnell's *Turtle* after a description of the craft first appeared in print in 1795. Fulton, who was living in France at the time, offered the device to the French government in return for a bounty on any ships sunk or prizes taken. Fulton's proposal was contained in a letter to the Executive Directory in Paris.

The letter began with a declaration of his intention to construct a mechanical "Nautilus" that would annihilate the British fleet. In return he expected the French government to pay his company a bounty for each British ship destroyed. All prizes taken by the *Nautilus* would become the company's property. As a citizen of the "American States," Fulton still considered himself a patriot as evidenced by his stipulating that the invention "not be used by the Government of France against the American States."[1]

Although the French Minister of Marine was initially inclined to favor Fulton's proposal and a draft decree of acceptance was drawn up, it was never issued. Submarine warfare was not considered a civilized means of combat, and the Minister of Marine changed his mind in the belief that the agents of such heinous methods of warfare were little better than pirates.

Fulton continued to negotiate with the ruling Directory without success until November 1799, when Napoleon Bonaparte became the First Consul. Fulton lost no time in calling upon the new Minister of Marine, Pierre Alexandre Laurent Forfait. Forfait, who had been a member of a commission of experts appointed by the previous minister in 1798 to examine Fulton's proposal, had already commented favorably on the idea. With Forfait's approval secured, Fulton began to build a submarine to be called *Nautilus*. Official documents indicate that the boat was completed in Rouen, although construction may have been started in Paris.

The boat was designed in the shape of an imperfect ellipsoid approximately 21 feet long and 6½ feet in diameter. A metal ballast tank that could be evacuated by a lever-actuated suction pump was installed at the bottom of the main wooden hull. There was a metal dome, or conning tower, at the bow, pierced with thick glass viewing ports and a manhole that served as a means of ingress for the three-man crew. A hand-actuated propeller was installed for underwater propulsion along

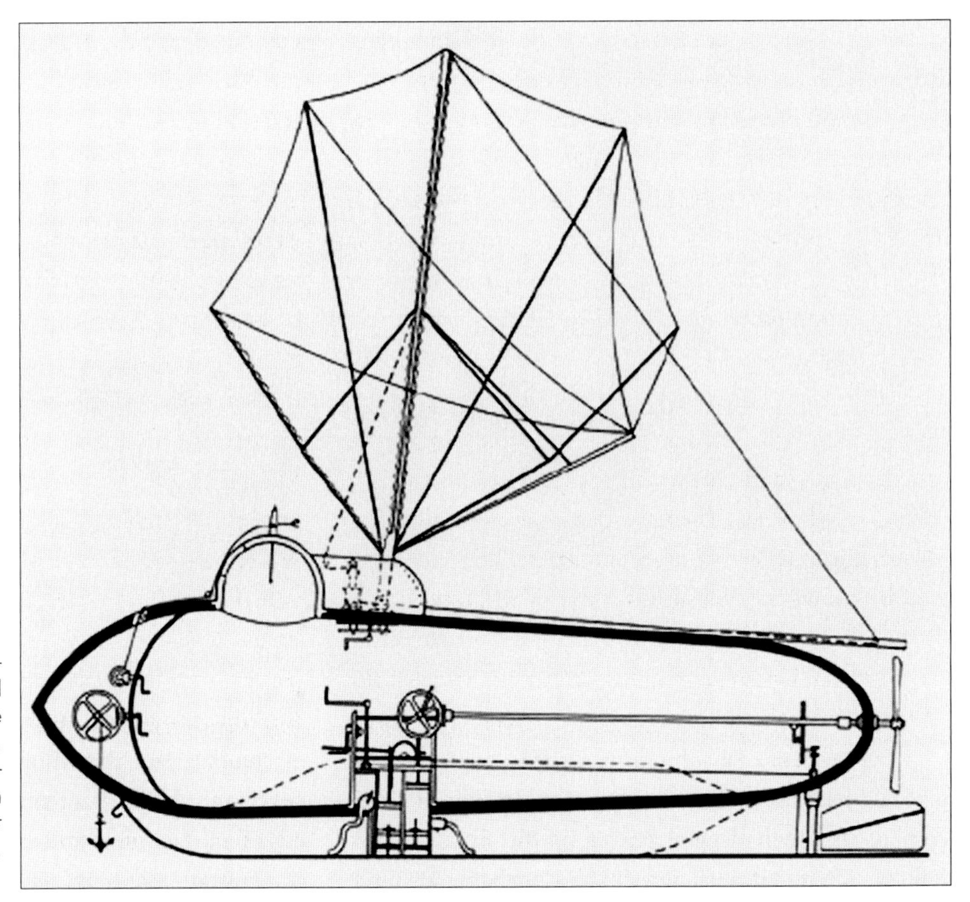

Robert Fulton, a highly successful artist-turned-inventor, offered his submarine *Nautilus* to France and, subsequently, to England. But his efforts to sell his submarines were unsuccessful, although the craft aptly demonstrated their potential against target ships. (Library of Congress)

with a horizontal rudder for depth control. When surfaced, the boat depended on a fan-like set of sails that were bent on a hinged mast located about one-third of the length from the bow. Before submerging, the sails would be furled on the ribs, lowered against the hull, and covered by envelopes shaped like the sheath wings of a fly.

The purpose of the *Nautilus* was to covertly place a copper barrel containing 100 pounds of gunpowder against the hull of an enemy warship. This torpedo, or mine as it was termed by Fulton, was detonated by a gunlock triggered by a lanyard pulled by one of the crew after the *Nautilus* had retreated to a safe distance. The mine or torpedo was placed into position by means of a tow rope that passed through a screw eye attached to the front end of a pole that projected through the conning tower. The screw eye, which Fulton called "the horn of the *Nautilus*," was to be embedded in the enemy ship's planking with a few quick blows to the inside end of the shaft projecting from the conning tower. The *Nautilus* would then set off, leaving the horn behind, until the tow rope brought the torpedo into contact with the ship's bottom.

The *Nautilus* was launched on 24 July 1800. Fulton commenced testing the boat five days later when he made two "plunges" in 25 feet of water. The first of these lasted 8 minutes and the second 17. This first test cruise lasted for three hours, during which the boat changed her depth frequently. The current, however, caused Fulton considerable trouble, and he resolved to proceed to Le Havre, where he could test the boat in the open sea. The first sea trial was conducted on 24 August, when he plunged the boat to a depth of 15 feet with two passengers on board. They remained below the surface for one hour without experiencing the slightest inconvenience. Trials to test the boat's maneuverability on the surface as well as his ability to use a compass underwater continued for two more days. The next day Fulton proceeded to test a torpedo containing 30 pounds of powder—the craft's weapon system.

Having successfully tested the *Nautilus* and its "torpedo," Fulton was now ready to conduct his first operation against the British. He sailed on the surface from Le Havre on 12 September and put into the little harbor of Growan three days later. During the next month Fulton twice approached two English brigs that were anchored nearby, but both times the ships set sail before he could close in for his attack. During one of these attempts he remained underwater for six hours, taking in air through a small tube, a harbinger of the later snorkel device. The onset of winter forestalled further operations, with Fulton retiring to Paris.

Unknown to Fulton, his attempt to blow up the English brigs off the French coast was frustrated not by accident, but by design. The British Admiralty generally knew of Fulton's activities and movements.

When Fulton returned to Le Havre the next summer he was disconcerted to find the *Nautilus* was full of rust, having been assembled using iron bolts. It took nearly two months to repair the damage before the boat was ready for sea. Although the boat performed in a most satisfactory manner, it appears that Fulton became fixated on the development of the submarine mine. He considered the *Nautilus* to be "an imperfect engine" and chose to dissemble her to keep the ideas embodied in his invention secret.

When the French government refused to provide further assistance to Fulton, he turned his attentions to the steam engine. He remained in France until Napoleon Bonaparte declared himself Emperor in the spring of 1803. To a republican like Fulton, the form that the government of France was assuming under Napoleon was utterly repugnant, causing him to change his allegiance to the British, who were anxious to secure his services. He arrived in London on 19 May 1804, using the assumed name of Francis.

On 15 October 1805, Fulton demonstrated his mine's effectiveness, sinking the *Dorothea* with his "torpedoes," which were rowed into position and allowed to drift into the target. Within a week, however, the combined French and Spanish fleets were defeated at the Battle of Trafalgar, undermining further British interest in Fulton and his submarines and torpedoes.

Fulton returned to the United States in 1806 after a 17-year hiatus abroad. He continued to conduct mine experiments for the next four years in an effort to convince Congress of their value as an inexpensive means of protecting American ports and harbors. Although Congress had appropriated $5,000 for "trying the practical use of the torpedo," only $1,500 was expended and his system for harbor defense was never implemented.[2] In 1810 Fulton published his book *Torpedo War and Submarine Explosions*. The knowledge disseminated in Fulton's book was put to good use by various individuals during the War of 1812, who used mines to harass and annoy British ships from Lake Ontario to the Virginia Capes. Their failure to sink a single ship caused the "torpedo" to drop out of favor in the United States. It would lay dormant for five decades until the Confederacy, with an insignificant navy, was forced to seek new methods of keeping Union ships out of its rivers and away from its ports and harbors.

When the American Civil War erupted in April 1861, most of the nation's existing warships were in the hands of the Union, which also possessed most of the nation's major shipyards. As soon as war was declared, Union naval leaders began to develop plans for blockading Southern ports to deny the Confederate States access to desperately needed arms and munitions that could only be obtained from Europe, and to prevent the export of cotton to earn funds needed for the execution of war. The South's answer to the Union blockade was the torpedo-mine: inexpensive, readily produced devices that could create havoc among the Union Navy's inshore squadrons. Matthew Fontaine Maury was one of the first Confederate naval officers to conduct mine experiments. The celebrated naval scientist, like many of his fellow officers, had resigned from the U.S. Navy after the South withdrew from the Union. Under Maury's guidance, the Naval Submarine Battery Service was established in Richmond in 1862 to develop mines and other underwater explosives.

This engraving from Fulton's *Torpedo Warfare and Submarine Explosions* (1810) shows the sinking of the target ship *Dorothea* after being attacked by his submarine *Nautilus*. More than an artist or submarine developer, Fulton is remembered primarily for his steamboat developments.

The development of "torpedoes"—essentially underwater mines—was spurred on by a Confederate law providing that "the inventor of a device by which a vessel of the enemy should be destroyed should receive 50 percent of the value of the vessel and armament." Before long Confederate ordnance officers were sowing mines wherever they were likely to encounter a Union steamer or riverboat. The mines were designed and manufactured locally, using whatever materials were at hand, contained in any kind of barrel, float, or keg. Although a few were detonated by wire using an electric charge, most were set off by a trigger mechanism attached to a lanyard snagged by some part of the enemy ship. Some mines were set adrift in a river current or on a rising tide to strike a random ship's hull. Others were anchored by grapnels or weights, to float just beneath the surface. But no matter how they were constructed, each design had to overcome three basic problems: (1) how to deliver the torpedo to the target, (2) how to keep the powder dry, and (3) how to detonate the charge.

Torpedoes remained passive devices until Captain Francis D. Lee, an engineer officer in the Confederate Army, invented the spar torpedo in 1862.[3] Lee was serving on the staff of General Pierre G. T. Beauregard, the commander of the defenses of the Carolina and Georgia coasts with his headquarters in Charleston, when he conceived of the idea of mounting a light-weight copper barrel containing at least 60 pounds of gunpowder at the end of a long wooden spar. Lee planned to attach the spar to the bow of a low-silhouette, high-speed steamer, which could approach an enemy warship under cover of darkness and ram the torpedo into the enemy's side. To explode the charge, Lee designed a new type of fuse that would detonate the charge upon contact.

Lee's detonator consisted of a small lead tube approximately three inches in diameter that was screwed into the front end of the mine or torpedo. The detonator's front end was covered by a thin piece of metal that was easily dented. Behind this was a hermetically sealed glass vial containing sulfuric acid. The rest of the tube was packed with a mixture of potassium chlorate, powered sugar, and gunpowder. Striking an object would dent the cap, thus breaking the glass vial to release the acid, igniting the mixture behind it, which would quickly burn through the tube to detonate the main explosive charge.

The design of Lee's "torpedo ram" called for a steam-driven craft that would lay so low in the water that only a small compartment at the stern for the pilot and the smokestack would be exposed. Protruding from the bow would be the spar torpedo that could be lowered or raised from within by a winch.

Convinced of the scheme's viability, General Beauregard sent Lee to Richmond to obtain the funds and resources necessary for the craft's construction. Although Secretary of War James A. Seddon heartily approved of Lee's plan, it was a subject that pertained to the Navy. The Secretary of War would do nothing, and neither would the Secretary of the Navy, Stephen R. Mallory, because Lee was not a naval officer under his command. Rebuffed by both men, Lee returned to Charleston.

After a lapse of a few months Lee was sent back to Richmond. This time he carried a complete set of plans and a scale model, as well as the endorsement of senior naval officers. His efforts resulted in the transfer of an unfinished hull on the stocks in Charleston to his project. Beauregard made the project official on 31 October 1862, when he ordered Lee to take charge of the craft's construction, giving Lee full power to supervise all accounts and to spend the $50,000 appropriated for the project.

As the Union's siege of the port of Charleston ground on, it was becoming evident that Lee's torpedo ram would take much longer to complete than originally planned. With this reality in mind, more immediate and less material-extravagant means of attacking Union ships were proposed, including the use of numerous small boats. When Lee and Beauregard learned of these plans, they proposed that they be equipped with spar torpedoes in lieu of the floating mines that were being prepared for them to tow.

Before this work could be undertaken, however, it was necessary to remove any doubts about a small boat or skiff's ability to survive the blast of a torpedo just a few feet away. Lee was sure that the force of the explosion would be expended through the side of the target ship's hull, leaving the small boat unharmed. He proposed to prove his theory by a trial test using the abandoned hulk of an old gunboat as the target.

The experiment began late in the afternoon of 12 March 1863, using a 20-foot canoe. A long spar was suspended 6 feet beneath the boat's keel by ropes that were attached to the boat's bow and stern, the spar being fitted with 30 pounds of gunpowder. The spar extended 8 to 10 feet beyond the boat's bow. For the test the canoe was pulled toward the target by a line affixed to the bow that was threaded through a pulley on the stern of the intended target; this line was then run back through another pulley on the stern of the canoe and was taken up by a nearby rowboat. Lee ordered the rowboat to pull away and watched as the little canoe rushed toward the waiting hulk. It struck with a dull thud, but there was no explosion.

Lee waited for an hour before rowing out to investigate. He quickly determined that the spar torpedo had passed underneath the hull, barely missing the target. With darkness approaching, he decided to postpone continuing the test until the next day. The test was repeated early the next morning. Everything worked perfectly; the spar torpedo exploded and the hulk was sunk without damaging the fragile canoe.

Lee now elicited the support of William T. Glassell, the first lieutenant of the Confederate ironclad *Chicora*, based in Charleston. Glassell, who was present during the spar torpedo tests, was greatly impressed by the weapon's effectiveness. He immediately approached his immediate superior, Captain Duncan N. Ingraham, for permission to equip 40 boats with Lee's spar torpedo and to be allowed to lead them against the Union blockade. Although Ingraham scoffed at the idea, he allowed Glassell to mount a single attack using a cutter borrowed from the *Chicora* fitted with one of Lee's spar torpedoes containing a 50-pound charge.

The cutter, with six volunteers manning the oars and Glassell in command, set off in the early hours of 18 March 1863, to attack the Union steamer *Powhatan* at the harbor mouth. Although the Union ship's deck watch discovered them, the cutter's crew managed to close to within 40 feet of the *Powhatan* when one of the seamen suddenly backed his oar, stopping the boat's headway. The other crewmen stopped rowing as the cutter drifted past the Union ship's stern with the tide. When the *Powhatan* began lowering a boat into the water, Glassell ordered the torpedo cut loose and directed his men to pull away at full speed. No shots were fired and the Confederate craft reached Charleston safely. The abortive attempt convinced Glassell that steam, not oars, was the "only reliable motive power" to be used in the future.[4] Although Glassell retained his enthusiasm for torpedo boats, he would not participate in anther torpedo action until he returned from Wilmington, North Carolina, where he was assigned as the executive officer of the newly completed ironclad *North Carolina*.

The failure of Glassell's cutter and the flotilla of similar craft that attempted to attack the blockading Union ships renewed interest in Lee's torpedo ram, whose construction had languished for lack of money and iron for her armor. Construction resumed in July 1863, after a team of Confederate naval engineers sent out to survey the unfinished ship reported favorably. The unfinished vessel—she still lacked the armor plating included in her original design—was launched on 1 August 1863 and christened the CSS *Torch*.

The *Torch* was sent out on 20 August 1863, with a volunteer crew under the command of James Carlin, a successful blockade-runner who had offered to buy the *Torch* before she had been launched. Her target was the largest, most heavily armored warship in the U.S. Navy, the casemated ironclad steamer *New Ironsides*. To improve the chances of his success and to eliminate any possibility of failure, Lee attached three 100-pound torpedoes to the end of her bow-mounted spar, which could be lowered into position by a series of ropes as she neared the target.

Carlin had lowered the spar and its three torpedoes. The *Torch* was within 40 yards of the 3,486-ton *New Ironsides* when he ordered the engines stopped and the helm put over so that he could get into the proper position to attack. The craft did not answer the helm properly and she drifted past the *New Ironsides*, which swung at her mooring with the ebb tide. Carlin had to abandon the attack when the *Torch*'s engines failed to start, and the Confederates were lucky to get away unharmed after the *New Ironsides* opened fire on them. In his after-action report to General Beauregard, Carlin condemned the *Torch* as being unfit for the purposes it was intended.

While the finishing touches were still being put on the *Torch* in Charleston, the wooden hull of a small torpedo steamer was nearing completion at Stoney Landing on the Cooper River, 30 miles northeast of Charleston. The vessel would later be christened the *David*, the name apparently derived from the biblical tale of David and Goliath. She was like no other warship ever seen before. She had a cylindrical hull 48½ feet long and 5 feet in diameter that was tapered at both ends. Her hull was heavily ballasted, allowing her to lay low in the water, and she may even have been equipped with a rudimentary pair of diving planes. A steam engine could propel the craft at a speed of five to seven knots while partly submerged. She could be trimmed to float with her decks awash and only the pilot's conning tower and her smokestack above water, making her difficult to see or hit with gunfire.

The finished hull was launched in the Cooper River and taken to Charleston, were she was hoisted out of the water, placed on a railroad car, and moved to the Northeastern Railroad slip, where her boiler and steam engine were installed. She was moved to a wharf at the end of Broad Street, relaunched, and officially named *David*. Although the *David*'s construction had been funded by a group of private investors, the little steamer was immediately taken over by the Confederate Navy. Lee wasted no time in affixing a 14-foot piece of three-inch boiler tubing to her stem to serve as the spar for the 70-pound explosive charge that he fitted.

Shortly after her launching, Glassell, who had just returned to Charleston, was offered and accepted command of the *David*. After examining the craft and recruiting a three-man crew—consisting of engineer, fireman, and pilot—Glassell began an intense training period, during which he found the *David* handled well and was reasonably fast under a full head of steam.

The *David* saw action for the first time on the night of 5–6 October 1863, when Glassell with a crew of three—assistant engineer J. H. Toombs, fireman James Stuart, and pilot T. W. Canon—steamed across Charleston Harbor on the ebb tide, just after dark. Their target was the Union ironclad *New Ironsides* moored just outside the harbor on blockade duty. Glassell intended to strike the Union ironclad just under the gangway, but the tide carried the *David* toward the enemy's starboard quarter. Glassell was 300 yards from the target when a sentry on board the *New Ironsides* hailed the approaching craft. "We were moving towards them at all speed and I made no answer, but locked my gun," explained Glassell.[5] The officer of the deck made his appearance and loudly demanded, "What boat is this?" The *David* was within 40 yards of the *New Ironsides* by now, close enough for Glassell to fire his gun, mortally wounding the Union officer, who fell back. Glassell immediately ordered Toombs to stop the engine. The *David* was making plenty of headway; within moments the torpedo struck the starboard quarter of the Union ironclad, detonating under the ship's starboard quarter, throwing up a high column of water that rained back over the *David*, extinguished her boiler fires. With her engine dead, the Confederate craft hung under the quarter of the *New Ironsides* and was pelted with a hail of small-arms fire from the Union warship.

Believing that the *David* was sinking, the crew, with the exception of the pilot, who could not swim, abandoned ship. Toombs swam back a short time later and reboarded the immobilized craft. Rebuilding the fires, he succeeded in getting *David*'s engines working again, and with Canon at the wheel, steamed back to safety in Charleston Harbor. Glassell and the fourth crewman, fireman Stuart, were captured by Union forces.

Though the *David* did not sink the *New Ironsides*, the ship was leaking so badly from the damage caused by the attack that it was kept out of action until the last months of 1864. The successful attack fueled further interest in cigar-shaped, steam-driven torpedo boats. Before the war was over, no fewer than 15 additional "Davids" had been constructed.[6] While several attempts to attack Union ships were undertaken with such vessels, no damage was inflicted on any of the target ships. Nevertheless, the threat of the Davids and the multitude of small steamers fitted with spar torpedoes forced the Union ships on blockade duty, especially those off Charleston, to remain anchored within protective nets with their crews on special alert at night and during periods of fog. This allowed the Confederate blockade-runners much greater freedom of movement.

While the crews of the *Torch* and the *David* were preparing to do battle with the Union ironclads, an entirely new type of vessel appeared on the scene that was destined to make maritime history. This craft, which would later come to be known as the CSS *Hunley*, arrived in Charleston on 12 August 1863, cradled on two railroad flatcars. The craft was designed to attack while fully submerged. She was privately built in the spring of 1863 in the machine shop of Park and Lyons, Mobile, Alabama, from plans furnished by Horace L. Hunley, James R. McClintock, and Baxter Watson.

The *Hunley*'s 40-foot hull had a four-foot beam and was constructed of iron. To save time in constructing the submersible, her builders used an existing ship's boiler that was 25 feet long and 4 feet in diameter as the basis for the hull. The boiler was cut in half lengthwise and reinforcing iron bars were added to the inside of each half. The upper and lower halves were then rejoined by riveting two 12-inch strips of iron between them on either side, and all seams were tightly caulked and sealed. Tapered sections were bolted to both ends to form the bow and stern, while on the exterior, a 12-inch strip of iron was bolted to the top to form a deck.

Access to the interior was provided by two raised hatches at either end that had heavy glass deadlights for viewing ports. When the vessel was on the surface, the viewing ports were only a foot or so above the level of the water. The motive power was a hand crank operated by eight of her nine-man crew. She had two ballast

The Confederate submarine *Hunley* is shown at Charleston, South Carolina, on 6 December 1863. The *Hunley* killed several of its crews, but earned a place in naval history as the first submersible to sink an enemy warship. (U.S. Navy)

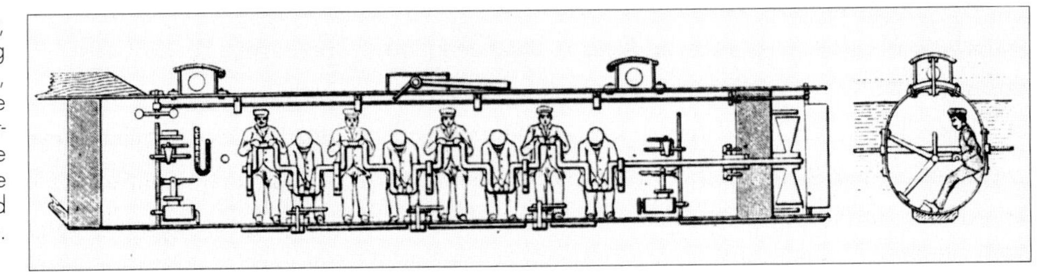

The interior plan of the *Hunley*, showing eight of her crew turning the propeller crank. In addition, an officer would command the craft, standing under the observation copula in the bow. The *Hunley* carried the name of one of her financial backers—and victims, Horrace L. Hunley. (U.S. Navy)

tanks for diving and surfacing, each equipped with a seacock and pump. A pair of diving planes could be used to adjust her depth while underwater. She carried a crude mercury depth gauge and a compass for navigation.

The submersible had been designed to sail above the water with her hatches projecting just above the surface until she was close enough to her target to submerge. She would then pass under the target towing a "torpedo" designed to explode on contact. The *Hunley* had been built in Mobile, Alabama, where her effectiveness had been successfully demonstrated against a coal-hauling flatboat anchored in the Mobile River.

After trials under Lieutenant George E. Dixon in Mobile Bay, General Beauregard ordered railway agents on 7 August 1863 to expedite *Hunley* to Charleston for the defense of that city.[7] She arrived in Charleston on two flatcars and under the management of part-owners B. A. Whitney, J. R. McClintock, B. Watson, and other persons who remain unidentified. B. A. Whitney was a member of the Secret Service Corps of the Confederate Army, his compensation to be half the value of any Union property destroyed by torpedoes or submarine devices.

During the third week in August 1863, the submarine made at least three nocturnal excursions aimed at striking the Union fleet at the mouth of Charleston Harbor. None was successful, although the volunteer crew (which had come with the boat from Mobile) gained experience handling the boat and learned important information about the harbor and its currents. The crew's failure to strike back after a group of Union monitors steamed into the harbor to shell Fort Sumter on the night of 23 August led to her seizure by the Confederate Army. Within days she was turned over to the Confederate Navy and placed under the command of Lieutenant John A. Payne.

Payne and the rest of his crew were volunteers who had been drawn from the *Chicora*. Payne rushed the crew through several training sessions, aided by advice from a few holdovers from the original civilian crew. Within a day or two of his taking command, Payne was ready, or so he thought, to make an evening attack on the Union ships outside of the harbor. He was putting the submersible through a series of practice dives on the afternoon of 29 August when disaster struck. She had just left the dock under tow of the gunboat *Ettawan* when Payne became entangled in the towline while climbing into the forward hatch. As he struggled to free himself, he knocked over the supporting prop that had been set to hold the diving planes in the "up" position. The boat immediately began to dive with both hatches open, causing her to fill rapidly with water. Payne and three members of the crew managed to scramble out of the boat before she sank, but five others went down with the boat and were drowned.

Following days of strenuous effort in the murky water, the *Hunley* was raised and her water-filled hull pumped dry. Horace Hunley monitored the repairs and cleanup himself. Three weeks after the tragedy he wrote to General Beauregard requesting that the craft be placed under his control and that he be allowed to select the crew from those who had built and tested her in Mobile. Hunley's request was rejected.

By the beginning of October 1863, a newly formed crew under the command of Lieutenant Dixon of the Confederate Navy was hard at work conducting practice dives in the Cooper River, using the Confederate receiving ship *Indian Chief* as a target. The goal was to draw the dummy torpedo-mine it was towing against the target's hull. On 15 October, Hunley, in Dixon's absence, decided to take the boat out for another routine practice run on the *Indian Chief*. Everything looked normal as the *Hunley* dove beneath the water when 200 feet from the starboard side of the ship. Sailors on the *Indian Chief* had become accustomed to this routine and expected to see the *Hunley* surface on the ship's port side. But as the minutes went by it became apparent that the submarine had taken her occupants on a fatal dive; this time none would survive.

It was later determined that the accident had probably been caused by an oversight on Hunley's part when he failed to close the seacock leading to the forward ballast tank. Once again the *Hunley* was raised to the surface, overhauled, and yet another crew of volunteers selected and trained. By mid-December the crew was ordered to proceed to the mouth of the harbor and to sink any Union ship contacted. The *Hunley* was towed to the mouth of the harbor several times that month, but none of her nocturnal expeditions was successful.

By January 1864 evidence of the *Hunley*'s existence was verified to Union officials by a Confederate deserter from the *Indian Chief*, who provided Rear Admiral John A. Dahlgren, the Union officer in charge of the squadron blockading Charleston, with a fairly accurate account of the submersible's troubled history. Dahlgren was already aware of the danger posed by the *David* and the other semi-submersible torpedo boats. To protect the ships of his squadron from the dangers of torpedo attack, Dahlgren ordered increased vigilance and instructed the ships to anchor in shallow water and to put out nets and chain booms.

Instead of targeting the ironclads and monitors lying near the harbor's entrance, the *Hunley*'s crew turned their attention to the main squadron of wooden warships anchored offshore. It was around this time that the *Hunley* was moved to Breach Inlet and fitted with a spar torpedo. "Its front was terminated by a sharp and barbed lance-head so that when the boat was driven end on [*sic*] against a ship's sides, the lance head would be forced deep into the timbers below the water line, and would fasten the torpedo firmly against the ship. Then the torpedo boat would back off and explode it by lanyard."[8] The torpedo itself was a copper cylinder holding a charge of 90 pounds of explosive fitted with a percussion primer set off by triggers.

In early February 1864, a new and inviting target appeared off Sullivan's Island: the screw sloop *Housatonic*. The ship had been a familiar sight around Charleston for 17 months and had recently been moved to her new location under strict orders to capture or destroy any blockade runner attempting to reach Charleston. Heavy seas prevented Dixon and his crew from venturing out to attack her until the evening of 17 February 1864. That night the *Hunley* made naval history when she became the first submarine to sink an enemy ship.

The 30-foot Union torpedo launch used by Lieutenant William B. Cushing to sink the Confederate ironclad *Albemarle* under way with its spar torpedo lashed to the starboard side. This engraving is based on a painting by R. G. Skerrett. (U.S. Navy)

As the *Hunley* neared the *Housatonic*, she had to surface to get her bearings, revealing her presence to an alert lookout who saw the candlelight showing through her viewing ports. The *Hunley* was so close to the *Housatonic* that the warship's heavy guns could not be depressed sufficiently to strike the *Hunley*, which appeared to be immune from the small arms that were fired at the low-lying craft. Dixon pressed home his attack, delivering the torpedo, which exploded beneath the *Housatonic*, quickly sinking the Union warship. Although it took only three minutes for the *Housatonic* to sink, just five lives were lost. The *Hunley* never made it back to safety, and her crew was lost when they went down with the submersible for reasons that remain unknown to this date.

THE UNION EFFORTS

While the Confederates were the first to make use of the spar torpedo, its utility as a weapon of war was not lost upon the Union. On 9 July 1864, William B. Cushing, a young U.S. naval officer who had already distinguished himself as an outstanding officer of exceptional initiative and courage, wrote Rear Admiral S. P. Lee, the commander of the North Atlantic Blockading Squadron, requesting permission to attack the Confederate ironclad *Albemarle* with spar torpedoes. The *Albemarle*, which had already survived a near-perilous collision with the Union ship *Southfield*, was on the Roanoke River at Plymouth, North Carolina, in shallow waters that no Union ironclad could navigate. Cushing was certain that he could destroy the troublesome ironclad using steam launches fitted with spar torpedoes.

Cushing found two suitable, 30-foot picket boats being built in New York, which he fitted with a spar torpedo invented by John L. Lay. The torpedo contained an air chamber that allowed it to float in a vertical position. The explosive device was mounted at the end of a 14-foot spar that could be raised or lowered by a halyard. It was released by pulling a second lanyard when the torpedo was under the enemy ship's hull. A third lanyard attached to the detonator set the charge off. One of Cushing's boats was lost en route to Norfolk, but he took the other, with a crew of seven officers and enlisted men, to the Union ships waiting in the sound off the mouth of the Roanoke.

On the night of 27 October 1864, Lieutenant Cushing and a crew of 14 took *Picket Boat Number 1* up river to Plymouth, North Carolina. A cutter in tow with two officers and ten men had been added to the force to attack and capture whatever Confederate picket boats were guarding the approaches to the ram *Albemarle*, which was moored along the riverbank and surrounded by a cordon of cypress logs chained together. Cushing wanted to board the *Albemarle* and try and take her out into the stream, but as he approached the ship a hail of gunfire met his boat. Cushing briefly studied the protective boom, steering parallel to it for a few moments before sheering off so that he could make

The Confederate ironclad *Albemarle* is struck by a spar torpedo and sunk by Lieutenant Cushing's torpedo launch at Plymouth, North Carolina, on 27 October 1864. Both the Union and Confederacy employed spar torpedo boats during the Civil War. (U.S. Navy)

Lieutenant William B. Cushing, USN. (U.S. Navy)

a full-speed run at it. After long immersion in the water, the logs were covered in slime and the steam launch easily rode over them.

Standing in the bow, Cushing ordered the spar torpedo lowered until the forward motion of the launch carried the torpedo under the ram's overhang. He pulled the detaching line with his left hand, waited a moment for the torpedo to rise under the hull before yanking the explosive line held in his right hand. A huge explosion ensued just as a canister of grape struck the boat. Crying "Men, save yourselves!" he plunged into the icy river and swam to shore, where he hid until paddling out to a Union ship in a stolen skiff. Miraculously, Cushing and one sailor survived the attack. The other men perished or were captured.

But the ironclad *Albemarle* sank. Upon hearing of this daring exploit, Secretary of the Navy Gideon Welles called William Cushing "the hero of the war." This action made him a national celebrity, and he was quickly promoted to the rank of lieutenant commander. The significance of Cushing's action, the first successful "torpedo" attack in the U.S. Navy, was not lost on the service when, in 1890, it named its first torpedo boat the USS *Cushing*.

CHAPTER TWO

False Starts

The First Self-Propelled Torpedo (1869–1890)

The first Whitehead torpedo was tested by the Austrian Navy in 1868. The U.S. Navy did not procure a Whitehead torpedo until 1891, more than two decades later. During that interval most of the major naval powers in the world acquired Whitehead torpedoes.

The origins of Whitehead's "infernal device"—as the weapon was subsequently called—began with the introduction of the Der Küstenbrander, or coastal fire ship, a self-propelled mine designed for harbor defense that was tested along the shores of the Austrian harbor of Cattaro in 1866. The Der Küstenbrander was the brainchild of Giovanni Luppis, a retired Austrian naval officer, who derived the idea from the drawings of an unknown officer in the Austrian Marine Artillery for a small, steam-propelled boat filled with explosives that could be guided from shore using tiller ropes.[1] Luppis added a percussion pistol in the bow to detonate the explosive charge. He set about making a clock work-powered model of the boat that he is purported to have presented to the naval authorities in Vienna in 1864.

What transpired in Vienna is unclear. Perhaps, as claimed in one account of events, Luppis was advised to seek technical assistance to perfect a workable full-scale prototype. In the event, Luppis soon solicited the help of Robert Whitehead, who was then managing director of a marine engineering works located at Fiume on the Adriatic coast.

Robert Whitehead was born into a family of well-to-do industrialists in 1823, at Bolton, England. From his early days he showed an interest in the new steam engines that were then coming into use in many factories around his hometown. At the age of 16 he was apprenticed to Richard Ormerod and Son, Engineers, of Manchester. This gave Whitehead a valuable grounding in practical engineering and he began to study mechanical drawing and pattern design, soon gaining a reputation as an exquisite draftsman. He left England in 1840 to seek his fortune abroad, where the talents of well-trained English engineers were in high demand.

After several years in France, Whitehead moved to Milan, Italy, where he established himself as an independent engineering consultant. In 1856 he was invited to become the chief engineer to Stabilimeno Technico Fiumano, a marine engineering firm located at Fiume, which is not far from Milan. By 1864 Whitehead, who had joined the company at its inception eight years earlier, had built the firm into one of Europe's leading producers of marine engines and had established himself as a talented engineer with demonstrated mechanical skill and ingenuity.

Luppis induced Whitehead to enter into an agreement to develop a surface torpedo.[2] The two men set to building the full-scale model of the Der Küstenbrander, which they tested at Cattaro. Although successful in principle, the slow speed and limited range of the device, coupled with the need to control the tiller ropes from shore, limited its effectiveness as a military weapon.

Realizing these shortcomings, the two men abandoned the Der Küstenbrander and set out to ascertain whether it was possible to build a locomotive or self-propelled torpedo that could be entirely independent of outside influence once it had been launched on its course. With this concept as a basis, Luppis and Whitehead, aided by Whitehead's son and a trustworthy workman, set about to build the weapon. Within two years they had succeeded in constructing the first "fish" torpedo, a fully submerged, self-propelled, screw-driven device that could "swim" to its target like a fish.

From the fragmentary evidence that has survived, it appears that the first Whitehead-Luppis torpedo was completed sometime in 1867, if not earlier. This prototype—of what would become the first in a long line of Whitehead torpedoes—was 11 feet long and 14 inches in diameter, weighed 300 pounds, and had a speed of about 6 knots. It was constructed of wrought iron in the shape of a cigar and was fitted with a pair of vertical fins that ran the full length of the body to prevent the torpedo from rolling on its axis while travelling through the water. The warhead, which could be separated from the body of the torpedo for storage, contained 18 pounds of dynamite, which was detonated on contact with the target by means of a simple firing pin.

The first "fish" was a marvel of ingenuity made possible by Whitehead's clever use of several innovative engineering concepts. Among them was the introduction of compressed air for propulsive power and the use of a hydrostatic valve connected to a pair of horizontal rudders for depth control. The exact layout and specifications for the internal details remain shrouded in mystery, although it appears likely the air flask was charged to 370 pounds per square inch (psi) and connected to a rotary-type pneumatic engine of Whitehead's own design.

On 26 May 1867, Whitehead, in company with other prize-winning exhibitors from the Paris Exhibition of 1867, was received in Vienna by Emperor Franz Josef. This was his second meeting with the Austrian monarch and served, no doubt, to enhance Whitehead's reputation and opened the door for the initial demonstration of the prototype torpedo, which appears to have taken place at Fiume in October. Although exact details of this event also remain unknown, the performance of the Whitehead-Luppis torpedo was sufficiently impressive to warrant serious consideration by the Austrian naval observers, who recommended that further tests be conducted to evaluate the potential capabilities of this new weapon.

Robert Whitehead.

Naval trials of the Whitehead-Luppis torpedo were conducted before a board of Austrian naval officers assembled at Fiume starting in May 1868. In addition to the first prototype, Whitehead prepared a larger model patterned after the original that was intended for actual service. Dubbed the "normal torpedo," the latter was 14 feet, 1 inch long, 16 inches in diameter, and was designed to carry 60 pounds of guncotton explosive. A quantity of junk equal in weight to the explosive charge was placed in the torpedoes for the trials.[3]

The first set of tests was conducted from the Austrian gunboat *Gemse*, which was fitted with an underwater launching tube, also designed by Whitehead. Constructed of bronze, the tube was located in the ship's bow and had a watertight outer door three feet below the waterline. Two other watertight doors at the rear of the tube allowed the torpedo to be loaded and to admit seawater into the tube after the torpedo was loaded.

The target was a large net, 200 feet long by 24 feet deep, attached to the yacht *Fantasie*, which was moored approximately 700 yards from the *Gemse*, providing an effective area for capturing torpedo hits.

The torpedo supplied for this test was the small prototype that had been demonstrated in October. The weapon was expelled by compressed air, its engine being started at the same time by the withdrawal of a small retaining pin. Whitehead had intended that it be launched 12 feet below the surface, the depth that it had been adjusted to run. The *Gemse*'s draft was too small to allow for the torpedo tube's installation at that depth. Thus the torpedo, once launched, was forced to seek its own level. This caused huge oscillations in the torpedo's run, varying in depth from as little as 8 feet to as much as 40. So great were these gyrations in depth that at times the torpedo rose to the surface, and at times it dove to the bottom, where it stuck in the mud. Not surprisingly, only 8 of the 54 shots hit the net. Sixteen passed below the net and the rest missed altogether.

The trials were suspended for three weeks while Whitehead determined a means to solve the depth-keeping problem. The answer was relatively simple, but extremely clever: a device consisting of a pendulum weight, which Whitehead added to the depth-keeping mechanism. It provided control of the horizontal rudders whenever the torpedo strayed from a horizontal attitude. Whitehead referred to this feature as the "Secret" and refused to reveal it for many years, fearing that it might be stolen by those interested in making their own fish torpedoes.

A second series of trials was conducted after this improvement had been added to both torpedoes. More than 50 percent of the shots in this series hit the net, while the torpedo speed was increased to 6.8 knots. The commission of naval officers convened to view the trials was so impressed with the performance of the Whitehead torpedo that the vote was unanimous to acquire the "Secret" of Whitehead's torpedo. The Austrian government reportedly paid Whitehead £20,000 for the nonexclusive rights for the depth-keeping secret while conceding the right for Whitehead to manufacture the deadly device and agreeing to pay him £600 for each of the 11-inch diameter models and £1,000 for the 13-inch versions.[4]

As word of Whitehead's successful torpedo spread through Europe, representatives from the various naval powers, including officers from the U.S. Navy's Mediterranean Squadron, went to Fiume to view the device. The Americans, according to *The Times* of London, were "greatly interested in these torpedoes and entertain a serious idea of making them the base, so to speak, of naval warfare, not only for defense, but for attack."[5] One of those Americans was Lieutenant Commander J. D. Marvin. His report on the visit to Whitehead's factory was forwarded to Vice Admiral David D. Porter, the Assistant Secretary of the Navy.

THE TORPEDO STATION

Admiral Porter was greatly interested in the development of torpedoes, which he believed would be the great weapon of naval defense for the United States. He believed that torpedoes would be militarily effective and relatively economical. He had earlier argued that half a dozen tugs armed with spar torpedoes could destroy any blockading fleet. Porter's enthusiasm for these weapons had led him to conduct his own experiments with spar torpedoes on the Severn River in 1866 while he was superintendent of the Naval Academy.

Ulysses S. Grant's election to the presidency in 1868 gave Porter an opportunity to turn his "hobby" into one of the major naval innovation programs of the century. Were it not for the policy of appointing civilians as cabinet secretaries, Grant would have undoubtedly appointed Porter to be Secretary of the Navy. Instead, he named Adolph Borie of Philadelphia, who agreed to serve as figurehead while his assistant, Porter, executed actual control over the Navy Department.

In June 1869, Admiral Porter issued orders to the Chief of the Bureau of Ordnance to organize a Torpedo

Corps. The Bureau of Ordnance was responsible for developing and maintaining weapons for the Navy. Porter's plan included the establishment of a torpedo station for administration and technical development of the weapon. The idea for such a station was not new: Captain Henry A. Wise had proposed it in 1865. Unlike Porter, Wise had neither the authority nor the persuasiveness to obtain funds for such "extravagances" in the face of the severe budget cutbacks that plagued the Navy after the Civil War. To command the new corps, Porter selected Lieutenant Commander Edmund O. Matthews, the head of the Department of Gunnery at the Naval Academy, who had helped Porter with his torpedo experiments on the Severn River. Matthews' first task was to select an appropriate site for a torpedo station. After checking several possible sites, Matthews recommended Goat Island at Newport, Rhode Island. It was approved and occupied in the fall of 1869 with a cadre of 23 officers and enlisted men, and one civilian chemist.

From its beginning the Torpedo Station was an exception to the normal pattern of naval stations. The Torpedo Station was a bold attempt to develop new technology. Unlike the work at navy yards and the earlier experimental batteries, the work at the Torpedo Station was to be "confidential," which was equivalent to "secret" or "top secret" in modern security classifications. Besides serving as the headquarters for the Torpedo Corps, the station became a combination of torpedo school for officers, a laboratory for torpedo development, and a center for torpedo experimentation. Thus, like a modern research and development center, it would have laboratories for exploring new ideas and experimental facilities for testing them in a realistic environment. Its mission included a heavy emphasis on officer instruction. And, with a few notable exceptions, the new implements of war produced at the Torpedo Station were to be the products of the creative minds of naval officers. Moreover, most of the officers serving as station commanders and instructors and in technical capacities were line officers assigned to the station between tours of sea duty.

THE STATION "FISH" TORPEDO

In the meantime, Admiral Porter appears to have forwarded Commander Marvin's report to Matthews along with a note asking if it would be possible for the station to build an automobile torpedo similar to the one Matthews had observed at Fiume. Although Marvin's report and its accompanying drawing provided considerable detail about the external shape and construction of the torpedo, and how it performed, no information was available about the design of its internal components and how the fish worked.[6] After requesting authority to conduct experiments with the "Austrian Torpedo," Matthews, assisted by Lieutenant Francis M. Barber, began the arduous task of designing from scratch a self-propelled torpedo, which they undertook during the winter of 1869–1870. The torpedo they designed had a length of 12 feet, 6 inches and was 14 inches in diameter. It was propelled by a liquid carbonic gas generator that powered a reciprocating rotary engine connected to a single propeller.

Fabrication got under way as soon as components began arriving at Newport in May 1870. There were problems galore: the castings were full of flaws, the rotary engine leaked and did not function properly, and gases from the carbonic acid generator were severely corrosive. A number of unsuccessful attempts were made to repair the engine and reduce the leakage, but the engine could not be used.

In March 1871, Matthews gave up on the original engine and began to design a new one. The new design was a two-cylinder horizontal cam engine that was referred to as a "diamond" engine. The drive mechanism, which included a 3.5:1 gear reduction, was similar to that used for whip drills and was designed to drive the propeller at 250 revolutions-per-minute for 1,300 yards. The new engine worked fine on steam, but the severe corrosion problems continued when carbonic acid gas was used. These problems led to a decision to use compressed air for power, requiring the use of a high-pressure air compressor. A special order had to be placed for the compressor, causing further delays in the project.

The Naval Torpedo Station at Newport, Rhode Island, continued to experiment with spar torpedoes well after the end of the Civil War. Here a station boat is about to test an electrically detonated explosive charge, one of several variations of the original spar torpedo. (U.S. Navy)

In the interim, Commander Matthews proceeded to redesign the depth control mechanism, replacing the static bellows concept with a set of balanced fins that were activated by the depth bellows. If the torpedo went deeper than the set depth, the bellows gave the fins an "up-elevator" signal; if the torpedo were running shallow, the bellows transmitted a "down-elevator" signal to increase the torpedo's running depth. Although the concept was similar to the depth-control mechanism used by Whitehead, it lacked the pendulum needed to dampen the controller's tendency to over-shoot the depth setting, causing the torpedo to move up and down and thus providing only marginal depth control.

In June, Matthews, who was impatiently awaiting the arrival of the special air compressor, decided to test the torpedo using the carbonic acid gas generator. Upon launching, the torpedo immediately dove to the bottom and promptly filled with water. The engine was so badly corroded that no further tests could be conducted until after the new compressor arrived in July.

By the end of the month, the "Station Fish Torpedo" was ready for its second test run. This time the torpedo ran for 200 yards, deviating slightly to port as it ran through the water. Although the torpedo was set to run at a ten-foot depth, the propeller occasionally broke the surface, indicating an erratic depth performance. After running 200 yards, the torpedo began to fill with water and sank. During the third test, conducted in August, the torpedo made a great curve to starboard and the propeller again broke the surface. Performance during the fourth test was no better.

A design review conducted after this final test trial led to a number of design changes. The size of the air tank was increased, the size of the fins was increased, the size of the vertical guards was reduced to save weight, and a governor was installed on the engine. Still the torpedo's performance continued to fall below expectations. It got tangled in the eelgrass during the fourth test, ran into the mud bank on the fifth, and scored a direct hit on a piling during the sixth and final test. By then Commander Matthews had concluded that a major redesign was needed and deferred all further work until a new design could be perfected.

Several dock trials took place in 1872 after improvements were made to the existing torpedo. From the performance data obtained during these static tests it was estimated that the torpedo would have a speed of eight knots and a run of 4,000 feet. This was theoretically better than the numbers attained by the Whitehead torpedo in actual practice. No further tests were conducted, and the station's "fish" torpedo was placed in storage in 1873.

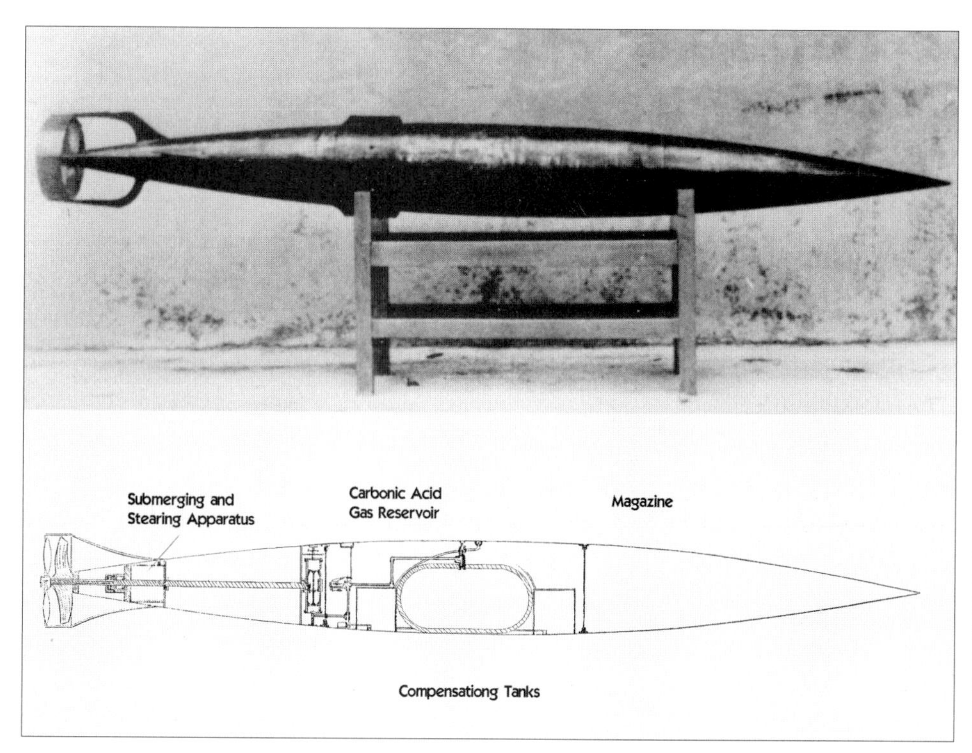

A contemporary photograph and drawing of the Newport "Fish Torpedo" of 1873. The term "magazine" was used for the warhead, which would have contained guncotton. Note the circular enclosure for the torpedo's propeller (U.S. Navy)

WHITEHEAD TORPEDOES

In November 1873, Robert Whitehead invited Commander William. A. Kirland, then serving with the U.S. Navy's Mediterranean Squadron, to Fiume to obtain first-hand information on the Whitehead-Luppis fish torpedo and to observe tests of the device. Kirland was present on 6 and 7 December 1873, to observe the tests, which were conducted in an underwater course laid out in the Gulf of Fiume. The course was marked with cork floats at 20-foot intervals, running from the launching tube to a series of target buoys 600 feet away. A series of eight-foot nets were set below the surface to catch any errant runs that were too deep or off course. Kirkland watched as three torpedoes were successfully launched on both days. As Whitehead had promised, the torpedoes ran a straight course and remained at proper depth setting.[7]

On 12 December Whitehead wrote to Commodore William N. Jeffers, Chief of the Bureau of Ordnance, offering to license the Whitehead-Luppis torpedo to the U.S. Navy. For $44,000 Whitehead would furnish a sample torpedo, supply working drawings, and provide instruction in its operation for up to six naval officers. An apparatus for compressed air charging and launching torpedoes from a ship for experimental purposes was also in the proposal. In return, the U.S. government had to agree to keep the secret of the torpedo and not sell it or any of the torpedoes manufactured under the license to any other nation.[8]

Within weeks of receiving Whitehead's offer the Navy received a letter from Henry Westbrook, a former employee of the Woolwich Arsenal in England, who claimed to have worked on the fish torpedoes being manufactured there. Westbrook offered to sketch "the whole of it" and provide all of the details if the U.S. Navy would give him employment. Although the record indicates that the Navy declined the offer, a set of drawings was obtained from him and turned over to Commodore Jeffers. The drawings were not exploited, but were the subject of a lengthy exchange between Commodore Jeffers and Whitehead's American agent.

Although many European countries had adopted the Whitehead torpedo, Jeffers declined Whitehead's offer. While the Whitehead torpedo performed well under favorable circumstances, Jeffers remained unconvinced of its suitability for use under operational conditions at sea. Moreover, the continuing downward slide in the Navy Department's budget and congressional reluctance to spend money for new weapons weighed heavily against any decision on his part to request the considerable sum that would be needed to purchase the Whitehead license. Whitehead's price was almost as much money as Congress budgeted for the annual operation of the Torpedo Corps—$50,000.

In the interim, the Torpedo Station continued to investigate and test a number of different types of self-propelled torpedoes that were designed and built by a variety of civilian and military inventors. The most prominent of these being:

- Lay Torpedo: A moveable surface torpedo powered by carbonic acid gas and controlled electrically by a trailing cable.
- Ericsson Torpedo: A completely submerged rectangular shaped torpedo powered by compressed air supplied from a trailing tube.
- Howell Torpedo: A flywheel propelled fish torpedo first submitted to the Bureau of Ordnance in 1870. The initial design utilized an iron flywheel weighing 100 pounds.

Other experimental torpedoes included the Barber Torpedo, the Sims-Edison Torpedo, the Patrick Torpedo, the Hall Torpedo, and the Cunningham Torpedo. Characteristics for all of the experimental torpedoes mentioned above can be found in Appendix C. Other types of automobile torpedoes were also tested by the Naval Experimental Battery at Annapolis, and by the inspector of ordnance at the Navy Yard in Washington, D.C.

THE HOWELL TORPEDO

After the Civil War, budgetary constraints placed upon the Navy Department by a frugal Congress led to a rapid decline in the material condition of the fleet. By 1880 the Navy's budget had plunged to a postwar low of $13.5 million. The condition of the Navy was so bad that Admiral Porter called it "nearly worthless" for war.[9] William Hunt, who became Secretary of the Navy in 1881, concurred with Porter's assessment of the deplorable state of the Navy. Hunt clearly stated his position when he wrote the following in the introduction of his annual report on the Navy for 1881: "The condition of the navy imperatively demands the prompt and earnest attention of Congress. Unless some action be had in its behalf it must soon dwindle into insignificance."[10] Although Hunt served as Secretary of the Navy for less than a year, he laid the groundwork for the emergence of the so-called "New Navy" by establishing the first Naval Advisory Board.

Hunt's installation as Secretary of the Navy coincided with the selection of Commodore Montgomery Sicard as the new Chief of the Bureau of Ordnance, appointed on 1 July 1881. Unlike his predecessor, Sicard regarded the introduction of the automobile torpedo as "a matter of the greatest importance and urgency."[11] He also knew that the development of such a torpedo was a long-term experimental process that would consume a great amount of time. Commodore Sicard was well aware that the Whitehead torpedo had been adopted by every naval power of consequence and was the only automobile torpedo that had shown itself capable of working with certainty over any reasonable range.[12] He wanted to acquire this weapon for the U.S. Navy and believed that the best way to achieve this was to purchase a limited number of torpedoes directly from Whitehead along with their working drawings, which could then be used to produce a domestic version of the weapon. Sicard was so convinced of the need to obtain the Whitehead torpedo that he included the purchase of 25 Whitehead torpedoes and a set of working drawings in his proposals for the next year's budget.

Commodore Sicard's pleadings were only partially answered by Congress, which added the following paragraph to the Naval Appropriations Act of 1883:

> For the purchase and manufacture, after full investigation and test in the United States under the direction of the Secretary of the Navy, of torpe-

The Lay-Haight torpedo photographed in March 1894. This experimental torpedo was driven on the surface by carbonic acid gas. Steering depended upon an electric current generated from the shore that was sent over a cable payed out from the torpedo. (U.S. Navy)

The Cunningham rocket torpedo and its inventor, a shoemaker, at New Bedford, Massachusetts, in July 1893. The after end of the launching tube appears at right. (U.S. Navy)

does adapted to naval warfare, or of the right to manufacture the same and for the fixtures and machinery necessary for operating the same, one hundred thousand dollars: *Provided*, That no part of said money shall be expended for the right to manufacture the same until the same shall have been approved by the Secretary of the Navy, after a favorable report to be made to him by a board of naval officers to be created by him to examine and test said torpedoes and inventions.[13]

Although the $100,000 appropriated was a considerable sum, no money could be spent without the satisfactory approval of the Torpedo Board, which was subsequently established under the able leadership of Captain George E. Belknap. In the meantime, the Navy Department—most probably via the Bureau of Ordnance—issued a circular letter addressed to various torpedo manufacturers soliciting their participation in the trials that would be used to determine the best torpedo design. Included in the list of addressees were foreign manufactures Whitehead, Swartzkoff, and Berdan. The Americans were John Ericsson, Asa Weeks, A. Rowe, Thomas F. Rowland, John A. Howell, the United States Torpedo Company, and the Dynamite Projectile Company. Each was asked to provide a torpedo and to demonstrate its performance at his own expense before the Torpedo Board. Public notices of the meetings of the Board were also given by advertisements in leading journals, and every opportunity possible was given to inventors and manufacturers to display the merits of their torpedoes.[14]

None of the European manufactures were willing to participate under the conditions stipulated, and only three of the domestic manufactures completed the program of trials specified by Commodore Sicard in his instructions to Captain Belknap dated 3 December 1883.[15] These were:

- An electrically steered surface-running torpedo presented by the American Torpedo Company similar to the Lay-Haight Torpedo that was powered by carbonic-acid gas acting upon a Brotherhood engine.
- The Weeks surface rocket-torpedo presented by Asa Weeks.
- The flywheel powered subaqueous torpedo presented by Captain John A. Howell.

A fourth company, the Sim's Electrical Fish Torpedo Company of New York, also submitted an electrically controlled torpedo, but was unable to supply a firing apparatus and thus failed to participate in all of the required tests.

Inspection and testing of these torpedoes took place in the spring of 1884, during various sessions of the Torpedo Board that were held at Milford, Connecticut; Norfolk, Virginia; Washington, D.C.; and in the waters off Hampton Roads, Virginia. The only torpedo not rejected by the board was Captain Howell's, which was recommended for further consideration. The torpedo submitted by Howell was an improved version of the torpedo that had been tested at the Torpedo Station on two previous occasions. Howell had first applied for and received a patent on his self-propelled torpedo in 1871. He was then a lieutenant commander serving as head of the Department of Astronomy at the Naval Academy in Annapolis. The key innovative feature of Howell's patent was the heavy flywheel, which was spun up before launching to provide the propulsive energy needed to rotate the propeller shaft. In the ensuing years Howell had continued to test and modify the design of his torpedo and had made many improvements to the device.

THE FIRST TORPEDO ATTACK

The first launch of a self-propelled torpedo in combat occurred in 1877, when a Whitehead torpedo was fired by the British cruiser *Shah* in her fight with the Peruvian monitor *Huscar* off the Bay of Ilo, in southern Peru. Both ships were armed with Whitehead torpedoes as well as guns. The *Shah*'s torpedo attack failed, as did the first attempt to use Whitehead torpedoes by the Russians in their war with Turkey the same year. After an earlier failure, the Russian torpedo launches *Chesma* and *Sinop* attacked the Turkish steam frigate *Intibakeh* with torpedoes off Batoum on the Black Sea. The two craft, previously used with towed torpedoes, were too small to carry the Whitehead torpedoes and their launching apparatus. The *Chesma* had a torpedo tube lashed under her hull with the intention of cutting it away as soon as the torpedo was fired; the *Sinop*'s torpedo tube was secured to a raft that was lashed alongside her. Their torpedoes had warheads of 60 pounds of guncotton. The two torpedo launches made their approach undetected on 26 January 1878. Both fired at the large gunboat *Intibakeh*. She was struck and quickly sank—the first vessel known to be sunk by a self-propelled torpedo.

Commodore Sicard wrote to Captain Howell in August 1884 to advise him that the Bureau was favorably impressed with his torpedo and would pay for the manufacture of three, but as a serving naval officer he would receive no royalty. Work on a torpedo design based on Howell's concept began in the ordnance shops of the Washington Navy Yard under the Inspector of Ordnance, Commander Casper F. Goodrich. The torpedo Goodrich produced was 9 feet long and approximately 14 inches in diameter. The torpedo's casing was hand wrought from 1/32-inch galvanized iron. It weighed about 284 pounds and had a 112-pound flywheel for motive force and gyro stabilization. In place of the explosive charge the test torpedoes carried 40 pounds of ballast placed in the nose.

The flywheel, which was located in the horizontal plane set at right angles to the axis of the torpedo, was spun up by an auxiliary steam engine on the launching ship. The flywheel was connected to two separate propeller shafts through a set of beveled gears attached to each side of the flywheel's axle. The twin shafts each turning a three-blade propeller rotated in opposing directions. As the flywheel slowed down, the amount of energy delivered to the propellers fell off markedly. To

A rear view of the Howell torpedo, showing its twin, variable-pitch propellers and steering gear. It was driven by a 132-pound flywheel that was spun to 10,000 revolutions per minute prior to launch by a steam turbine mounted on the torpedo tube. (U.S. Navy)

compensate for this effect, a set of sliding cams was installed on each propeller that increased the pitch of the propeller blades so that they cut more deeply into the water as the flywheel decelerated.

The gyroscopic action of the flywheel has been credited with providing the excellent directional stability exhibited by the Howell torpedo. Any force tending to deflect a Howell torpedo from its course would cause it to heel over. Captain Howell's design took advantage of this phenomenon to control the torpedo's direction by installing a pendulum suspended on the fore-and-aft axis. As the pendulum swung from one side or the other, it caused the rudders to move in an opposite direction, acting as an anti-rolling device that served to keep the torpedo on course.

In the summer of 1885, Commander Goodrich sent the three experimental Howell torpedoes that he had produced to the Torpedo Station at Newport for testing and evaluation. The test program did not start out auspiciously. The first torpedo launched from the deck of the station tug *Triana* on 5 August 1885 hit the water and immediately sank to the bottom of Newport Harbor. Five days later, Torpedo No. 2 was successfully launched, but after a brief run it too sank. It took a number of months to correct all of the problems. By July 1886, the Howell torpedo was consistently making successful in-water runs.

The success of the Howell torpedo vindicated the Bureau of Ordnance's long-standing objections to the Whitehead torpedo. By all accounts the model 1888 Howell torpedo was equal or superior to the Whitehead in every category (see table 2-1). Although both torpedoes had about the same speed and range, the Howell's warhead could carry 20 percent more explosives. It was much lighter than the Whitehead, took up less space, and was a much simpler design, making it less costly to manufacture and easier to operate and maintain and it cost one-half that of the Whitehead. The principal disadvantage of the Howell torpedo was time needed to run-up the flywheel and the need to provide an auxiliary power source. But it eliminated the high-pressure air flask, which was considered a hazard, especially in the face of hostile gunfire. The Howell had one more advantage. It did not leave a wake, making it impossible to detect as it streaked toward its target!

TABLE 2-1

WHITEHEAD AND HOWELL TORPEDOES

MODEL	WHITEHEAD MARK IV	HOWELL PROTOTYPE
Produced	1885–1886	1886–1888
Diameter	14 in	13.2 in
Length	14 ft, 9 in	8 ft
Weight	660 lb	325 lb
Explosive charge	58 lb	70 lb
Propulsion	Compressed air @ 1250* psi	110 lb flywheel @ 10,000 rpm (max.)
Speed	23 knots @ 600 yds	24 knots, 1st 200 yds
Range	600 yds	600-700 yds

Source: Murray F. Sueter. The Evolution of the Submarine Boat, Mine and Torpedo, from the Sixteenth Century to the Present Time *(Portsmouth, England: J. Griffin and Co., 1907); Hotchkiss Ordnance Company.* The Howell Torpedo: General Description and Notes *(Washington, D.C.: Hotchkiss Ordnance Co., 1888).*

**Estimated from other sources.*

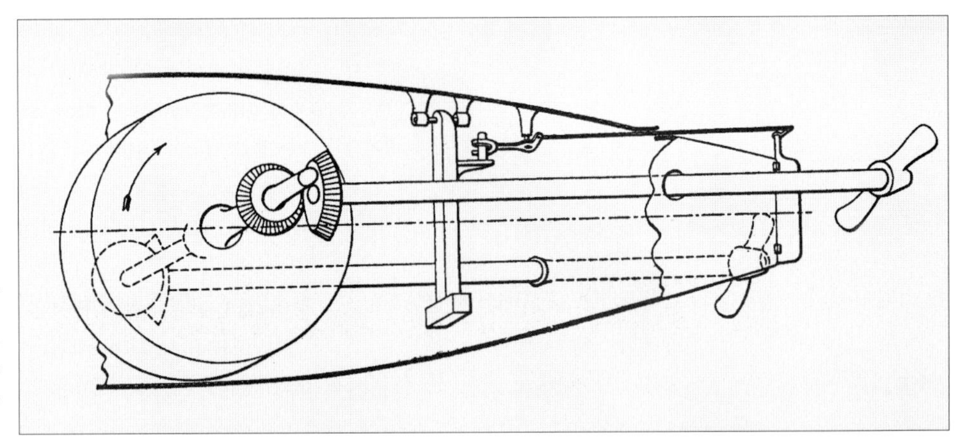

A schematic view of the Howell's flywheel motor and steering gear. The rotating flywheel (left) created a gyroscopic effect with deviations in azimuth being adjusted by a pendulum coupled to the rudder that sensed the heel of the torpedo when it deviated from its course. (U.S. Navy)

The final torpedo design developed by Lt. Com. J. A. Howell on the deck of the USS *Stiletto*, the first U.S. Navy torpedo boat capable of launching self-propelled torpedoes. Despite depth-keeping limitations, Howell torpedoes were used by the Navy until 1898. (U.S. Navy)

Also important, the gyroscopic action of the Howell flywheel provided greater accuracy, especially when discharged from the broadside of a warship moving at high speed. This was later confirmed when repeated trials with Whitehead and Howell torpedoes fired from a torpedo boat running at a speed of 14 knots showed the Whitehead torpedo to have a mean irregular deflection of 10½ degrees at 400 yards, while the Howell's deflection was only 5 degrees. Test data compiled during the same period showed the Howell much more reliable with a mean rate failure of 9.1 percent versus 28.6 percent for the Whitehead.[16]

The Navy Department did not place its first production contract for the Howell torpedo until 5 January 1889.[17] By then Howell, now a Navy captain, had sold all of his rights to the Hotchkiss Company, which undertook the development of the production model at its facilities in Providence, Rhode Island. The loss of several torpedoes during preliminary trials and other difficulties extended the time needed to produce a reliable torpedo that could meet the critical performance characteristics specified by the Navy. As a result, the first 10 of the 30 Howell torpedoes contracted for were not delivered until December 1891—16 months late. In the interim, the Bureau of Ordnance sought another supplier of torpedoes.

CHAPTER THREE

The Torpedo Perfected

Bliss-Leavitt Torpedoes (1890–1913)

The Hotchkiss Ordnance Company's inability to deliver a workable torpedo on a timely basis created a minor crisis in the Bureau of Ordnance, which, because of the delay, was unable to provide torpedoes for the experimental torpedo boat *Stiletto*, purchased in 1887, or the newly launched *Cushing*. The latter was the first torpedo boat built specifically for the U.S. Navy. The craft was named for William B. Cushing, the officer who commanded the first successful use of the spar torpedo in the Navy when he attacked and sank the Confederate ironclad *Albemarle* in 1864.

When the *Cushing* was commissioned on 22 April 1890, the Navy did not possess a single torpedo. Although ten Howell Torpedoes were scheduled for delivery by 1 July of that year, continuing development problems forced Hotchkiss to request an extension to 20 November 1891. Although the company assured the Navy that its torpedo would soon be ready for trials, the delay in delivering operational versions of the torpedo and the uncertainty of its ultimate success fostered the urgent need for the Navy to find another source of torpedoes.

A month after the *Cushing* was commissioned, the Chief of the Bureau of Ordnance, Commodore William M. Folger—who had relieved Montgomery Sicard on 12 February 1890—took the first step in a process that would lead to the Navy's procurement of Whitehead torpedoes. On 12 August 1890, Commodore Folger cabled the Whitehead factory in Fiume, Austria, asking if the firm would agree to arrange for an American company to manufacture torpedoes for the U.S. Navy. Folger's cable was forwarded to Robert Whitehead's home in Sussex, England, the following day. Whitehead did not like the idea of putting another torpedo manufacturer in business but was well aware that he would never receive an order from the U.S. Navy unless the torpedoes were manufactured in the United States.[1] Thus he agreed to license the manufacture of his torpedoes provided that a royalty of £50 be paid on each torpedo manufactured for the next 15 years. As part of the agreement, Whitehead stipulated that they could only be manufactured for the U.S. government and that he could benefit from any improvements introduced by the manufacturer. Whitehead sent this proposal to Folger in a letter the day after he received Folger's cable.

Commodore Folger agreed to Whitehead's terms and immediately began looking for a company that was capable of manufacturing Whitehead torpedoes. It did not take him long to convince Eliphalet W. Bliss to send Frank McDowell Leavitt from his company, the E. W. Bliss Company of Brooklyn, New York, to the Whitehead Factory at Fiume to investigate the feasibility of manufacturing Whitehead torpedoes for the U.S. Navy.

Leavitt was the mechanical genius behind the company, which had been formed in 1871 and had undergone several changes in name and ownership over the years until 1881, when Bliss took over as its sole proprietor. Leavitt, who joined the company five years earlier, had previously invented an automatic can-making machine, which had revolutionized the tin can business, and had invented the toggle drawing press for producing pails, cooking utensils, and hollow dishes of all kinds. The Bliss Company was one of the largest suppliers of metalworking machinery in the United States, and when it was incorporated in 1890, Leavitt was given a partial interest in the new firm.

Bliss was also president of the U.S. Projectile Company, which supplied shells to the Navy. He was obviously familiar with the ordnance business and was clearly interested in expanding his firm's dealings with the Navy when he sent Leavitt to Fiume in 1890. (By that time the U.S. Projectile Company had been acquired by the E. W. Bliss Company.)

Leavitt arrived in Fiume in November. On the 27th of that month he cabled the home office with an estimate of what it would cost to build Whitehead torpedoes in their factory in Brooklyn. Leavitt was certain that he could handle the task. None of the machining or assembly work seemed insurmountable. The most difficult pieces of the torpedo to manufacture were the forward and aft shells, which were hammered up from flat blanks and brazed. Although this required a considerable amount of skill on the part of the man handling the hammer, Leavitt believed that they could be subcontracted to a metalworking shop located in Providence, Rhode Island.[2]

Leavitt was not enamored of Fiume, which he called a "Godforsaken" place. His disposition was not helped by the difficulties he was having in getting the information he needed from Whitehead's technical experts. "Getting their secrets from them," he wrote, "is like drawing their heart's blood and it takes an immense amount of pumping to get at it." In his cable, Leavitt estimated it would take him another three weeks to finish up, although he thought that things would move much faster "now that these people are waking up to the fact that we mean to know it all."

Leavitt's mission was successful, and Secretary of the Navy Benjamin F. Tracy awarded the Bliss Company a contract on 19 May 1891, to produce 100 Whitehead torpedoes. The model ordered was similar to the 11 foot, 8 inch "short" Whitehead torpedo with an 18-inch diameter that was also being offered to the British. This was an improved version of the Whitehead torpedo, being faster, with a greater range, and carrying more explosives than the 14-inch model that had been in production for some time. The version manufactured by the Bliss Company was designated according its metric dimensions—it was 3.55 meters long and with a 45-centimeter diameter. This became the "3.55m x 45cm Whitehead Torpedo," in U.S. Navy parlance. Three versions of this torpedo—identified as the Marks I, II, and III—were subsequently manufactured, each incorporating slight mechanical improvements to the basic design.

As the delivery of torpedoes neared, Commodore Folger appointed Commander George A. Converse, then assigned to the Bureau of Ordnance, as the president of a board of officers to determine "all questions relating to the torpedo outfit for the vessels presently

THE FIRST MAJOR WARSHIP SUNK BY A TORPEDO

The first successful self-propelled torpedo attack ever on an enemy warship occurred during the Chilean Civil War, when the 3,500-ton *Blanco Encalada* was sunk in Caldera Bay on the night of 23 April 1891. The ironclad warship was at anchor when the 710-ton torpedo gunboats *Almirante Lynch* and *Almirante Condell* attacked her just before dawn. Of the seven—some reports state five—14-inch Whitehead torpedoes launched, only one found its mark, striking the *Blanco Encalada* abreast of her engine room. This was considered good torpedo marksmanship at the time, because target practice with the Whitehead torpedo—then supplied without gyroscopic control—often resulted in erratic torpedo runs that could take the underwater missile in any direction once it entered the water. When the torpedo struck the *Blanco Encalada* the ensuing explosion was so severe that every man on the gun deck was thrown off his feet. One of the ship's 8-inch guns was hurled from its carriage. Water poured into the hull, sinking the *Blanco Encalada* in about five minutes, with some 300 of her crew going down with her.

The experimental torpedo boat *Stiletto* launching a flywheel-driven Howell torpedo from her bow tube during tests on Narragansett Bay circa 1893. The *Stiletto*, built as a private venture, was launched in 1885, purchased by the Navy in 1887, and entered service in July of that year. She was assigned to the Newport Torpedo Station. (U.S. Navy)

under construction or completed."[3] On 4 September 1891, the Torpedo Board, which included Lieutenants F. J. Drake, T. C. McLean, and C. H. Bradbury, examined the *Stiletto* and recommended that she be fitted with a bow tube for the Howell torpedo under the main deck and extending through the bow. The board examined drawings of the *Cushing* at the same time, recommending that she be fitted with tubes for launching the Whitehead torpedo: one bow tube under the deck, passing through the stem, and two deck-mounted tubes fitted on a pivoting plate to permit launch over either side of the craft.

During the next two years, the Torpedo Board conducted extensive trials and acceptance tests of the Howell and Whitehead torpedoes and their respective launching gear. The board members visited the Bliss works to inspect air flasks; witnessed numerous in-water trials at Mackeral Cove on Narragansett Bay, near the Torpedo Station; were present for the official trials at the Hotchkiss firing station at Triverton, Rhode Island; and were present on board the *Stiletto* and the *Cushing* to witness launchings from those ships.

Testing continued through the fall of 1894, by which time both torpedoes were fully qualified for service use. By then all 100 of the Whitehead torpedoes called for in the original contract had been delivered and the Navy had entered into a follow-on agreement with Bliss to manufacture 50 additional torpedoes of an improved type, designated as the Mark II. Although only two of the 30 Howell torpedoes already on order had been accepted, the Navy increased its order by 20, based on the fact that 21 of the Howell torpedoes were undergoing trials and were close to acceptance. Thus by the end of 1894, the Navy had 200 torpedoes ready for service or on order.

The first torpedoes, other than those carried on board the *Stiletto* and *Cushing*, the only torpedo boats in the U.S. Navy, did not enter service until 1895, when torpedo outfits—above-water launching tubes, torpedo directors, and air plants or auxiliary engines in the case of the Howell—were installed on seven protected cruisers. Most of these cruisers received six tubes located on the berthing deck, two to a side, plus one tube installed in the bow and one in the stern. A few ships received the side-mounted tubes only. The cruisers *San Francisco* (No. 5), *Olympia* (No. 6), *Raleigh* (No. 8), *Montgomery* (No. 9), *Minneapolis* (No. 13), and *Columbia* (No. 12) were armed with Whitehead torpedoes and launching tubes; only the cruisers *Cincinnati* (No. 7) and *Detroit* (No. 10) received Howell torpedoes and their launching gear.

The *Stiletto* was fitted with bow and stern tubes. Here she launches a torpedo from her stern tube. The 31-ton, 94-foot wood-hull *Stiletto* was built at the Herreshoff Manufacturing Co., Bristol, Rhode Island. She was on the Navy List from 1887 until January 1911. (U.S. Navy)

The Navy expected eventually to arm every battleship, cruiser, and large gunboat with torpedoes, which was by then considered an essential naval weapon. The typical naval engagement was expected to take place when the opposing ships were less than 1,000 yards apart. Under such conditions, the torpedo, which then had a nominal range of 800 yards and carried about 100 pounds of explosive, was considered to be a formidable weapon that could easily "hole" the underwater hull of an enemy warship.

The first U.S. battleships were outfitted with torpedoes in 1896. The *Maine*, which had been commissioned the year before, received four 18-inch, trainable torpedo tubes on her berthing deck, two on each side of the ship, to take the 3.55m x 45cm Whitehead torpedo. A similar installation was installed on the *Texas*, but with just two torpedo tubes, one on each side. Whitehead torpedoes were also installed on the coast battleships *Indiana* (No. 1), *Massachusetts* (No. 2), and *Oregon* (No. 3). The first had six 18-inch torpedo tubes mounted on the berthing deck: two on either beam, one in the bow, and one in the stern. The two later ships were similarly outfitted, except that one broadside tube on each ship was omitted to provide more berthing space for the crew.

By the end of 1896 the Bureau of Ordnance was projecting that it would need 186 additional torpedoes to arm all of the new ships under construction or under contract. To fulfill this obligation it placed another order with the Bliss Company for 100 "long" Whitehead torpedoes, which were capable of carrying 220 pounds of explosive. The new torpedo had the same 18-inch diameter as the first Whitehead torpedoes manufactured by the Bliss Company, but was almost five feet longer. When it entered service in 1898 it was designated as the 5m x 45cm Whitehead Mark I torpedo.[4]

Although Bureau of Ordnance officials believed that it was prudent to have two manufacturers of torpedoes, a board appointed to investigate and report on the relative merits of the Howell and Whitehead torpedoes in 1897 concluded that Navy should discontinue manufacturing Howell torpedoes.[5] Of the 50 Howell Mark I torpedoes that had finally been delivered, 8 had been lost, 2 were on the *Stiletto* for instructional purpose, 30 were in reserve, and 10 were on board the new battleship *Iowa* (battleship No. 4), the only ship with Howell torpedoes in 1897.

Another event took place that year that would greatly enhance the future value of the torpedo as an effective naval weapon: the inclusion of a gyroscope for directional stability. In 1897 the Torpedo Station installed and tested a gyroscopically controlled steering mechanism invented by Ludwig Obry of Austria, which had been patented in the United States a year earlier.[6] The device, which quickly became known as the Obry gear, was so effective that it was immediately added to the remaining order for 40 5m x 4cm Whitehead torpedoes that had yet to be delivered. With the Obry gear

The first Whitehead torpedo was ordered by the U.S. Navy in 1891 as the Whitehead 3.55m x 45cm torpedo. The Mark I—shown here—entered U.S. service in 1895, followed by the Mark II in 1896, and the Mark III in 1898. Whitehead torpedoes were used by many other navies. (U.S. Navy)

A Mark III Whitehead torpedo being test fired from East Dock at Goat Island—the Newport Torpedo Station—in 1894. A torpedo boat is seen beyond the torpedo. (U.S. Navy)

installed, the torpedo's accuracy was increased by *300 percent* as evidenced by the revised specification issued by the Navy. The original specification for the 5m x 45cm torpedo, issued in 1896, called for a maximum horizontal deviation (left or right) of 24 yards at a range of 800 yards. With the Obry gear installed, the deviation was reduced to just 8 yards.[7]

The scientific principle behind the Obry gyroscope was fairly straightforward. Adapting it to the complexities of a self-propelled torpedo and getting it to work right was quite another matter. The Obry device was based on the conservation of angular momentum. Once a spinning mass (usually in the shape of a disk or wheel) is set in motion, it tends to resist any change in the plane of its rotation. In a torpedo the gyroscope is installed so that the spinning disk rotates in a vertical plane parallel to the centerline of the torpedo. At the time of launch—and this is the tricky part—the gyroscope is freed from its normally locked position and a rapid impulse of en-

The Brotherhood reciprocating engine of a Whitehead torpedo. The engine, which was powered by compressed air stored in the air flask, was connected to the torpedo's propeller by a drive shaft. Torpedoes, beginning with the earliest models, have been among the most complex weapons employed by navies. (U.S. Navy)

Charging the air flask of a Whitehead torpedo. (National Archives)

One of three 18-inch, above-water tubes that were fitted in the battleship *Oregon* (BB 3) as completed in 1896. U.S. battleships were fitted with torpedo tubes until the 1930s. There are two Whitehead torpedoes stowed against the bulkhead in this circa 1900 photograph. (Library of Congress)

ergy is imparted to the disk in order to get it to a high speed of rotation in less than a second. When the torpedo is discharged from the launching tube, the vertical plane of the disk is in alignment with the initial line of fire. If the torpedo begins to deviate from this line, the gyroscope, which remains fixed in position, will begin to exert a small force on an arm that connects the gyroscope to the steering engine slide-rod driving. Any movement away from the spinning plane of the gyroscope causes the slide-rod to move backwards or forwards depending upon the direction in which the torpedo deviates from the line of fire (see Figure 3). The other end of the slide-rod is connected to a servomotor that amplifies the correcting force transmitted from the gyro

The gyroscope used in the steering mechanism of most U.S. torpedoes was patented in the United States by Ludwig Obry in 1896. The directional stability provided by the addition of a gyroscope increased the torpedo's accuracy by 300 percent.

to the rudders causing them to move in a direction that will bring the torpedo back in line.

The most complicated part of the Obry gyroscope was the unlocking mechanism, which freed the gyroscope and released the impulse spring to start the wheel spinning. The Obry gyroscope, and its future derivatives, were intricate precision devices that had to be carefully adjusted and maintained to work properly. Although the gyroscope had greatly increased the accuracy of the torpedo, it had one great disadvantage, according to Lieutenant Lloyd H. Chandler, who discussed the practical limitations on its use in the March 1900 issue of the U.S. Naval Institute *Proceedings*. Chandler wrote,

> If not in perfect adjustment, this gear is worse than nothing, as it renders a miss certain. No gyroscope can be adjusted or even examined to see whether it is in adjustment or not, except upon the most stable platforms. It is doubtful if it could be done on board even the largest ship.[8]

This problem was addressed by the Torpedo Station in 1899, when it commenced development of a "more satisfying steering device" that would be more reliable and require less adjustments. Several experiments were conducted in 1901 on a number of mechanisms intended to improve the gyroscope's performance, including two devices invented by Lieutenant Commander Washington I. Chambers that would enable the gyroscope to be set for curved fire—an important feature that would greatly enhance the tactical value of the broadside mounted underwater torpedo tubes intended for the Navy's newest battleships.

In 1901 the Bliss Company began a concerted effort to improve the torpedo's propulsion system when Leavitt installed an experimental air-heating device within the air flask of a 5m Whitehead Mark II torpedo. The experimental torpedo attained a speed of 35

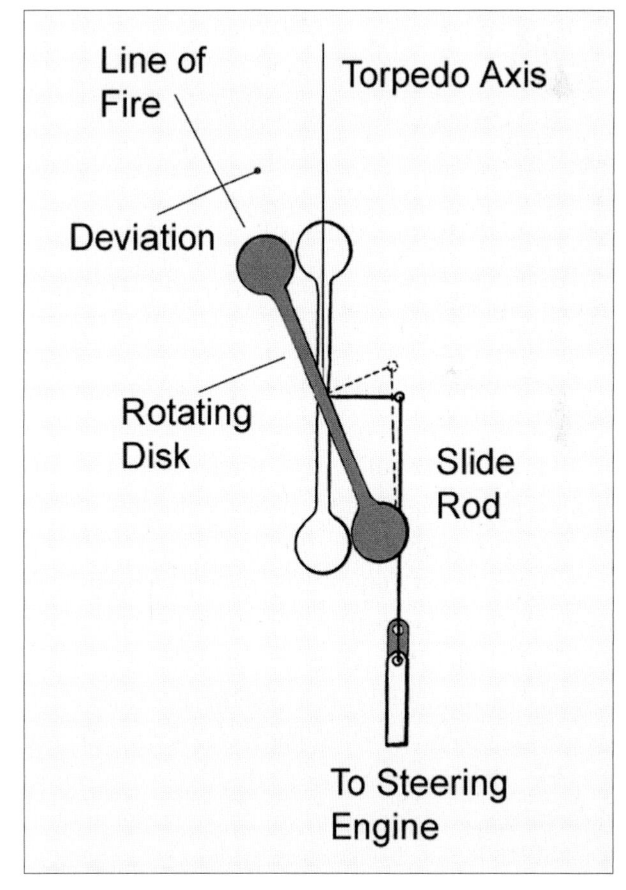

A schematic drawing of gyroscopic operation. If the torpedo begins to deviate from the line of fire, the gyroscope, which remains fixed in position, will begin to exert a small force on an arm that connects the gyroscope to the steering engine slide-rod driving. Any movement away from the spinning plane of the gyroscope causes the slide-rod to move backward or forward, depending upon the direction in which the torpedo deviates from the line of fire.

knots over a 1,200-yard range and 24.5 knots for 3,000 yards. This was a significant improvement over the 28-knot speed and 1,200 yard range that could normally be expected from the 5m Mark II Whitehead torpedo.[9]

Because the high temperature of the heated air was considered to be highly detrimental to the life of the Brotherhood engine, which was constructed totally of bronze, Leavitt replaced the reciprocating engine with a single-stage turbine rotating about the torpedo's longitudinal axis connected to the propeller shaft.[10] The turbine was tested in an experimental torpedo at the Torpedo Station in 1903 using unheated compressed air as the power source.[11] The successful demonstration of the turbine led to a contract for two prototype torpedoes 21 inches in diameter designated as the Bliss-Leavitt 5m x 21-inch Mark I torpedo.[12] The Mark I was the first torpedo produced in the United States that used the principle of "superheating" to impart additional energy to the air charge. The addition of superheating substantially increased the torpedo's performance without adding any significant weight to the torpedo or its air charge. The term is a misnomer, for no superheating of air occurs, just heating, but the term is used here in accord with the terminology used in the ordnance pamphlets of the era.

Orders for several versions to the superheated Bliss-Leavitt torpedo were placed before a major defect was identified: the unbalanced torque created by the single-wheel turbine created a tendency for the torpedo to roll.[13] The solution was the development of a two-stage turbine that consisted of two balanced wheels turning in opposite directions. The development of the two-stage balanced turbine is credited to Lieutenant Gregory Davison. The Bureau or Ordnance authorized its use on 2 February 1906, but it was not tested until October 1906, when it was installed in an experimental torpedo at the Torpedo Station. In the meantime, the Bliss Company had come up with its own balanced turbine design, which it began adding to its latest torpedoes in 1907.

Another innovation introduced by Leavitt after 1907 was the outside superheater. Leavitt's design was based on a modified version of the so-called "Elswick" heater that had been introduced in 1904. The Bliss Company purchased a license for this technology from the Armstrong Company in 1905.[14]

The continued installation of torpedoes on the Navy's large warships was not greeted with approval from all quarters. The damage suffered by Spanish cruisers during the Spanish-American War of 1898 had demonstrated the extreme vulnerability of torpedoes contained in above-water launch tubes. The hazard was considered so great that in 1898 the Navy Department ordered the removal of all such installations from unarmored cruisers and gunboats. The predominance of naval gunfire during the war—no torpedoes were fired during either of the two major battles except to sink disabled hulks—caused U.S. naval leaders to question the wisdom of placing torpedoes on its major combatants.

Rear Admiral Charles O'Neil was one of the most outspoken opponents of installing torpedoes on major combatants. O'Neil was Chief of the Bureau of Ordnance in 1903, when he included the following statement in the Bureau's section of that year's annual report of the Secretary of the Navy:

> It is the opinion of this Bureau that the most effective use of torpedoes is most likely to be obtained when they are used on vessels designed especially to use them, such as torpedo boats and submarine boats. Their value on board large vessels is, to say the least, problematical. The gun, with its great power and accuracy, is and must be the main factor in all decisive naval combats. The torpedo may occasionally get in its deadly work and may have an important effect upon the battle tactics of the fleet, but it is not likely to decide an action.[15]

In O'Neil's view, the torpedo was "essentially a weapon of opportunity, and probably of remote opportunity." The controversy continued until the Navy's leadership reached a consensus in January 1904 to install torpedo tubes below the waterline or behind armor in all battleships and armored cruisers then under con-

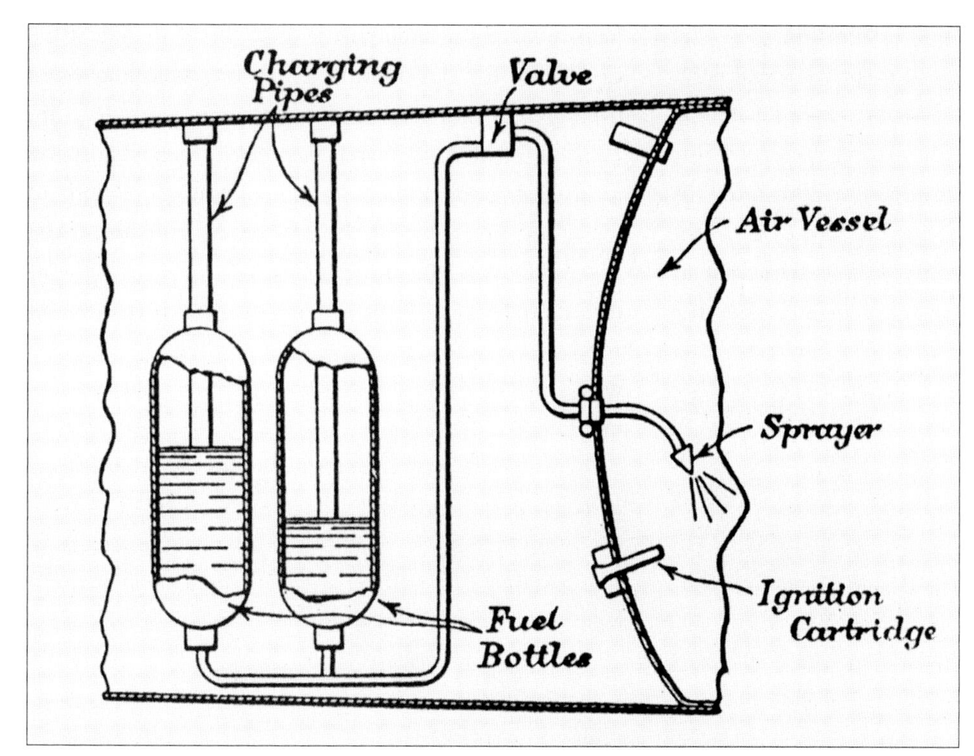

The Elswick heater, patented by the William Armstrong Company of England in 1904, was licensed to the E. W. Bliss Company of Brooklyn, New York, for use in U.S. torpedoes in 1905. Frank McDowell Leavitt's design for the Mark III torpedo was based on this device. Bliss-Leavitt torpedoes were used by the U.S. Navy through World War I.

struction or authorized. The weapon chosen for this purpose was the new, 21-inch-diameter Bliss-Leavitt Mark II torpedo, which was showing much promise. Unfortunately, Bliss encountered the same technical difficulties that Hotchkiss experienced 14 years earlier when that firm undertook the development of the Howell torpedo. Bliss's problems were exacerbated by the introduction of two unproven technologies—air heating and the turbine drive—at a time when the company had simultaneously committed to the production of four different torpedoes (the 18-inch Mark III and Mark IV ordered respectively in 1904 and 1905, and the 21-inch Mark I and Mark II, which was also ordered in 1905).

Once again the failure of the only domestic manufacturer to deliver reliable torpedoes as scheduled forced the Bureau of Ordnance to turn to the Whitehead Company. Several factors entered into this decision. First and foremost was the inability of Bliss to deliver enough of the torpedoes to arm the 26 torpedo boats that the Bureau wanted to refit with new, superheated torpedoes. Bliss had enough capacity to manufacture about 250 torpedoes per year, which was not enough to supply the 13 battleships, 4 armored cruisers, 16 destroyers, 5 submarines, and 27 torpedo boats that required them.[16] Second was a reliability problem with the Bliss-Leavitt 5m x 45cm Mark III torpedo. Although the Bliss Company had made several modifications in an effort to improve its reliability, it still lacked the reliability that the Navy desired. Finally, there was the matter of the 5m x 45cm Mark IV, and 5m x 21-inch Mark I torpedoes. Although models of both torpedoes had been accepted by the Navy for experimentation, they had numerous defects that the Navy wanted corrected before it would accept them for service. As an expedient, the Navy asked for and received approval to purchase 130 torpedoes from the Whitehead Torpedo Works Company of Weymouth, England. A contract for those weapons, designated as the Whitehead 5.2m x 45cm Mark V torpedo, was issued on 7 July 1908. Twenty more were ordered to be manufactured under license by the Navy's Torpedo Station at Newport, which had received authorization from Congress to establish a torpedo factory as part of the Naval Appropriation Bill passed on 2 March 1907. They were the only U.S. torpedoes in which kerosene was used as fuel.

THE "WET" HEATER OR "STEAM" TORPEDO

The next major breakthrough in torpedo design came from the original Whitehead works, still located in Fiume, which had been building torpedoes for more than 40 years. The high temperatures produced by the Elswick hot-air heaters played havoc with the Brotherhood engine. To solve this problem the engineers at Fiume injected water into the combustion generator. This reduced the temperature of the preheated air while producing copious amounts of steam, which provided even greater motive power to the engine. The development of the so-called "wet" heater, or "steam" torpedo as it would come to be known, would have a profound effect on naval warfare, changing the torpedoes role from a weapon of coastal defense to an offensive threat that would greatly influence future naval tactics.

A torpedo being launched from the torpedo boat *Morris* (TB 14) in July 1906. The 105-ton, 139 ½-foot craft was commissioned in 1898, one of a series of 23-knot torpedo boats. The craft had three 18-inch torpedo tubes. Note the conning tower forward; there was a second one aft. (U.S. Navy)

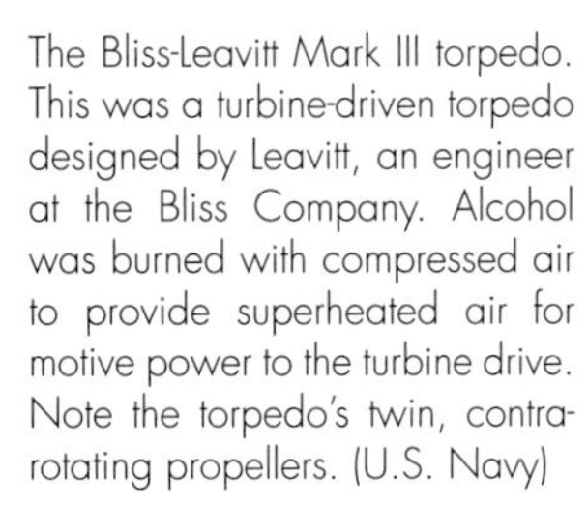

The Bliss-Leavitt Mark III torpedo. This was a turbine-driven torpedo designed by Leavitt, an engineer at the Bliss Company. Alcohol was burned with compressed air to provide superheated air for motive power to the turbine drive. Note the torpedo's twin, contra-rotating propellers. (U.S. Navy)

A 21-inch Mark III torpedo being hoisted aboard the battleship *Kansas* (BB 21) during the around-the-world cruise by U.S. battleships—the "Great White Fleet" —in 1907–1908. The *Kansas*, completed in 1907, was fitted with four 21-inch, below-waterline torpedo tubes. (U.S. Navy)

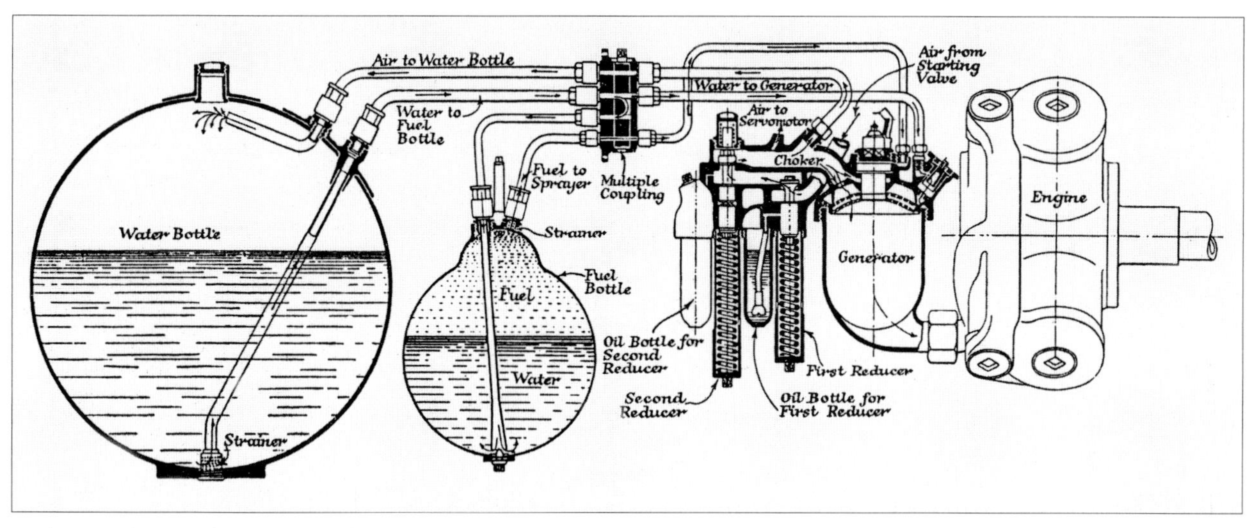

A schematic drawing of a Fiume "wet" heating system, circa 1909.

The first naval power to produce a torpedo with a "wet" heater was the Royal Navy. In 1908, the 18-inch diameter Mark VII torpedo manufactured at the Woolwich Arsenal had a wet heater that required less piping than the Fiume heater (see illustration on page 36). Design work on a steam torpedo for the U.S. Navy did not begin until 1910, when the Bureau of Ordnance began the design of a long-range 21-inch diameter torpedo to have a speed not less than 30 knots and an ultimate range, if possible, of 10,000 yards.[17] By that time the Bureau was well aware that the 6.2m x 21-inch torpedoes in foreign service had greater range than the existing 21-inch torpedoes in the U.S. Navy's inventory.

Thus, in the U.S. Navy the torpedo was then seen as a weapon to be used on essentially all types of warships, from battleships, to torpedo boats and submarines. The long-range 21-inch torpedo, which had just begun development, was viewed as a weapon for use in the line-of-battle where an engagement range was expected of between 8,000 to 10,000 yards, and a minimum speed not less than 30 knots. It was to be fired from submerged tubes on surface ships and was capable of being used as a defensive and offensive weapon. It was thought at the time that the weight and length of such a long-range torpedo precluded its use on all but the largest vessels in the Navy, namely battleships and cruisers.[19]

For smaller ships, a second class of torpedoes was more suitable, which, according to the Bureau of Ordnance, should be of high speed and short range, not

TABLE 3-1

TORPEDOES IN U.S. NAVY SERVICE (CIRCA 1912)[18]

TORPEDO	LAUNCH PLATFORM	SPEED KNOTS	EFFECTIVE RANGE/YDS	INVENTORY
3.55m x 45cm Whitehead Mk III	A-class submarines	28	800	50
5m x 45cm Whitehead Mk I	torpedo boats	27.5	800	30
5m x 45cm Whitehead Mk II	battleship *Missouri*	28	1,500	5
5m x 45cm Bliss-Leavitt Mk III*	—	28	3,500	45
5m x 45cm Bliss-Leavitt Mk IV	D-class submarines	28	3,500	33
5.2m x 45cm Whitehead Mk V	*Paterson*-class destroyer	29	4,000	170
5.2m x 45cm Bliss-Leavitt Mk VI	G-class submarines	36	2,000	100
5m x 21-inch Bliss-Leavitt Mk II	battleships	28	3,500	316

** to be converted into B-L Mk IVs*

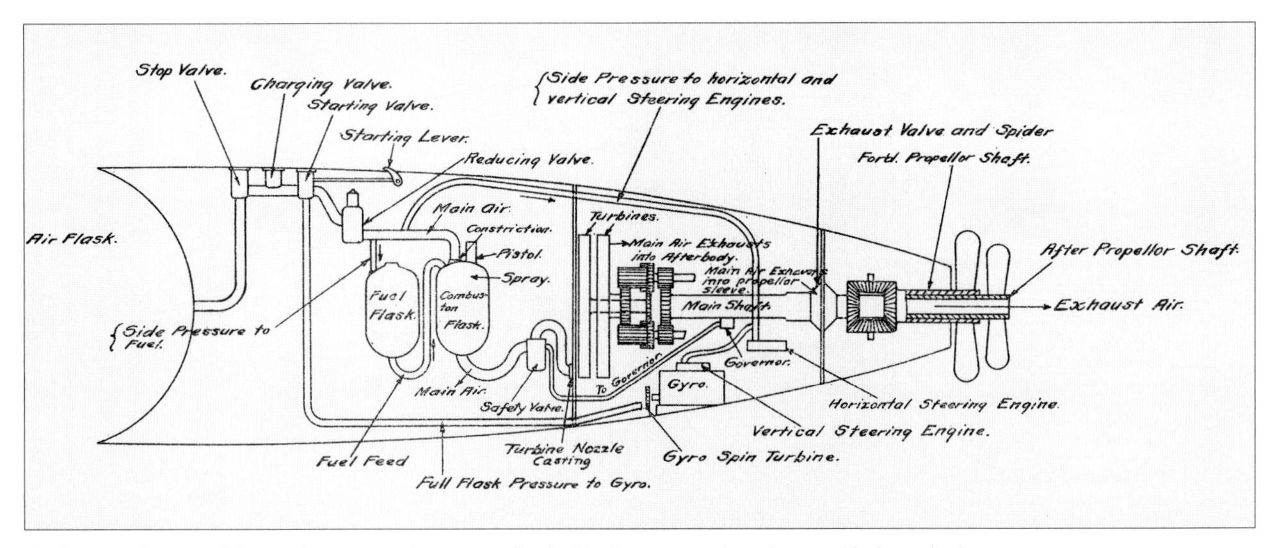

A schematic drawing of the wet heater propulsion system for the Bliss-Leavitt torpedo with vertical balanced turbine.

TABLE 3-2

U.S. TORPEDO DESIGNATIONS (C. 1913)[21]

FORMER DESIGNATION	NEW DESIGNATION
Whitehead 3.55m x 45cm Mk III	Type A
Whitehead 5m x 45cm Mk I	Type B
Whitehead 5m x 45cm Mk II	Type C
Bliss-Leavitt 5m x 21-inch Mk I Mod 2	Mark I Mod 2
Bliss-Leavitt 5m x 21-inch Mk II	Mark II
Bliss-Leavitt 5m x 21-inch Mk III	Mark III
Bliss-Leavitt 5m x 45cm Mark IV	Mark IV
Whitehead 5.2m x 45cm Mark V	Mark V
Bliss-Leavitt 5.2m x 45cm Mark VI	Mark VI
Bliss-Leavitt 5.2m x 45cm Mark VII	Mark VII
Bliss-Leavitt 21-foot x 21-inch Mark IV	Mark VIII
Bliss-Leavitt 5m x 21-inch Mark III Mod. 1	Mark IX

less than 4,000 yards extreme range with a speed of not less than 35 knots. It would be used by destroyers, torpedo boats, and submarines. The torpedo would be no longer than 17 feet and not greater in diameter than 18 inches.

By 1911 the Torpedo Station had issued contracts for two new types of long-range torpedoes: The first torpedo to emerge from this process was the 5.2m x 45cm Bliss-Leavitt Mark 7.[20] It was the first torpedo accepted by the U.S. Navy that sprayed water and alcohol into the combustion pot of the superheater. The resulting mixture of steam and combustion products was much more efficient than heated air and dramatically improved the speed and/or range of the torpedo. Torpedoes of this type came to be known as "steam" torpedoes. The configuration of the Mark 7, i.e., the installation of a "wet" superheater in conjunction with a horizontal balance turbine, would form the basic design format for every U.S. torpedo developed prior to World War II. The Mark 7 was the first to use TNT (trinitrotoluene), which had twice the explosive force of the guncotton (nitrocellulose) previously used in torpedo warheads.

Thus by 1913 the U.S. Navy inventory of torpedoes included both "hot" and "cold" running Whitehead and Bliss-Leavitt design torpedoes, with some identified by the same mark number. Consequently, new designations were formulated, as shown in table 3-2.

All other torpedoes in the inventory, i.e. Howell, Whitehead Mark 1, Whitehead Mark 2 (3.55m versions), and the Schwartzkopff torpedoes of foreign manufacture that were purchased or captured during the Spanish-American War were condemned against further service use.[22]

CHAPTER FOUR

Submarines

The Ultimate Torpedo Boat (1900–1918)

The U.S. Navy's interest in the torpedo was initially for the main armament for the torpedo boat, which was regarded as an inexpensive method of providing coastal defense. These lightly built craft relied upon their high speed to provide the tactical advantage and protection that would enable them to close with the enemy warship to within the limited range of the early Whitehead torpedoes.

The designers of these state-of-the-art boats had to depend on extreme weight reduction in hull, coal supply, and secondary armament to achieve this high speed. Their limited endurance and light construction made these craft unsuited for use at sea with the battle fleet. They were soon replaced by "destroyers," which were larger, had better sea-keeping features, and had a greater steaming range. Between 1890 and 1902 the U.S. Navy commissioned 34 torpedo boats. They were built to a variety of designs by several shipyards with lengths varying from 100 feet to 228 feet. The first two boats, the *Cushing* (TB 1) and *Ericcson* (TB 2), both commissioned 18 February 1897, were initially armed with a single, bow-mounted 18-inch torpedo tube to launch the 3.55m x 45cm Whitehead torpedo.[1] All of the boats that followed were armed with three 18-inch torpedo tubes, one on either beam facing forward and one mounted at the stern facing aft. The early boats—those ordered before 1898—were armed with 3.55m Whitehead torpedoes; the 15 boats constructed after that date, according to most records, were armed with the 5m x 45cm Whitehead Mark I.

In December 1899 a new form of torpedo boat appeared at the Washington Navy Yard for inspection, a submarine designed by Irish schoolteacher-immigrant, John P. Holland. The *Holland IV*, as the craft was called, arrived for official trials held on 14 March 1900 that included a demonstration before Admiral of the Navy George Dewey, the bureau chiefs, and Assistant Secretary of the Navy Charles H. Allen. The boat was just short of 54 feet long, was powered by a gasoline engine for running on the surface and by a battery-fed electric motor while submerged. It was armed with a single 18-inch torpedo tube in the bow and two dynamite guns, one facing forward and the other aft (the latter being removed when the boat was modified for naval service).

Admiral Dewey, who had gained fame during the Spanish-American War for his victory over the Spanish fleet at Manila Bay, now headed the newly established General Board of the Navy. A principle function of the board was to advise the Secretary of the Navy on characteristics that should be incorporated in future warships. Because of Dewey's reputation and his unique rank, he exerted a great deal of influence over the board. (Dewey, who was a commodore when he engaged the Spanish in Manila Bay, was promoted to admiral the following year. In March 1903 Congress

The torpedo boat *Ericsson* (TB 2) mounted two 18-inch torpedo tubes—one to port and one starboard in this photo taken in late 1900. Completed three years earlier, the *Ericsson* displaced 120 tons and was 149½ feet long, with a rated speed of 24 knots. At one point she carried three torpedo tubes. (U.S. Navy)

commissioned Dewey in the specially created rank of Admiral of the Navy, unrivaled in the Navy's history, to date from March 1899.)

Dewey was so impressed with Holland's submarine that he would later tell a congressional committee that if the Spanish had possessed such submarines he could never have taken Manila. The submarine, he noted, was "infinitely superior to mines or torpedoes or anything of the kind."

With Dewey's approval assured on 11 April 1900, Secretary of the Navy John D. Long authorized the purchase of Holland's submarine for $150,000. The boat, which became the USS *Holland*, was the Navy's first submarine. A few weeks before her formal commissioning, on 12 October 1900, the *Holland* (SS-1), under the command of Lieutenant Harry H. Caldwell, demonstrated the devastating effect that could be achieved by a submarine under the proper circumstances. The occasion was an exercise in blockade conducted by ships of the U.S. Fleet that had surrounded the entrance to Narragansett Bay. The *Holland*, which was assigned to the defending forces, sortied at sundown, trimmed so that her conning tower was just awash. The submarine was not equipped with a periscope and Caldwell kept the submarine awash under cover of darkness. He maneuvered toward the battleship *Kearsarge* (BB 5) without being detected until he was close enough to hail the flagship of the blockading force. "Hello, *Kearsarge*," he yelled, "you are blown to atoms. This is the *Holland*!"[2] Caldwell later claimed that he could have torpedoed three blockading ships that night, one for each of the three 3.55m x 45cm Whitehead torpedoes that the *Holland* could carry.

By then contracts for six more submarines of an improved *Holland* type had been awarded. A seventh boat was added to the contract when the Holland Torpedo Boat Company agreed to refund the $93,000 previously paid for an unfinished submarine ordered five years earlier. The six new boats would be called the *Adder*-class after the lead ship, the USS *Adder* (SS 3), which was commissioned on 12 January 1903.

The *Adder*s were the first submarines designed by the Electric Boat Company, which had acquired Holland's financially troubled torpedo boat firm in 1899. The

The *Holland* (SS 1), the U.S. Navy's first submarine, was armed with a single 18-inch torpedo tube in the bow and two dynamite guns, one facing forward and the other aft. Note the protruding "nose," consisting of a set of clamshell doors to protect the torpedo tube. (U.S. Navy)

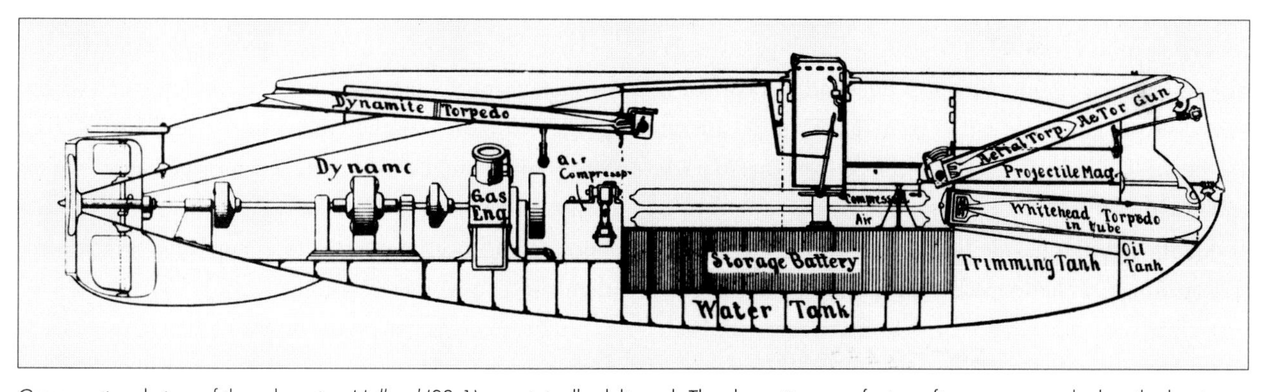

Cross sectional view of the submarine *Holland* (SS 1) as originally delivered. The dynamite gun—facing aft—was removed when the boat was modified for naval service. The Navy also built the cruiser *Vesuvius* (completed in 1890), which was armed with three 15-inch dynamite guns employing compressed air to fire their shells.

design of the new boats was a radical advancement over the *Holland*, and Electric Boat decided to build a full-scale prototype at the firm's own expense. This proved to be a wise decision, for the construction of the prototype, launched in 1901 as the *Fulton*, revealed a number of serious flaws in its design that required major changes to correct. Instead of carrying the five long, 5m x 45cm Whitehead torpedoes originally planned for the new boats, weight constraints limited the type and number of torpedoes that could be accommodated. As a result, the *Adder*-class submarines, redesignated as the A class, carried three short, 3.55m x 45cm Whitehead torpedoes

The U.S. Navy's second submarine, the *Plunger* (SS 2), cruises at full speed across Oyster Bay, Long Island, in 1905. She was later renamed *A-1* when the Navy assigned letter-number names to all submarines on 17 November 1911. (Library of Congress)

that were fired from a single 18-inch bow torpedo tube. One torpedo was loaded into the firing tube, with the two others resting on wooden loading skids that could be slid across the deck so that they would align with the empty tube for reloading. The single torpedo tube was closed by an outer cap raised by a bell-crank. The tube had to be flooded before the torpedo could be fired.

The six *Adder*-class boats were followed by the three submarines of the *Viper* (SS 10) class and five of the *Octopus* (SS 9) class, all commissioned in 1907–1908.[3] Like their predecessors, these submarines were armed with the 3.55m x 45cm Whitehead torpedo with two 18-inch launching tubes mounted side by side in the bow.[4] Each boat carried four torpedoes, one in each tube, with a pair of reloads mounted on wooden skids inside the submarine, as in the *Adder* class. The external openings of the torpedo tubes in both classes were closed by a single rotating cap. To fire a torpedo, both tubes had to be flooded. The caps would be unseated and rotated, making it impossible to fire only a single torpedo and leave the other one dry. This prevented the rapid reloading of a tube in the event that only one torpedo was fired.

Because the underwater endurance of these boats was extremely limited, they were suitable only for use in coastal and harbor defense. Their seven-man crews lived ashore, or aboard whatever ship was available to be designated as a tender. The commanding officers of the submarines were ensigns, a couple of years out of the Naval Academy, whose submarine training consisted of a few practice dives with their predecessors.

In wartime, according to the doctrine of the time, the boats would remain with the tender, inside of a harbor, until they were alerted that enemy ships were approaching. The boats would then proceed to the harbor entrance, where they would anchor in the "awash" condition, radio aerials hoisted up, and keeping a lookout for the enemy. As soon as smoke appeared on the horizon they were to up their anchors, lay down their masts, and submerge. Once the enemy was within range—in clear weather these subs could easily see a large ship from a distance of seven or eight miles with a moderate amount of periscope exposed—the submarines would head toward the enemy and attack with their torpedoes.[5]

The three boats of the D class, which began entering service at the end of 1909, were the first of six classes (D, E, F, G, H, K) of submarines built between 1909 and 1914 for coastal defense. Their greater endurance, higher speeds, and better habitability conditions permitted them to operate farther away from harbors. For the pre-1914 Navy, coastal defense meant attacking battleships. All but two of these boats (*Seal* and *Thrasher*) had four 18-inch bow torpedo tubes, because it was assumed that four hits would disable or sink any capital ship.[6]

The bow compartment of the submarine *Adder* (SS 3) at Manila Bay in 1909, showing the interior door to her single torpedo tube and stowed torpedoes. U.S. submarines were based in Manila Bay until early 1942, when the Philippines were seized by the Japanese. (U.S. Navy)

The submarines *Shark* (SS 8), at left, and *Porpoise* (SS 7) on cradles at the New York Navy Yard, circa 1915. Note the single torpedo port and its covering door on each submarine, with a sailor in the *Shark*'s torpedo tube.

The *Adder* loading a torpedo at the Cavite Navy Yard in the Philippines, circa 1912. She has been renamed *A-2*, as evidenced by her new name on the hull below the torpedo. (U.S. Navy)

The D class, starting with the *Narwhal* (SS 17), was armed with the new Bliss-Leavitt Mark IV torpedo, which had a range of 2,000 yards and a speed of 29 knots. The Mark IV was the first U.S. torpedo to be designed specifically for submarine use. The four torpedoes carried by these boats were stowed in the four forward tubes. As in the B and C classes, they shared a single bow cap that had to be rotated before each pair of tubes could launch their torpedo. The extended range of these 5-meter-long torpedoes provided the impetus to develop a torpedo director for use with a periscope that could solve the problem of how much lead was needed (the torpedo deflection problem) to hit a moving target.

The D boats were the first to be internally subdivided with bulkheads placed foreword and aft of the control space amidships, as well as between the torpedo room and officers quarters and, aft, between the crew quarters and engine room. Because officers in the control room could no longer see or speak into the torpedo room, a signal buzzer from the commander's periscope was installed in the torpedo room to signal when to fire torpedoes.

To make room for a walk-around periscope installed in the control room of the *Salmon* (SS 19), the last of the D boats, two of her bulkheads had to be eliminated. Although this compromised the boat's survivability in case of a collision, the added space in the control room simplified the periscope arrangement and enabled the addition of a torpedo director specifically designed for use with the periscope. The director, engineered by the Electric Boat Company, was a modified form of the standard three-bar analog calculator that served as the standard torpedo director in surface ships.[7] The sight bar was aligned with the view through the periscope so that the operator had only to adjust the target bar and enter the torpedo's speed. Once the data was entered, the submarine commander had to swing his boat until the cross hairs of the periscope were on the target at which point he would depress the switch connected to the firing buzzer signaling the torpedo room to fire.

The E, F, G, and H boats carried 5.2m x 45cm Bliss-Leavitt Mark VI torpedoes. This "18-inch" torpedo was eight inches longer than the 5-meter Bliss-Leavitt Mark IV, requiring slightly longer torpedo tubes. These torpedoes were six knots faster and had 60 pounds more explosives than the Mark IV, with a warhead of 200 pounds. Although the gyros in Bliss-Leavitt torpedoes could be set for angled fire, the torpedo tubes on these boats had no provision for setting gyro angles. Instead, the typical firing spread would be set with two torpedoes running straight, one torpedo set to run 5 degrees to the right, and one torpedo set to run 5 degrees to the left. Each of the boats (except for *G-1* and *G-4*) carried eight torpedoes, four in the tubes and four reloads on racks in the torpedo room.

The last of the so-called coast defense boats—the K class—were armed with the Bliss-Leavitt Mark 7 torpedo. The Mark 7 was the first "steam" torpedo to enter service in the U.S. Navy. It had the same dimensions as the Mark 6, with a length of 17 feet and a diameter of 17.7 inches. Its warhead contained more than 300 pounds of explosives and was capable of maintaining at least 33 knots for a distance of 3,000 yards.[8] The Mark 7 also saw service with the first so-called fleet boats (L, M, N, and O classes) that were commissioned between 1916 and 1918. The number of torpedoes carried and the arrangement of the torpedo tubes in these boats was identical to that used in the K boats—four bow tubes with eight torpedoes. The Mark 7 was the last 17.7-inch diameter torpedo developed by the U.S. Navy. It underwent many modifications and remained in service with the older submarines throughout World War II.

A NEW DIMENSION IN NAVAL WARFARE

The German submarine offensive of World War I opened on 6 August 1914 when the commander of German U-boats ordered a flotilla of ten small submarines to attack the British Grand Fleet at its base in Scapa Flow in the Orkney Islands off the northern coast of Scotland. The U-boats, all gasoline powered, were units of the first class of submarines built in Germany between 1906 and 1912 and were composed of *U-5*,

A Mark 7 torpedo outside of the torpedo factory on Goat Island in August 1913. The Mark 7 was the first "steam" torpedo to enter service in the U.S. Navy. Its warhead contained over 300 pounds of explosives and was capable of maintaining at least 33 knots for a distance of 3,000 yards. It was the last 18-inch diameter torpedo developed by the U.S. Navy and remained in service with the older submarines throughout World War II. (U.S. Navy)

U-7 to *U-9*, and *U 13* to *U 18*. These were small submarines, ranging in size from 489 tons surface displacement to—the last two units—564 tons. All had a tow bow and two stern tubes for 17.7-inch torpedoes, with most carrying a 2-inch deck gun.

The displacements of these first German U-boats were:

U-5, 7, 8	505 tons
U-9	493 tons
U-13 to *15*	516 tons
U-16	489 tons
U-17, 18	564 tons

The campaign started rather unpromisingly when the *U-5* and *U-9* had to turn back because of engine troubles. Two days later the *U-15* launched a torpedo at the battleship *Monarch*, without success. This was the first time that a self-propelled torpedo had been launched against an enemy ship from a submarine. It was also the first submarine attack upon a ship under way. On the following day the *U-15* was rammed and sunk by the cruiser *Birmingham* as the submarine was trying to close in for the attack. The *U-13* disappeared on 12 August, probably sunk by a mine. The remaining boats arrived back at Heligoland on the same day, ending the first war cruise of the German U-boat force. Although two boats had been lost without damage to the enemy, the campaign caused great uneasiness to the commander-in-chief of the British Grand Fleet, for it revealed the vulnerability of the Grand Fleet's anchorage at Scapa Flow.

During the next few weeks German U-boats, concentrating on purely naval targets, chalked up a number of remarkable successes against British warships. On 1 September 1914, the entire British battle fleet was chased out to sea by an alarm over a submarine that had supposedly been sighted within the anchorage at Scapa Flow, causing the fleet to move to Loch Ewe, Scotland, which was considered a more secure anchorage. Five days later, the *U-21* drew the first blood of the undersea campaign when she torpedoed the British light cruiser *Pathfinder* off the Firth of Forth on a calm, sunlit day. The cruiser sank within minutes, with heavy loss of life: 259 from her crew of 296. This striking success by a small German U-boat was surpassed on 22 September when the *U-9* sank three armored cruisers—HMS *Aboukir*, *Hogue*, and *Cressy*—off the Hook of Holland in about an hour, killing 1,460 British seamen.

On 7 October, a German submarine discovered in Loch Ewe forced the British fleet to make another hasty exit to sea. After various undignified excursions between Loch Ewe and Scapa Flow with extended stays at sea, the Grand Fleet finally took up temporary anchorage at Loch Swilly, Northern Ireland, until the submarine defenses at the two fleet bases could be completed. There the fleet suffered the loss of one of

A Mark 7 torpedo being loaded in the submarine *O-15* (SS 76). She is moored outboard of her sister ship *O-2* (SS 63). Note the laundry drying on the hand lines and the T-shaped hydrophone antenna mounted forward of the torpedo loading hatch. (U.S. Navy)

its newest dreadnoughts, HMS *Audacious*, which sank on 27 October after striking a mine laid by a German submarine. The battleship went down with no loss of life, the first battleship to be lost in the war.

The German U-boat campaign against Allied warships continued, but on 20 October a seemingly insignificant event occurred that was to profoundly affect the whole course of the war. On that date the *Glitra*, a small steamer of under 1,000 tons, was stopped, searched, and scuttled by the *U-17*, initiating the start of what would later evolve into Germany's campaign of unrestricted submarine warfare.

FLEET SUBMARINES

In the fall of 1915, while the first of the fleet boats was under construction, the General Board of the Navy, with Admiral Dewey still its president, met to discuss the torpedoes that would be needed for the next submarine building program.[9] The tactical employment of the fleet submarine, in the board's opinion, was quite different from that of the coastal submarine. The latter was chiefly intended to lie in wait in an assigned area and usually act independently. For success it relied upon stealth and surprise, and was well served by a 35-knot torpedo with a range of at least 2,000 yards. A fleet submarine was more like a destroyer. It had to be ready to deliver an aggressive attack in a fleet action.

Like all submarines of the period, fleet boats were slow under water and had a limited radius of action. They depended upon their faster surface speed to close rapidly on the enemy to attain a favorable firing position for attack. The General Board expected that in a fleet action these submarines would fire at long- or medium-range (4,000 to 10,000 yards) before submerging to fire a second round at shorter range (2,000 to 4,000 yards) before making their escape.

Even then, the General Board felt that it would be difficult for a submarine in the submerged condition to get within short range of an enemy. The Board believed that the slow approach would be less advantageous than an earlier action of the same craft with a long-range torpedo. A Board report of 16 November 1915 stated, "A submarine which submerges for attack and finds herself thrown out by the course of the battle can scarcely hope to make a second approach. Time and submerged radius will both be lacking."[10]

To facilitate the tactical employment of fleet submarines with the main body of the fleet, the Board recommended that future submarines be fitted with tubes for long-range torpedoes similar to the 21-foot long, 21-inch diameter Mark 8 torpedoes being developed for battleships and destroyers. The torpedo that emerged from the Board's recommendation was the 21-inch diameter Mark 10. It weighed 2,215 pounds, was 16 feet,

The *L-1* (SS 40) was one of the flotilla of U.S. submarines sent to bases in Ireland and the Azores to hunt (unsuccessfully) for German U-boats. Shown off Bantry Bay, Ireland, in 1918, most of the *L-1*'s crew is on deck; there is a canvas cover on the upper portion of her conning tower. The prefix letter "A" for American was added to U.S. submarines in European waters, in this instance to distinguish her from the British submarine named *L-1*. (U.S. Navy)

4 inches long, and was initially configured to run 5,000 yards at 30 knots.[11] It was rushed into service for the R boats and entered service with the *R-18* (SS 95), which was commissioned on 27 July 1918.

By then, the knowledge gained from the actual battle experience of World War I had dramatically changed the tactical view of submarine warfare. Gone by then was the idea of a long-range surface attack by submarines. Rather, they would remain submerged, relying upon stealth to gain a favorable firing position. A high-speed torpedo of moderate range was the ideal weapon for such tactics. Experiments at the Naval Torpedo Station at Newport had shown that the new Mark 10 was capable of 36 knots.[12] This led to the development of the Mark 10 Mod 1, a 36-knot, 3,500-yard torpedo. This torpedo, with minor modification, became the standard torpedo issued to all subsequent classes of U.S. submarines until the introduction of the Mark 14 in the late 1930s.

U.S. SUBMARINES AT WAR

At the outbreak of the war the United States was lagging behind the great powers of Europe in submarine development. During early years of World War I the U.S. Navy built the *G-3* (SS 31)and the K, L, and O classes of submarines, but failed to keep pace with European technology improvements. When the United States declared war on Germany on 6 April 1917, the U.S. Navy wanted to get its submarines involved in the conflict. The 392-ton K-boats were not meant for blue-water operations and were obsolete, but in October 1917 the Navy decided to tow four submarines across the Atlantic later to conduct patrol cruises off the Azores: *K-1* (SS 32), *K-2* (SS 33), *K-5* (SS 36), and *K-6* (SS 37).

Two months later, the L boats of Submarine Division 5—plus *E-1* (SS 24)—left Newport under tow by the *Bushnell* (AS 2) and two ocean-going tugs bound for Ponta Delgada in the Azores. This group ran headlong into a hurricane and was forced to divert toward Bermuda. Although the flotilla was badly scattered, with one tug and a submarine actually returning to Boston, the other tug and four submarines eventually reached their destination. Then, after several more straggled in, *Bushnell*, a tug, and four submarines completed the remaining 1,000 miles to Bantry Bay on 27 January 1918, with three more boats to follow. They were promptly redesignated the "AL" class to avoid confusion with British L-class submarines and under the tutelage of the Royal Navy began preparing for their role in the antisubmarine warfare effort off southern Ireland.

Eight O-class boats were also on their way across the Atlantic when the war ended.

TABLE 4-1

U.S. SUBMARINES BUILT DURING WORLD WAR I

	NAME COMMISSIONED	BUILT	SURFACE DISPLACEMENT
G-3	1915	1	400 tons
K-1 to *8*	1914	8	392 tons
L-1 to *11*	1916–1917	11	450 tons
N-1 to *7*	1917–1918	7	340 tons
O-1 to *16*	1918	16	521 tons

CHAPTER FIVE

Battleships Made Vulnerable

Torpedo Boats and Destroyers (1893–1919)

The proliferation of torpedo boats in world navies in the late 1880s and early 1890s led to the development of a new type of seagoing warship to protect the battleships that had become the mainstay of the world's battle fleets. These "torpedo boat catchers"—as they were originally called—were introduced by the Royal Navy to chase down and destroy torpedo boats before they could come within launching range of the main battle fleet. The first of these "catchers" was HMS *Rattlesnake*, completed in 1886.[1] Although the *Rattlesnake*'s ability to catch torpedo boats proved limited—her margin of speed was no more than two knots—the British Admiralty ordered 13 more "catchers" of an improved design. By the time the last of these craft was launched in 1892, the ships were outclassed by the newest torpedo boats, whose 24-knot speed was four knots faster than the "catchers."

To meet the continuing threat of the torpedo boats the Admiralty placed an order for two ships of a new class having a minimum speed of 27 knots with a displacement of 250 tons. These ships were to be armed with one 12-pound gun, three 6-pounders, and *three torpedo tubes*. The first of these ships, HMS *Havoc*, was launched in 1893. The generic name of this class was changed to "torpedo boat destroyers." These "destroyers," with three Whitehead torpedo tubes, in theory at least, after successfully destroying the enemy's torpedo boats, would carry out torpedo attacks against enemy battleships.

Although the U.S. Navy was well informed on the development of torpedo boat destroyers in the Royal Navy and in other European navies, it saw no need to acquire its own destroyers until it began planning for war with Spain in 1898. The profusion of torpedo boats in Europe did not pose a menace to the U.S. Navy, as these small craft did not have sufficient range to cross the Atlantic while the U.S. Fleet was charged primarily with coastal defense. But attached to the Spanish squadron, then known to be assembling in the Cape Verde Islands, were three new torpedo boat destroyers—the *Furor*, *Pluton*, and *Terro*—which were armed with torpedoes. The Spanish torpedo boats were considered a major danger to U.S. ships according to a special war plans board convened in March 1898 to assess the Spanish threat. The report, issued by the board on 16 March 1898, emphasized the urgent need to obtain torpedo boat destroyers as soon as possible.

Congress responded on 4 May 1898, authorizing the Navy to build 16 torpedo boat destroyers. Construction did not begin until 15 August 1899, when the keel for the first destroyer, the *Bainbridge* (DD 1), was laid down. In the interim, the Navy requisitioned a number of pleasure yachts—including J. Pierpont Morgan's *Corsair*—that were quickly converted into naval auxiliaries. The *Corsair* was fitted with four 6-pounders and renamed the *Gloucester*. She was commissioned on 16 May 1898, although her potential role in determining Spanish torpedo attacks was highly questionable.

The USS *Bainbridge* (DD 1) was the first torpedo boat "destroyer" acquired by the U.S. Navy. The ship is shown here in 1915. The 420-ton ship had an armament of two 18-inch torpedo tubes plus two 3-inch guns and lighter weapons. (U.S. Navy)

Sixteen days earlier, on 30 April, the Spanish squadron, under the command of Admiral Pascal Cervera, sailed from the Cape Verde Islands en route to the Caribbean. The squadron, which was made up of the most modern ships in the Spanish Navy, included the destroyers *Pluton* and *Furor*, two of the most feared torpedo-armed warships in the world at that time. Cervera's squadron, which remained undetected for almost a month, caused great consternation with the American public, with fears of a Spanish attack on U.S. coastal cities. It was discovered in Santiago Harbor, Cuba, on the morning of 29 May 1898, by elements of the Commodore Winfield S. Schley's Flying Squadron. The squadron was soon reinforced by the arrival of Commodore William T. Sampson's North American Blockading Squadron.

For more than a month the two fleets faced off, with a few inconclusive skirmishes the only result. For his part, Admiral Cervera was content to wait, hoping for bad weather to scatter the Americans ships so that he could make a run to a position more favorable for engaging. By the end of June, Cervera, realizing that his forces could no longer remain safely in the harbor, decided to break out into the open seas. He planned the breakout for Sunday morning, when the Americans would be at religious services.

When Cervera's force sortied from Santiago Harbor on 3 July 1898, the *Furor* and *Pluton* were the last ships out of the harbor. Before the Spanish destroyers were within effective torpedo range they were engaged at short range by the converted yacht *Gloucester*, commanded by Lieutenant Commander Richard Wainwright (the former first officer of the battleship *Maine*). The Spanish destroyers then came under heavy fire from the secondary batteries of the U.S. battleships *Indiana* (BB 1) and *Iowa* (BB 4). Both Spanish ships were soon riddled by gunfire, the *Furor* plunging to the bottom and the *Pluton* being run up on rocks and later blowing up.

While the *Furor* and *Pluton* were being destroyed, the *Oregon* (BB 3) and the *Brooklyn* (ACR 3) commenced the 3½ hour chase that would result in the destruction of the remainder of the Spanish squadron. At the end of the day the U.S. Navy had won a tremendous victory, totally defeating the Spanish Navy and ending the threat of further naval action against the United States. *No torpedoes were launched by either side.*

The Navy's success in the Battle of Santiago eliminated any urgency in constructing the 16 destroyers authorized by Congress. Without a prototype to rely on, the Bureau of Construction and Repair took a considerable amount of time to come up with its own design. The first to join the fleet, the *Bainbridge* (DD 1), was not commissioned until 24 November 1902; the last three were not commissioned until a year later. The 16 destroyers bristled with a main battery of rapid-fire guns to combat the torpedo boat threat; all but two of the ships mounted two 3-inch guns and five or six 6-pounders. As secondary armament each destroyer carried two 18-inch torpedo tubes for the long Type 5m x 45cm Whitehead torpedo. One tube was placed amidships between the two pairs of funnels, the other on the fantail.

Torpedomen pose at the 18-inch stern torpedo tube of the destroyer *Perry* (DD 11), circa 1905. The torpedo appears to be a Whitehead type. Destroyers 1 through 16 were 420-ton, 250-foot, coal-burning ships completed in 1901–1902. These ships were authorized during the Spanish-American War in response to the perceived threat of torpedo boats. (U.S. Navy)

No additional destroyers were authorized until 1906. By then improvements in torpedo technology, as well as the success of the Japanese torpedoes against Russian ships at the Battle of Tsushima in May 1904, had begun to affect Navy thinking about the destroyer's function. In addition to protecting the battleships from attack by hostile torpedo craft, whether at sea or harbor, the torpedo boat destroyers had to be capable of attacking offensively with torpedoes. In its report of 7 January 1905, the Converse Board, which had been convened by President Theodore Roosevelt, under the chairmanship of Rear Admiral George A. Converse, "to consider the types and qualities of torpedo vessels," specified that a destroyer's armament should consist of 3-inch guns and 21-inch torpedo tubes. The board suggested that future destroyers be armed with at least four 3-inch guns and four torpedoes carried in twin central-pivot torpedo tubes.

In June 1906 Congress authorized the construction of another three destroyers, which would become the *Smith* class (DD 17–21). Two more were added in March of the following year. The U.S. Navy was now undertaking a sustained program to produce significant numbers of destroyers. The designs drawn up by the Bureau of Construction and Repair were greatly enlarged versions of the *Bainbridge*. The new ships, displacing 700 tons, were almost 300 feet in length, and were to have a trial speed of at least 28 knots. Their main battery consisted of five 3-inch guns and three single 18-inch torpedo tubes, i.e., one less than recommended by the Converse Board.

By the time the last two ships were authorized and the design completed in March of 1907, the situation with regard to the supply of torpedoes had reached crisis proportions. The Bliss Company was already heavily committed to the production of 300 21-inch torpedoes that were urgently needed to outfit the torpedo batteries of 15 battleships and 7 cruisers. But the company's manufacturing capacity was only 250 torpedoes per year. In addition to 21-inch torpedoes, it was also constructing smaller numbers of Bliss-Leavitt 5m x 45cm Mark III and IV torpedoes for submarines and destroyer use. It was thus prudent for the Bureau of Construction and Repair, which had cognizance over the construction of the Navy's ships, to stay with the 18-inch torpedo for destroyers.

An unusual design feature of the torpedo battery for the new destroyers was the addition of enclosed stowage for one reload located on deck near each of the three 18-inch torpedo tubes. This relatively cumbersome system was replaced in 1916 with the installation of twin 18-inch tubes, similar to those used on the *Paulding* class (DD 22–31) that followed. These ships carried no reloads.

The destroyer *Smith* (DD 17), shown in 1910, was the first destroyer to carry the new Bliss-Leavitt Mark III torpedo. Its 18-inch diameter was designed to fit into existing torpedo tubes. Destroyers grew rapidly in size; the DD 17–19 were 700-ton, 294-foot ships carrying three 18-inch tubes and five 3-inch guns. (U.S. Navy)

The single 18-inch torpedo tube was the standard torpedo mounting in early U.S. destroyers. Shown here is the starboard mount of the USS *Flusser* (DD 20). The 21-inch torpedo tube was introduced (in twin mounts) with the destroyer *O'Brien* (DD 51), commissioned in 1915. She was a 1,050-ton, 305¼-foot warship; eight torpedo tubes and four 4-inch/50-caliber guns were fitted in the *O'Brien*. (U.S. Navy)

The design data for the ships of the *Smith* class made no mention of torpedo. The most likely candidate to be installed on these destroyers was the 5m x 45cm Bliss-Leavitt Mark III torpedo. The first of these torpedoes was delivered in 1907, well in advance of the anticipated completion date of the destroyers ordered that year. Although 45 torpedoes of this type were ultimately accepted by the Navy, questions about their reliability kept them out of service. They were replaced by the 5.2m x 45cm Mark V Whitehead torpedo, which became the standard torpedo for all 45 of the destroyers laid down by the U.S. Navy between 1906 and 1912 (DD 17–50).

The maximum range of the Mark V Whitehead torpedo was 4,000 yards at a time when the battle fleet expected to engage in gunnery duels at 10,000 yards or farther. Thus, the torpedo, because of its limited range, was regarded solely as a secondary weapon for use at close range.

This tactical situation changed with the advent of the Mark 8 steam torpedo in 1912, which could travel 10,000 yards at 27 knots. Arming the fleet's destroyers with the Mark 8 would place the torpedo on a par with battleship gunnery. Instead of being a secondary weapon, the torpedo could now be considered a primary weapon for fleet use. Rear Admiral Nathan C. Twining, Chief of the Bureau of Ordnance, wrote to the General Board on 7 November 1912.

Sailors on the *Walke* (DD 34) hoist an 18-inch Bliss-Leavitt torpedo aboard ship in preparation for torpedo practice, circa 1914. Practice torpedoes had water-filled warheads that approximated the weight of high-explosive warheads. Training with armed torpedoes was rare because of their relatively high cost; unarmed practice torpedoes could be retrieved for reuse or arming. (U.S. Navy)

Sailors pose with the port-side twin 18-inch torpedo tubes aboard a destroyer. The torpedoes are partially withdrawn from the tubes for maintenance. The twin-tube configuration was the standard U.S. destroyer torpedo mount until 1916, when the triple tube was introduced with the destroyer *Sampson* (DD 63). (U.S. Navy)

> The adoption of a long-range torpedo places this weapon in an entirely different position from that which it has hitherto held as a battleship weapon. It must now be reckoned with as an important factor in a day action between battleships and its possibilities as a night weapon have been immensely increased since it will be possible to discharge it successfully at ranges beyond the reach of searchlights and anti-torpedo battery guns.[2]

The Board had already recommended the development of the 21-foot long, 21-inch diameter Mark 8 Bliss-Leavitt torpedo. According to the characteristics specified by the Board, it was to carry an explosive charge of 316 pounds of trinitrotoluene (TNT), sufficient to inflict major damage upon a capital ship, have a minimum range of 4,000 yards at 36 knots, and a range of 10,000 yards with the best speed obtainable. Two changes were needed to convert the 36-knot, short-range version of the torpedo into the slower, longer-range torpedo that would be a potent weapon against enemy battleships. First, it was necessary to replace the high-speed propellers with ones designed for lower speed; and, second, it was necessary to reduce the engine speed by lowering

A torpedoman holds a gyro for the 18-inch Bliss-Leavitt torpedo. (U.S. Navy)

the amount of alcohol sprayed into the combustion pot. The latter was accomplished by modifying or replacing the combustion pot and piping.

Earlier studies of the 10,000-yard torpedo conducted at the Naval War College showed that the primary mission of the destroyer armed with such weapons would be changed from that of torpedo boat catcher to that of attacking the enemy battle line with torpedoes. Because all of the torpedoes delivered in a single attack were those in the tubes ready for firing, it was essential that every destroyer carry the maximum number of tubes.

The Commander-in-Chief Atlantic Fleet, Rear Admiral Hugo Osterhaus, agreed that the torpedo armament should be as heavy as possible and that the torpedo was the principle weapon of the destroyer. However, since destroyers might be called upon to repel or attack an enemy's torpedo-armed destroyers, it was necessary that their armament be consistent with that mission too.[3] The solution was found in a compromise: the next class of destroyers, the "Thousand Tonners"

An 18-inch Bliss-Leavitt torpedo is launched by the destroyer *Walke* (DD 34) during firing practice. Note the chief petty officer sighting through the telescope of the Mark 4 director mounted atop the torpedo tubes. The *Walke*, displacing 787 tons and 294 feet long, was fitted with six torpedo tubes and carried five 3-inch guns. (U.S. Navy)

(DD 43–67) would have one less gun mount (albeit size of the guns were increased to 4-inch) than their predecessors in favor of an additional twin torpedo tube, thereby providing a main armament of four 4-inch guns and four twin-mount torpedo tubes. The first of the "Thousand Tonners" were completed with 18-inch tubes that were designed to fire a 5.2m x 45cm torpedo. It was not until the *McDougal* (DD 54) was commissioned on 16 June 1914, that their torpedo battery was upgraded to fire the Mark 8 torpedo.

A further increase in torpedo armament occurred the following year when the triple-tube mount was introduced in the U.S. Navy. Beginning with the *Sampson* (DD 63), four of these mounts were provided, giving destroyers a battery of 12 21-inch tubes, which could be fired six per side. This 12-tube torpedo battery became a hallmark of the flush-deck, four-stack destroyers that would be mass-produced for the U.S. Navy during World War I. Almost 300 destroyers were built to this configuration.

WORLD AT WAR

Although no U.S. destroyers nor any other U.S. surface ship fired a torpedo in anger during World War I, the torpedo was recognized as an important weapon that could change the course of naval battle. The Battle of Jutland between the British Grand Fleet and the German High Seas Fleet in 1916 held lessons for the future of the torpedo in the eyes of most admirals and naval analysts of the time. The battle, which took place in the North Sea off of Denmark's Jutland Peninsula, began in the afternoon of 31 May 1916 and continued through the early hours of the following day. It was the most important naval battle of World War I and the largest in history in terms of the number of battleships (44) and battle cruisers (14) that participated in what was essentially a gunnery duel between the capital ships of the opposing fleets.

The battle pitted a much superior British force against the inferior German fleet. Both sides attempted to ambush and annihilate a portion of the other's fleet. The battle quickly turned into a running gunnery duel conducted at ranges varying from 10,000 to 18,000 yards. Although the capital ships of both fleets were accompanied by large numbers of screening destroyers, the torpedo did not come into play until the out-gunned High Seas Fleet had executed its Gefechtskehrwendung (Battle Turn Away) maneuver for the second time at approximately 7:21 PM. In Gefechtskehrwendung, all ships turned at once instead of in succession, as was normally the case. This special maneuver was perfected by the German Navy to permit it to rapidly disengage when confronted by a superior force. It was implemented at Jutland in a desperate effort to avoid the devastating fire of the 33 British battleships and battle cruisers that were pounding the German battle line.

Only a handful of German destroyers were in position to cover the retreating battleships, but the 31 torpedoes they unleashed were more than enough to dissuade the Commander-in-Chief of the Grand Fleet, Admiral Sir John Jellico, from pursuing the fleeing German ships. This turn away from the German fleet because of the torpedo threat became one of the most controversial aspects of the battle.

The standard tactical response to a massed torpedo attack was to turn away, so that the torpedoes could be more easily avoided by outdistancing them. "The [torpedo] tracks were quite clear to us aloft and could be picked out when nearly a mile away," reported one British officer stationed in the foretop of the battleship *Marlborough* during the torpedo attack.[4] Eight British battleships narrowly escaped being hit, some of them only by putting the helm hard over and swerving out of line. One torpedo passed under the *Marlborough* without exploding. According to one historian, "It was the alertness of the lookouts and the skill of the helmsmen that saved the British fleet from serious losses."[5]

When the United States joined the war against Germany on 6 April 1917 the E. W. Bliss Company was the primary supplier of the Navy's torpedoes. Although the Navy had taken steps in 1914 to increase the output of torpedoes at the Naval Torpedo Station in Newport, and had ordered 100 torpedoes from the Washington Navy Yard, the lion's share of the contracts for torpe-

The USS *Stevens* (DD 86) shows the arrangement of 12 torpedo tubes in four triple mounts that were long the standard for U.S. destroyers. The first destroyer with this torpedo battery was the *Sampson*, completed in 1916. The standard destroyer gun armament of four 4-inch/50-caliber guns is evident: one forward of the bridge (beneath the awning), two amidships (outboard of the funnels), and one aft, plus a few light guns. (National Archives)

The *Aylwin* (DD 47) and similar ships had 4-inch guns in place of the 3-inch weapons previously used, and an additional twin torpedo mount to provide eight 18-inch torpedo tubes. Two mounts can be seen on the starboard side in this photograph taken during World War I. These ships are not to be confused with the famed four-stack, flush deck destroyers produced in large numbers in wartime building programs. (U.S. Navy)

does continued to be awarded to the Bliss Company. By the time war was declared the company had outstanding contracts to supply 1,016 torpedoes to the Navy. Work on these contracts had almost come to a standstill in the months just prior to the declaration of war while the company concentrated on fulfilling its contracts to supply artillery shells to the British government.

Although the Bliss factory in Brooklyn, New York, had material on hand in various stages for the manufacture of 500 torpedoes, only 20 torpedoes were nearing completion. At the Navy's request, Bliss proceeded to close out its foreign shell contracts and then devote as much effort as possible to torpedo manufacture.

In addition to the torpedo contracts already pending, additional contracts for more than 2,000 torpedoes were signed soon after the United States declared war on Germany. To supply torpedoes in the quantities under contract, the Bliss Company would have to under-

Four-stack, flush deck destroyers, such as the *Hale* (DD 133), shown in 1936, were built during and immediately after World War I with 271 being completed. The *Hale* was one of 50 of the type transferred to Great Britain in 1940 in exchange for Western Hemisphere base rights. These were the last U.S. destroyers to mount triple torpedo tubes. (U.S. Navy)

A very unusual photo of two torpedoes being launched simultaneously by a U.S. destroyer. This unidentified ship is likely the *Blakeley* (DD 150), a four-stack, flush deck destroyer photographed firing torpedoes during fleet maneuvers in the 1920s. The *Blakeley*, commissioned in 1919, was one of 117 destroyers of this type in service in 1941, some modified to special roles. (U.S. Navy)

take a major expansion of its manufacturing facilities. The Navy entered into an agreement with the company for the construction of a new building entirely devoted to the torpedo production. The cost to expand the factory—approximately $2 million—was advanced by the U.S. government at an interest rate of 4 percent. The loan and the interest accrued would be repaid by proportional deductions from the final payments due on the torpedoes when they were delivered. Once the building was completed and tooled for production, the Navy expected production at the Bliss factory to reach 300 torpedoes per month. This goal was never accomplished. One hundred and fifty torpedoes was the largest number the company was able to produce in one month, which occurred in December 1918, after the war was over.

By the summer of 1918, the Navy was facing an extreme shortage of torpedoes due to the Bliss Company's failure to manufacture torpedoes in the quantities expected. In addition to the torpedoes ordered in 1917, the Navy had placed an order for 2,308 torpedoes in February 1918. Still, deliveries of torpedoes were averaging only 20 per week. Only 401 of the 5,901 torpedoes under contract had been delivered by 20 July 1918. The situation had become extremely acute because of the need to provide 2,600 long-range torpedoes needed to outfit the 150 destroyers that were then under construction.

To remedy the situation, the Bureau of Ordnance, under the direction of Rear Admiral Ralph Earle, recommended the erection of a Torpedo Assembly Plant in Alexandria, Virginia. This proposal was quickly approved by Secretary of the Navy Josephus Daniels. Steps were immediately taken to acquire the property and to prepare plans for the plant. The total cost to construct the facility, which included a heating plant, storehouse, dock, machinery and equipment, torpedo testing barge, and the land, was budgeted at $2,760,000. Work on the site began on 27 August 1918. Ironically the Navy began construction of the building on 11 November 1918—Armistice Day. The factory operated for five years before it was mothballed in 1923 as an economy measure.

Today the "Torpedo Factory" exists—as a center for artists, eateries, and a tourist center. An old torpedo mounted on a dolly adorns the lobby to recall the original purpose of the building at the foot of King Street in downtown Alexandria.

CHAPTER SIX

Between the Wars

(1919–1941)

The battleship *Maryland* (BB 46) on 23 September 1921, three months after being commissioned, loading torpedoes from the lighter *YF 49* while anchored off Newport, Rhode Island. The dreadnought, fitted with two underwater 21-inch torpedo tubes, took aboard 12 Mark 8 Mod 3 torpedoes that morning, the ship's full allowance. Note the mass of sailors on deck, probably waiting to board the lighter for liberty ashore. (National Archives)

CONVENTIONAL TORPEDOES

The massive reductions in naval expenditures that followed in the wake of the Washington Naval Conference of 1922 and the resulting treaty caused a precipitous reduction in the U.S. Navy's torpedo programs after World War I. All dealings with the Bliss Company were terminated upon completion of the Mark 9 project. Although disputes of patent rights and the Naval Torpedo Station's 15 years of experience in building torpedoes were cited as factors influencing the termination of work, economic savings was the primary factor. This was also the case for closure of the Naval Experimental Station at New London. Thus, on 1 July 1923, the Naval Torpedo Station at

Newport became the only Navy activity involved in the design and production of torpedoes. Budget restrictions and cutbacks in shipbuilding programs kept torpedo development and production in low gear for more than a decade. In the midst of the massive economic depression that followed, the National Industrial Recovery Act of 16 June 1933 rejuvenated naval construction. This act, along with the Vinson-Trammel Bill passed on 15 March 1934, allocated money for the construction of 126 Navy ships, including 77 destroyers and five light cruisers to be constructed over a period of eight years. The need to fill the torpedo tubes for these new ships and the eight *Farragut*-class (DD 348–355) destroyers already under construction required an expansion of both personnel and facilities at the Torpedo Station. During the next five years the government invested approximately $1.25 million in a new plant and equipment—enough to increase the station's manufacturing capacity to about 650 torpedoes per year. In spite of the expansion of manufacturing capacity at Newport, demand continued to run ahead of supply. By January 1938, the order backlog exceeded 300 torpedoes. Even the most conservative estimates of production—those that did not anticipate the outbreak of war—predicted a shortfall of some 2,425 torpedoes by 1 July 1942, unless a new facility to manufacture torpedoes was established.

A preliminary study of possible new production sources narrowed to three choices: an expansion of the facilities at Newport; reopening the factory in Alexandria, Virginia; or the construction of a new facility at San Diego. It was inadvisable from a military point of view to concentrate all torpedo production at Newport. Moreover, to increase the Torpedo Station's output by 50 percent would require an investment of $4.5 million and take one and a half years to achieve. San Diego was ruled out. While there were advantages to establishing a plant on the West Coast, the Navy could not wait the three years that were estimated to start production in a new location. The most expeditious and most economical solution was thus to reopen Alexandria.

No sooner was the plan announced than the full weight of New England opposition was felt. Political and labor leaders demanded an expansion of Newport. Letters—whose common inspiration was reflected in repetition of misstatements and fallacious reasoning—poured into Congress. There was real doubt that a congressional appropriation could be secured for reopening Alexandria. Fortunately, the Alexandria facility had been a part of Washington's Naval Gun Factory since its closure in 1923, so that funds to rehabilitate the torpedo station were simply included in the Naval Gun Factory's budget for 1939. The Alexandria factory reopened in 1940 and was in full operation by the fall of 1941. By then the combined production of Newport and Alexandria was about 35 torpedoes per day.

In 1922 the Torpedo Station began an extensive program to inventory the Navy's torpedoes left over from World War I. All torpedo types prior to the Mark 7 were declared obsolete and removed from service. The four torpedoes that remained in the Navy's inventory—the Mark 7, 8, 9, and 10—were all single-speed torpedoes that had nearly identical propulsion systems based on the original Bliss-Leavitt steam torpedo. They differed mainly in the capacity of the air flasks, which were greater for the long-range Mark 8 and 9 torpedoes used on battleships and destroyers than for the short-range Mark 7 and 10 used by submarines.

As a result of World War I experiences, the Navy considered the speed of its torpedoes insufficient and the amount of TNT carried—between 300 and 400 pounds—too small to be effective against a modern large warship. The "rundown-type" gyro was not accurate enough to provide the performance required at the long ranges now possible, and the standard depth mechanisms were both unreliable and took too long to get the torpedo to the correct depth setting if the initial dive was too deep.

To address these problems the Bureau of Ordnance in 1920 directed the Naval Gun Factory to begin development of a new 21-inch diameter torpedo designated the Mark 11. Following a precedent established by the Royal Navy during World War I, the bureau specified

The Naval Torpedo Station was established on Goat Island in 1869 and was greatly expanded over the next 100 years. It became the Navy's primary producer of torpedoes in the 1920s. (U.S. Navy)

a three-speed torpedo with a high-speed setting of 45 knots that would suffice for use on all categories of surface ships as well as submarines. A new depth-keeping mechanism—the Uhlan mechanism—had been developed and an air-sustained gyro introduced, and both devices were incorporated in the Mark 11 design, which was 22 feet, 7 inches long, weighed 3,500 pounds, and carried 500 pounds of TNT. It was the first multispeed torpedo introduced in the U.S. Navy. It had an adjustable speed that could be set by an external spindle just prior to firing to one of three settings: high speed at 42 knots, medium speed at 35 knots, and low speed at 27 knots; the corresponding range at these speeds was 7,000 yards, 9,500 yards, and 15,000 yards, respectively.

To achieve this performance within the dimensions specified in the Mark 11's design characteristics required a 100 percent increase in the power output of the torpedo's engine. This, combined with the necessary gearshift incident for a multispeed torpedo, necessitated the use of a much heavier and more complicated propulsion plant. The heavier plant, additional lubricating oil capacity (48 pounds versus 10 pounds in Mark 9 and 10 torpedoes), and a 500-pound warhead led to a heavy torpedo that had a negative buoyancy both before and after its run.

Consideration was initially given to running the Mark 11 torpedo with a "light" exercise head for exercises—as instituted on the Mark 7 Mod 8—to increase the margin of positive buoyancy at the end of its run to make recovery of peacetime shots more certain. Such an expedient was not adopted in the Mark 11 because it was considered bad policy to school torpedo personnel with torpedoes running under conditions that would not duplicate those of wartime conditions, thus leading to erroneous test results. Just such a problem surfaced in 1929 when it was discovered that the Mark 7 Mod 4 torpedo—the standard for submarines during World War I—would run satisfactorily with an exercise head, but could not be depended upon to perform satisfactorily with a warhead. The Mark 11 developers considered it preferable to accept occasional torpedo losses during peacetime exercises than to suddenly discover in wartime that the weapon could not be expected to

The first U.S. warships completed after World War I where the ten scout cruisers of the *Omaha* (CL 4) class. As launched they carried ten 21-inch torpedo tubes with a triple mount on each side of the after deckhouse and enclosed twin-tube mounts on each side of the main deck, below their aircraft catapults, as seen in this photograph of the *Omaha* taken on 6 December 1923. Once in service, it was found that the twin tubes on the main deck were easily flooded; those were removed and the openings plated over. (National Archives)

run properly with a warhead. Unfortunately, this policy was later reversed when "the previously established dangerous practice of lightening up the exercise head to gain positive buoyancy" was reinstituted.[1]

Toward the end of 1928, the General Board recommended that cruisers, destroyers, and submarines be provided with the Mark 11 torpedo. Rear Admiral Andrew T. Long, then the senior member of the General Board, explained the reasons in the endorsement addressed to the Secretary of the Navy:

> The tactical requirements as regards torpedoes for cruisers and destroyers are practically identical. Day attacks by cruisers and destroyers under average weather conditions will be delivered at ranges greater than the present high speed torpedo range of 6,000 yards, but less than 15,500 yards corresponding to a torpedo speed of 27 knots. The attacking speed of cruisers and destroyers in formation under the conditions mentioned will be approximately 27 knots and it is highly desirable to have the speed of the torpedo the same in order to simplify greatly the methods of attack. At night or with low visibility conditions at other times, cruisers and destroyers will often deliver their torpedo attacks at ranges not exceeding the high speed torpedo range of 6,000 yards. In such attacks cruisers and destroyers should be enabled to utilize the advantages inherent to high torpedo speed.
>
> Submarines will, in all probability, make torpedo attacks when submerged to at least periscope depth. Under such conditions the limited range of visibility through the periscope and the inherently crude methods of tracking preclude the possibility of accurate torpedo fire at ranges in excess of the high speed torpedo range of 6,000 yards. Upon rare occasions groups of enemy vessels may be sufficiently massed to warrant firing at greater ranges, the group target presented being sufficient to offset the inaccuracy of fire inherent to such ranges.[2]

Although production of the Mark 11 torpedo was started in 1927, it was soon superseded by the Mark 12, an improved version that had the same dimensions,

The Mark 11 torpedo was the first torpedo designed by the Naval Torpedo Factory at Newport, Rhode Island. (U.S. Navy)

but with an air flask shortened by three feet and ball bearings in the side gears to reduce the amount of lubricating oil required. Development of the Mark 12 was completed in 1928. Approximately 400 Mark 10 and 100 Mark 11 torpedoes were manufactured by the Naval Torpedo Station before production of the newer model began in the mid-1930s.

Submariners were never happy with either the Mark 10 or the Mark 11 torpedo, which they felt were excessively large for their needs because of the long ranges demanded for surface-launched torpedoes. The submariners wanted a shorter torpedo with a speed of 46 knots over a range to 4,500 yards and an explosive charge of at least 400 pounds of TNT. The torpedo they envisioned would not begin development until January 1931, when the Bureau of Ordnance instructed the Naval Torpedo Station to draft specifications and commence design work on the Mark 14, a two-speed torpedo having an overall length of less than 20 feet, with a 500-pound TNT charge. The maximum range of the torpedo at low speed was to be 8,000 yards; the high-speed range was 1,000 yards at 42 knots.

In February, the Bureau of Ordnance authorized $143,000 to undertake the development of the Mark 14 torpedo. Another $803,000 was allocated for the production of 76 torpedoes, although the actual authorization to manufacture the torpedoes was not received until May 1932. While the Torpedo Station was preoccupied with initial development work for the Mark 13 and Mark 14 torpedoes, the General Board was busy reviewing the characteristics for the new "treaty cruisers" and the eight *Farragut*-class destroyers recently authorized.[3] After conducting hearings on the design of the new cruisers the General Board unanimously recommended the omission of torpedoes on all cruisers

> on the grounds generally of expense; the extra weight . . . when tonnage is limited by treaty; from experience at the War College . . . that the torpedo was a secondary weapon on the cruiser; and in the service that it was a secondary weapon on the cruiser; and, lastly, and of special importance, that one warhead if struck by a shell or fragment would probably detonate and set off a group of warheads.[4]

As for the new destroyers, the Board's preference for a main battery of four dual-purpose 5-inch guns and three triple 21-inch torpedo tubes mounted on the centerline was overruled by the design changes suggested by Secretary of the Navy Charles F. Adams. His views resulted in the addition of a fifth gun and a torpedo armament of two quintuple 21-inch tubes. As was customary at the time, the design characteristics for the new destroyers were circulated throughout the highest level of command within the Fleet.

A sailor aboard the scout cruiser *Trenton* (CL 11) in the 1920s works on a Mark 8 torpedo being loaded into one of the ship's triple torpedo mounts. Light cruisers and destroyers were the only U.S. warships designed to carry torpedoes after this period. The next U.S. cruiser class to carry torpedoes was the *Atlanta* (CLAA 51) class, with the lead ship completed in December 1941. Eight so-called "treaty" heavy cruisers (CA 24–31), completed in 1929–1931, had their torpedo tubes removed before World War II. (U.S. Navy)

On 5 August 1931, Rear Admiral William H. Standley, a future CNO who was then commanding all U.S. destroyers, was highly critical of the decision to arm the new destroyers with the three-speed, 21-inch Mark 12 torpedoes. Standley expressed his misgivings about the torpedo to the Chief of Naval Operations: "The interests of national defense," he wrote, "*demand* that the new destroyers be equipped with torpedoes capable of making a speed of *at least* 35 knots for at least 14,000 yards."[5] The success of a destroyer's torpedo attack depended to a large extent on the speed differential between the torpedo and its target, and Standley considered the 27-knot torpedo (at 15,000 yards range) "to be entirely inadequate as a weapon for daylight attack on a modern battle line."[6] Standley assumed that the required criteria could only be obtained by increasing the diameter of the torpedo.

Standley's memorandum elicited a nine-page response from the Chief of the Bureau of Ordnance, Rear Admiral Edgar B. Larimer, on 19 November 1931. The first three pages contained a summary of foreign torpedo performance supplied by the Office of Naval Intelligence, which Larimer used to show how well the Mark 9 and Mark 12 performed in comparison with their foreign contemporaries. Next, Larimer pointed out the necessity of switching the 96 partially completed Mark 12 torpedoes originally intended for new cruiser construction for use by the first of the *Farragut*-class destroyers. Any major changes in the characteristics of torpedoes intended for destroyer use would have to be deferred until Congress authorized the manufacture of additional torpedoes needed to outfit the destroyers built subsequent to DD 351. Lastly, Larimer explained the technical difficulties in achieving the 35-knot/14,000-yard range characteristics desired by Standley in either the existing dimensions of the Mark 12 or even in a new, larger 24-inch diameter torpedo. As head of the Bureau of Ordnance, Larimer recommended the continued investigation of any or all means available, i.e., increased diameter, increased turbine speed, and increased oxygen in the air flask, for achieving the speed-range characteristics that Standley desired.

A month earlier, in October 1931, the Torpedo Station had conducted test runs on a Mark 7 Mod 4

The eight destroyers of the *Farragut* (DD 348) class were the first post–World War I destroyers built by the U.S. Navy. In a departure from previous practice, which provided four triple torpedo mounts in destroyers, these ships had two quadruple torpedo tube mounts. The *Farragut*, completed in 1934, and subsequent U.S. destroyers were fitted with adjustable torpedo tubes that could launch both Mark 8 and Mark 15 torpedoes. This class also introduced the 5-inch/38-caliber dual-purpose gun. (National Archives)

torpedo that had been modified by the Naval Research Laboratory in Washington, D.C., to run on oxygen instead of compressed air. From the limited success of this experiment it appeared that an oxygen torpedo could "just be around the corner."

The General Board was aware of the progress being made with the oxygen torpedo. The Board's members also know that the evolution of a successful design of a new and larger torpedo and its associated equipment was a lengthy and costly process that would take many years to bring to fruition. Under these conditions, the Board felt that it was unwise to undertake the design of a larger (24-inch) torpedo "before the possibilities resulting from the adaptation of new sources of power to existing torpedoes have been fully explored."[7] Instead of a new 24-inch torpedo, the General Board recommended that the Torpedo Station (1) vigorously pursue a new source of power, especially in the liquid-fuel field, and (2) continue experiments with Mark 11 and Mark 12 torpedoes enlarging the oxygen content of the air flask and increasing the turbine speed.

In the interim, the Naval Torpedo Station began work on a new torpedo for future destroyers designated the Mark 15—the last operational torpedo introduced into the U.S. fleet before hostilities began on 7 December 1941. The Mark 15 had the same dimensions as the Mark 12 with a similar power plant that developed about 40 percent more horsepower at the high-speed setting. The major change involved a new type turbine bearing and the use of an improved air flask developed by Bethlehem Steel, similar to that used in the Mark 14 torpedo. The new air flask weighed 307 pounds less than the air flask used in the Mark 12, restoring some of the lost buoyancy. The reduction in weight also permitted the development of an even larger warhead, the Mark 15 Mod 2, which contained an explosive charge of 989 pounds of TNT.

Both the Mark 14 and Mark 15 torpedoes were developed during a period of severe austerity for the U.S. Navy. The country was still in the midst of the Great Depression, various naval limitations treaties were still in effect, and the population as a whole viewed defense spending unfavorably during an era when the country's political sentiments favored isolationism. Under these conditions it was difficult to find the funds necessary to develop new naval weapons. Nevertheless, the Torpedo Station was able to undertake the design of three new torpedoes in the 1930s: the Mark 13, Mark 14, and Mark 15.

The biggest impediment to the development of these weapons was an aversion to conducting live firings that would lead to the destruction of a $10,000 torpedo (about $150,000 in 2010 dollars) and the reluctance of the Office of the Chief of Naval Operations to provide suitable targets. To overcome these difficulties, the engineers and ordnance officers at the Torpedo Station had to rely on an extensive system of bench tests and experimental firings on the test range. Studies of the torpedoes' performance under these "ideal conditions" were hampered by inadequate on-site testing and a reliance on inaccurate depth measurements. Having little if any wartime experience with these weapons—only 11 U.S. torpedoes were fired in anger during World War I—they failed to anticipate the complicated problems that would arise when these highly complex devices were used for the first time against real targets.[8] This would have dire consequences for the sailors who had to rely on these weapons in wartime.

THE MAGNETIC "INFLUENCE" EXPLODER

The first Whitehead torpedoes came equipped with detonators called a "war nose," designed by Whitehead's engineers. These early detonators were relatively simple devices that contained a firing pin protruding from the torpedo's nose that was held in place by a shear pin. When the target was struck, the pin would shear, allowing the firing pin to strike a percussion cap to ignite a primer of dry guncotton, which in turn set off the main explosive charge. As a safety measure, the exploder had an arming propeller that restrained the firing pin until the torpedo had traveled a minimum distance through the water.

Until the beginning of World War I, U.S. torpedo warheads employed a contact exploder similar to those used on the early Whitehead models first employed by the Navy. Shortly before World War I, the Bureau of Ordnance developed a more complex device that operated on the principle of inertia. Successive changes to improve the device were made during the war. The result was the Mark 3 exploder, which contained another safety device—an anti-countermining mechanism—that was designed to prevent the torpedo from detonating by an explosion close to it, such as might happen if two or more torpedoes struck a target in rapid succession.

Development work on the Mark 14 submarine torpedo (top) and the Mark 13 aerial torpedo (bottom) began in the early 1930s. Both torpedoes were used extensively in World War II, with both having major success after a myriad of problems were solved. (U.S. Navy)

While work on the Mark 3 exploder was nearing completion, the Bureau of Ordnance also began to consider the possibilities of developing a new type of exploder based on a device the Germans had developed during World War I to explode their mines called a "magnetic pistol." The magnetic pistol contained a magnetic compass that was connected to a detonator. When an enemy ship came near its magnetic field, it caused the needle to swing, touching off the detonator. Although the simple compass-type detonator devised by the Germans was totally unsuitable for use in a torpedo, the possibility of using the influence of a magnetic field to initiate detonation promised to greatly increase the effectiveness of the torpedo. Use of such an exploder would eliminate the need for a direct hit. Better still, torpedoes could be exploded directly under a ship's hull, breaking the target ship's keel.

The prevailing opinion at the time was that such an explosion was more desirable than a direct hit against the side of an enemy warship, especially those protected by the new torpedo "blisters" that were being added to the hulls of capital ships as they were modernized. These torpedo blisters, or "bulges" as they were sometimes called, were streamlined structures added to the outside of the ship's hull that acted as a cofferdam to keep the explosive forces of the torpedo away from the vitals of the ship. Such a scheme was tested on the decommissioned *South Carolina* (BB 26) in 1924 using 400 pounds of TNT—the standard explosive charge carried by the Mark 10 warhead—placed at a depth of 15 feet. This test, along with another carried out on the unfinished *Washington* (BB 47), caused naval planners to believe that it would be impossible to severely cripple or disable a major warship equipped with torpedo blisters unless some way could be found of exploding the warhead directly beneath the hull where no such protection existed. The solution was the magnetic influence exploder that was activated by small changes in the earth's magnetic field that occurred when the torpedo's warhead came within close proximity to a ship's steel hull.

In the summer of 1922 the Bureau of Ordnance instituted project G-53 by allocating $25,000 to the Naval Torpedo Station for the development of an "influence" exploder actuated by perturbations in the earth's magnetic field caused by a passing warship. Very little was known about the these perturbations, especially as seen from beneath the keel of a ship, and the device would have to compensate for the large variations in the intensity of the earth's magnetic field at various geographical locations. Changes in the magnetic field were detected by sensitive induction coils, which were amplified electronically by special vacuum tubes that had to be rugged enough to withstand the torpedo-launching rough run to the target without causing a premature explosion. Some method would have to be devised to arm the torpedo after it was well away from the magnetic field of the submarine, otherwise it might blow up its own launcher.

With help from the General Electric Company, which produced the generator and the thyratron tubes used in the electronics, the Torpedo Station was able to complete a prototype in less than two years. By the early part of 1924 the torpedo's developers were ready to conduct a live test with a real warhead. For the target they wanted to use one of the battleships that was being scrapped under the terms of the Washington Naval Treaty. Their request was denied by the Bureau of Ordnance, which believed that the magnetic exploder was not far enough along to "permit sufficient surety of safety to the firing vessel to warrant use of a live warhead."[9] In addition, the bureau believed that the forthcoming tests on the battleship *Washington*'s hull that included the detonation of a mine placed beneath the warship's bottom would demonstrate the same effects as a torpedo. The bureau also refused the developers' request for more extensive surveys of the magnetic fields of surface vessels. The Bureau of Ordnance, in cooperation with the Carnegie Institute, had already conducted magnetic field surveys on two battleships and further surveys were not considered essential.

Throughout the next year Captain Thomas Hart, the officer-in-charge of the Torpedo Station, bombarded the Bureau of Ordnance with repeated requests for a "target whose complete destruction can be tolerated as the only practical means of obtaining the concrete information desired."[10] The bureau offered an obsolete tugboat that was about to be scrapped, but the Torpedo Station considered this too small a target. Finally in December 1925 the Navy Department agreed to provide the submarine *L-8* (SS 48), which was also due for scrapping. Although the *L-8* was far from being the battleship-size target for which the exploder was intended, Hart realized that this was the best that could be achieved at the time.

The test was conducted amid great secrecy at the station's torpedo test range in Narragansett Bay on 8 May 1926. A Mark 10 torpedo with a live warhead containing one of the hand-built magnetic exploders that had been constructed at the Torpedo Station was fired from one of the station's barge-mounted torpedo tubes at the *L-8*. Captain Hart and other senior officials invited for the test watched as the first shot passed below the target without exploding. Apparently the torpedo, for reasons that have never been established, ran too deep. A second torpedo was prepared and launched. This time the results were perfect: the torpedo exploded as it passed underneath the target, sending a huge column of water into the air and sinking the *L-8*.

After several redesigns, 30 production units were ordered from General Electric. These were subjected to testing in the artificially created magnetic fields in the Torpedo Station's lab and in a limited field test with the light cruiser *Raleigh* (CL 7).

The Mark 6 was designed to sense the changes in the horizontal component of the earth's magnetic field caused by a ship as the torpedo approached the target. The variation in the magnetic field inducted a voltage in a sensing coil, which triggered a thyratron. The thyratron discharged a capacitor through a solenoid, which in turn operated a lever that displaced an inertial ring triggering the mechanical exploder. Frederick J. Milford, a physicist who has extensively studied the history of the torpedo, believes that the complex arrangement was to enable the Mark 5 exploder—without the magnetic exploder, but otherwise identical to Mark 6—to be produced and issued to the fleet in peacetime as a security measure.[11]

Although the initial testing had validated the basic principles employed in the Mark 6 exploder, additional testing at other locations was required before the Torpedo Station would recommend full production. The changing nature of the earth's magnetic field as one moves from the poles to the equator greatly complicated the nature of the testing that had to be done. The magnitude in the horizontal perturbation of the earth's magnetic field by a ship depends on the inclination of the earth's field to the horizontal. This inclination varies from zero at the magnetic equator to 90 degrees at the magnetic poles (Newport is located at about 60 degrees). Because of the ferrous metal in its structure, a ship causes both horizontal and vertical perturbations of the earth's magnetic field, which vary with the distance and direction from the ship. The closer the earth's field is to vertical the greater the rate of change of the horizontal perturbation. A device that senses the change of the horizontal component of the perturbed field works best where the earth's magnetic field has a large vertical component. A device that works well at high magnetic latitudes may not work well where the earth's field is nearly horizontal. Thus, the performance of the magnetic exploder depended on the latitude at which it operated.

The Mark 6 developers were aware of this problem. More data was needed on the performance of the exploder in the southern latitudes. After much urging, the Navy Department agreed to make a cruiser and several destroyers available for more comprehensive field tests. What transpired was later described by Clay Blair Jr., in *Silent Victory*, his widely acclaimed study of the U.S. Navy's submarine service in World War II:

> The new 10,000-ton cruiser *Indianapolis* (CA 35) and two destroyers reported to Newport. Amid greatest secrecy, the ships took on personnel, torpedoes, and Mark 6 exploders and sailed to the equator, off the coast of South America. There, over a hundred torpedo shots were fired beneath *Indianapolis* in various locations along the equator or between 10 degrees north and south lati-

A test torpedo shot passes under the submarine *L-8* (SS 48), employed as a target ship during tests of the magnetic exploder in 1926. The *L-8* was stricken the previous year and retained for torpedo testing. (U.S. Navy)

tude. The dummy warheads were ingeniously fitted with a new photoelectric device known as the electric eye. As the torpedo passed beneath the cruiser, the eye, mounted in a window in the warhead, caused the shadow of the ship to be recorded on film. The tripping of the magnetic exploder was indicated by burning a wisp of guncotton. During the tests, the technicians made over 7,000 readings on magnetic fields.[12]

After the *Indianapolis* field tests were completed, the Torpedo Station proclaimed the Mark 6 exploder ready for production. The station had wanted to perform additional live firing, but the Chief of Naval Operations would not permit the use of a live warhead. He did agree to provide the decommissioned destroyer *Ericsson* (DD 56) for additional testing, but insisted that the Torpedo Station be prepared to raise and repair her if she were sunk. The Torpedo Station declined the offer and no live fire test of the production version of the Mark 6 was ever conducted.

Production of the Mark 6 exploder was undertaken at the Torpedo Station amid the secrecy and the finished exploders were stored away and dummy contact exploders—the Mark 5—issued to the fleet. The Mark 6 would remain under strict lock and key until war came, when it could be substituted for the Mark 5. Even the appropriate manuals were kept in a locked safe. The result was a weapon that very few officers or torpedomen in the submarine service knew anything about.

The exceptional secrecy surrounding the Mark 6 exploder was maintained in the belief that the weapon's effectiveness might be destroyed through enemy countermeasures. The extreme caution turned out to be ill-advised. England, Germany, and Italy had all developed magnetic exploders of their own before the outbreak of World War II. Of the major naval powers, only Japan had failed to develop a magnetic exploder and then only because the Japanese Navy considered them impracticable.

Less than a year before the attack on Pearl Harbor, the extreme secrecy surrounding the Mark 6 exploder was relaxed by Rear Admiral William H. P. Blandy. As chief of the Bureau of Ordnance, Rear Admiral Blandy

The Mark 6 exploder triggers the warhead on a Mark 10 torpedo destroying the target—the submarine *L-8*. (U.S. Navy)

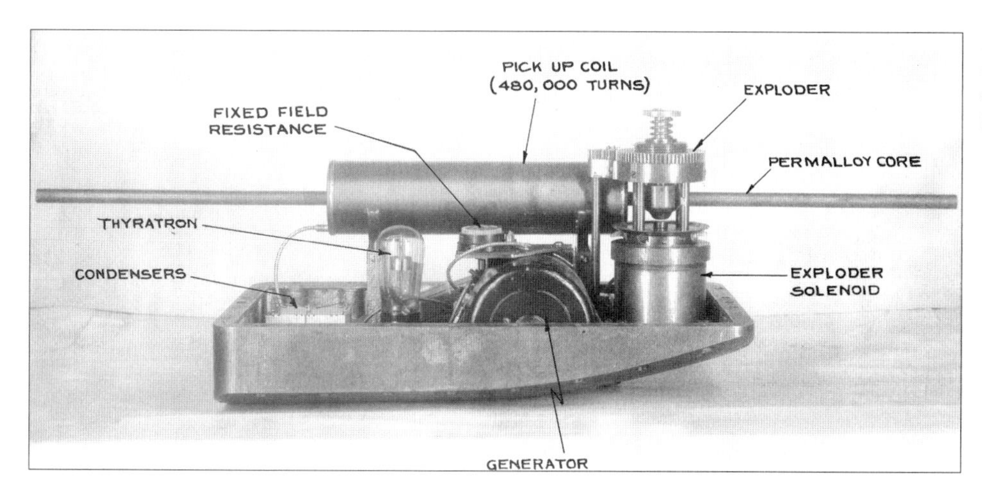

Internal components of the magnetic detection mechanism developed for the Mark 6 magnetic influence exploder. (U.S. Navy)

The Mark 6 Mod 1 exploder photographed at the Naval Torpedo Station, Newport, on 19 December 1941. This was the U.S. Navy's standard torpedo exploder at the beginning of World War II. (U.S. Navy)

realized that getting the fleet familiar with the weapon was more important than maintaining secrecy. During the summer of 1941 his ideas were implemented by the inauguration of a training program at Newport for selected officers, and by issuing a limited number of Mark 6 exploders to the fleet. Unfortunately, the move came too late to permit service testing and evaluation.

THE NAVOL TORPEDO

The development of the Navol Torpedo evolved out of the experimental work conducted on the oxygen torpedo, which was successfully tested for the first time in 1931 (see page 68). The origins of the oxygen torpedo tested in 1931 can be traced to the experimental work on exothermic (heat-producing) chemical reactions begun by the Westinghouse Electric and Manufacturing Company in 1915. These experiments, which were conducted under the direction of A. T. Kaisley, were initiated at the suggestion of Rear Admiral Joseph Strauss, chief of the Bureau of Ordnance, who saw the need to lay the scientific groundwork for a new propulsion system that would enable torpedoes to travel farther and faster. A major goal in torpedo design has always been to get the most possible energy out of the least possible weight of engine and fuel. Kaisley investigated the potential use of various liquid, solid, and gaseous compounds that could generate greater amounts of energy than could be obtained by the conventional compressed air, alcohol-burning engine in the conventional steam torpedo.

The cost of these experiments was borne entirely by Westinghouse until 1920 when the effort was placed under contract to the Navy. In 1927 the project was transferred to the Naval Research Laboratory in Washington, where it was assigned to Dr. Francis R. Bichowsky, the newly hired superintendent of the Division of Physical Chemistry. In his opinion the exothermic torpedo, at least in the form proposed by Westinghouse, was impractical. The primary reason for doubting its practicability was the difficulty of controlling the pressure and temperature of the combustion process between a solid and a gas. Instead, Bichowsky proposed replacing the solid oxygen carrier with a liquid carrier. One of the key issues to be solved before this could be accomplished was the problem of designing a combustion chamber to withstand the high temperatures and pressures involved. Bichowksy suggested that the first step to significantly improve the torpedo should be to develop an oxygen-powered torpedo. The experience gained from this effort could then be used as the basis for further work.

An oxygen torpedo, in theory, would provide more energy on a pound-for-pound basis by eliminating or reducing the amount of nitrogen carried in the air flask. In the steam torpedo the combustion agent is the oxygen that makes up 20 percent of the volume of gas contained in the air flask. The diluent is partly nitrogen and partly water. Although inert, the nitrogen, which was stored at high pressure in the air flask, was extremely undesirable from a weight standpoint, because it required nearly three times its own weight in steel to confine. Thus air flask in a typical steam torpedo accounted for a significant proportion of the torpedo's weight. (See Table 6-1.) Reducing the weight of the air flask would allow more fuel and oxygen to be carried extending the speed/range characteristics of the torpedo.

TABLE 6-1

WEIGHT DISTRIBUTION IN MARK 8-4 TORPEDO

COMPONENT	WEIGHT	N_2 (EST.)
Warhead	612 lb	
Air Flask	1,468 lb	1,000 lb
Air Charge	333 lb	90 lb
Fuel	41 lb	
Water	117 lb	
Other items	600 lb	
Total	3,171 lb	1,090 lb

With the approval of Secretary of the Navy Curtis D. Wilbur, the exothermic project was temporarily stopped and work was begun on the oxygen torpedo under the same job order. Dr. Bichowsky began by investing the combustion pot of the Mark 8-4. By the summer of 1929 he had constructed and successfully tested a burning combustion pot for the oxygen torpedo.

After successfully completing a series of dynamometer tests, the experimental torpedo was sent to Newport for the range trials to be conducted in October 1931. The major problem was the lack of high-pressure air needed to power the depth and steering control engines, which would not work properly on either pure oxygen or live steam bleed off of the combustion pot. Another cause of trouble was oxygen leakage from the supply pipe, which damaged the combustion pot and turbine rotors. The Torpedo Station was left with the problem of solving the control problem. Although the engineers at the Torpedo Station went to the trouble of installing a supplementary air supply for steering and depth control, there was little enthusiasm for a torpedo of this type, and little progress was made in developing an operational version of the oxygen torpedo.

The lack of enthusiasm for the oxygen torpedo was easy to understand. Pure oxygen reacts spontaneously and very energetically with many materials, particularly hydrocarbons (e.g., oils and greases), and the reaction could be explosive. Shipboard storage of such a volatile fuel was seen as extremely hazardous and was vigorously resisted within the Navy.

The confined spaces of a torpedo require small-radius bends in all piping, including pipes supplying oxygen to the combustion chamber. Such bends and any form of surface roughness on the interior of the oxygen tubes impede the flow. This impedance to the flow may cause compression heating of the oxygen, causing the oxygen to react with the surface of the pipe or with metal chips remaining from the machining process used to manufacture the torpedo. This problem could only be overcome by avoiding sharp bends and the meticulous cleaning of any residual oils and grease left from the machining process. This complicated the design process, increased manufacturing costs, and made it much more difficult to ensure reliable operation under the repeated trials and practice runs that all torpedoes were expected to experience during the weapon's lifetime.

While the Torpedo Station tried to correct the mechanical defects of the oxygen torpedo, the scientists at the Naval Research Laboratory in Washington returned to the exothermic project that had been suspended while they had worked on the oxygen torpedo. After a series of discouraging experiments the laboratory decided that a hydrogen-peroxide solution held the greatest promise for torpedo propulsion.

The virtue of hydrogen-peroxide is that it is a liquid containing over 90 percent oxygen by weight as compared to air, which is only 23 percent oxygen. Use of a hydrogen-peroxide solution offered several advantages that compensated for its greater weight compared to air. As a liquid it did not require the heavy flasks that were required to contain high-pressure oxygen or air. Hydrogen peroxide decomposes in an exothermic reaction that releases 48 percent of the oxygen present. The water resulting from the decomposition of the peroxide posed no problem, for it was no more than would have to be introduced to keep the engine temperatures at a working level. Lastly, there is practically no wake, because nitrogen, the principle component of torpedo wake, is not present and the combustion products themselves are soluble in the water. The particular solution of hydrogen peroxide selected by the Naval Research Laboratory was labeled "Navol" by the Bureau of Ordnance.

In June 1935 the Torpedo Station was directed to manufacture a special air flask for use with Navol and to modify a Mark 10 torpedo for its use. The experimental torpedo, which was designated as the Mark 10-3, was shipped to the Washington laboratory, where dynamometer tests were conducted. When everything was working properly the Mark 10-3 was returned to the Torpedo Station for tank and range tests, which were conducted in September 1937. Using Navol increased the range of the torpedo by about 300 percent. Success gave impetus to the work, and similar tests were later made with a converted Mark 14 submarine torpedo, designated the Mark 16, with a stainless steel Navol tank. On trials the torpedo ran 16,500 yards at 46 knots.

Encouraged by the prospects of obtaining a new, high-powered torpedo, the Bureau of Ordnance directed the Torpedo Station—in conjunction with the Naval Research Laboratory—to design a new surface-launched torpedo based on Navol, authorizing the manufacture of 50 Mark 17 torpedoes for the development effort. Specifications called for a 50-knot torpedo having a range of 17,000 yards with a 600-pound warhead. Six prototypes were constructed and tested by the fall of 1941.

By then the Torpedo Station was operating with three shifts on a seven-day-a-week basis in an effort to meet the huge demand for torpedoes needed to outfit the large number of destroyers and submarines that were being added to the fleet. Although the Naval Torpedo Station in Alexandria had been reopened and was manufacturing torpedoes, the two stations together were only able to make less than one-half of the torpedoes needed by the Navy. Under these circumstances, the General Board, which still had to be consulted before the Mark 17 could be assigned to the fleet, was reluctant to approve the production of such a radically new weapon.

The Board's decision was passed on to Secretary of the Navy Frank Knox on 3 November 1941.

> The General Board is aware that the problem of producing torpedoes in sufficient numbers under present conditions is already difficult and may be embarrassing if our Navy should be involved in active hostilities. The Board recognizes that the

> other most serious question is the present state of training of personnel in operations and upkeep which might be embarrassing if a radical change in type of torpedo were made at this time. Under the circumstances the Board does not recommend the adoption of this torpedo as a replacement for corresponding types at this particular time.[13]

Cognizant of the fact that the Navy would again be burdened with a large number of obsolete torpedoes at the end of the war, and with no money for future development, the Board went on to recommend that experiments with the Navol torpedo be continued with vigor. Thus, an opportunity to deploy a torpedo that could compete with the highly secret Japanese Type 93 oxygen torpedo—the so called Long-Lance torpedo—was forgone.

CHAPTER SEVEN

Attack from the Air

Aerial Torpedoes (1917–1945)

Although the idea for a torpedo that could be launched from an airplane was first raised in America by Rear Admiral Bradley A. Fiske in the summer of 1911, nothing was done about it until after the United States had entered World War I. By then Fiske, who had unsuccessfully lobbied Secretary of the Navy Josephus Daniels to provide a larger appropriation for the development of naval air, had been forced to retire because of age.

Fiske spoke out on the importance of the torpedo plane, for which he had received a patent in 1912, at the Pan American Aeronautical Exposition in New York City on 12 February 1917. Other nations were already way ahead of the United States in developing the aerial torpedo, causing Fiske to press for its development in our own Navy, stating:

> We can only hope to catch up with the other countries of the world by including in our naval program the most effective new inventions, and the torpedoplane, under favorable conditions, would make the $20,000 airplane a worthy match for a $20,000,000 battle cruiser. That the torpedoplane will become an important factor in naval warfare in the near future, many people have no doubt.[1]

When the Navy failed to provide any finds to develop a torpedo-carrying airplane, Fiske turned to the Aero Club of America for support. With $2,500 to defray expenses and three small torpedoes provided by the E. W. Bliss Company, Fiske was able to arrange a demonstration for the members of the Aero Club at Huntington Bay, Long Island.

On 14 August 1917, Lieutenant Edward O. McDonnell, an instructor assigned to the First Yale Unit, dropped a 180-pound, 10-foot long, 8-inch diameter dummy torpedo from an F-5L floatplane.[2] The dummy torpedo, which had been placed under one wing, struck the water at an unfavorable angle, ricocheted three times, nearly hit the plane. This test marked the beginning of the Navy's interest in launching torpedoes from aircraft.

Formal U.S. Navy development work did not begin until November of the following year when the Naval Aircraft Factory in Philadelphia began developing torpedo launching gear. Another year would pass before the first live torpedo, a 12-ft long, 18-inch diameter Type D was successfully launched from an R-6L Navy floatplane on 14 July 1919. Approximately 500 dummies and 43 Mark 7 Type D torpedoes were air launched during the next 12 months.

The small, obsolete, 1,036-pound Type D torpedo was soon replaced by the 1,650-pound Mark 7 aerial torpedo. The first of these was successfully launched on 12 May 1920. It became the standard aerial torpedo used throughout the 1920s and most of the 1930s. Two versions of this torpedo were modified for aerial use:

the Mark 7 Mod A and the Mod B. The latter had a larger warhead (319 vs. 205 pounds) and was 11 inches longer. Both torpedoes were strengthened to withstand the shock of water entry and were fitted with a breakaway nose drogue to limit the depth of the torpedo's initial dive once it entered the water.

While these torpedoes performed well when properly prepared and launched at a precise altitude at speeds below 80 knots, Lt. Comdr. Wadleigh Capehart, the commanding officer of Torpedo Squadron 9-S, considered it "far too fragile" for use as a war weapon. In his words, "it [was] a mechanical contrivance of extreme delicacy" that had to be gently placed in the water "otherwise, the governor link is out, the gyroscope tumbles, and the torpedo sinks or makes an erratic and ineffectual run."[3]

Work on a new torpedo that could withstand a launching speed of 140 mph from an altitude of at least 40 feet began in February 1925, but the project—designated G-6—was soon discontinued in favor of adapting the existing Mark 7 torpedoes. The project was revised in 1929 after the General Board recommended that a special torpedo be developed for launching from torpedo planes having the characteristics of:

Weight:	2,000 pounds
Length:	about 14 feet
Speed:	35 knots at a range of 4,000 yards

The Bureau of Ordnance objected to the length of the proposed torpedo, which it felt would adversely affect the design of future torpedo planes. The bureau was then in the process of studying how the performance of torpedo planes could be increased. These studies had shown that the over-all length of the torpedo should be reduced if the best in aircraft design was to be achieved. After much discussion and an exchange of correspondence between the various bureaus involved, the Chief of Naval Operations and the General Board changed the specifications to reflect the requirements for a shorter torpedo. The revised specifications called for a torpedo that was capable of being launched at 100 knots from an altitude of 50 feet having the following characteristics:

Length:	13 feet, 6 inches
Diameter:	23 inches
Weight:	1,700 pounds
Explosive:	400 pounds
Speed:	30 knots (minimum)
Range:	7,000 yards

The torpedo that evolved from these specifications was designated Mark 13 in August 1930. The first successful run of a Mark 13 torpedo was made the following year, in March 1932, when the No. 2 torpedo made 30 knots over 6,000 yards. Aerial testing of the Mark 13 torpedo began in 1935. At least 23 torpedoes fitted with bronze exercise heads were launched from aircraft between 27 May and 1 October. The Mark 13 torpedoes were dropped from altitudes of 46 to 105 feet at speeds varying from 85 to 114 knots.

Twenty additional drops were conducted the next year as part of the proof tests conducted on the first production order of 108 torpedoes. A number of these drops were made in excess of the 125-knot maximum launching speed now specified for the Mark 13 torpedo. None of the drops resulted in torpedoes diving in the mud at the start of the run due to abnormal initial dives. For the most part the damage sustained on the different torpedo tests was minor and did not adversely affect the torpedo's run.

Not surprisingly, the Torpedo Station believed that it had a good torpedo. The Mark 13 had exceeded the station's expectations in the matter of performance in the tank, in the proof work from the test barge, and in aircraft launches. The Inspector of Ordnance in Charge of the station, Captain Issac C. Johnson, considered "the Mark 13 torpedo an excellent service aircraft torpedo and far superior to anything we now have."[4]

Johnson's opinion was verified when the Mark 13 entered service in 1938. Torpedo Squadron 3 (VT-3) on board the USS *Saratoga* (CV 3), flying the new Douglas TBD-1 Devastator, dropped the Mark 13 during the squadron's gunnery practice of 26 September and 20 November. All drops were made at a speed of 100 knots or less and at altitudes from between 40 and 90 feet. The squadron rated the performance of the Mark 13

A dummy Mark 7D torpedo is released by a Curtiss R-6L during early U.S. Navy torpedo bombing trials at Hampton Roads, Virginia, in 1919. The Curtiss R-series aircraft of 1915–1918 were flown by the U.S. Army and Navy as well as the Royal Naval Air Service. The U.S. Navy's R-6 variant was the first American-built aircraft to serve with U.S. forces in Europe in World War I. It was used primarily for anti-submarine patrols. (U.S. Navy)

A Curtiss R-6L releasing a Mark 7 torpedo during experimental drops at Pensacola, Florida, on 2 April 1920. Note the position of the torpedo during what was obviously a failed attempt to launch the torpedo. Pensacola—the Navy's first major air station—is called the "cradle of naval aviation." (National Archives)

A Martin MBT releasing a torpedo into the Anacostia River at Washington, D.C., on 17 May 1920. The Army bought the twin-engine biplane as the MB-1 in 1918. The Navy procured two of the aircraft in 1920, designated MBT for Martin Bomber-Torpedo, and, subsequently, eight improved MT (Martin Torpedo) aircraft for the Marine Corps. The latter were used as bombers and to drop paratroops. (U.S. Navy)

The first U.S. Navy fleet exercise involving aerial torpedo attacks occurred in September 1922. Top: the battleship *Arkansas* (BB 33) takes a direct hit from a dummy warhead. Below: A miss astern of the *Arkansas*. The miss demonstrates the difficulty of hitting a large non-maneuvering ship steaming at 20 knots. The *Arkansas*, commissioned in 1912, was updated between the world wars and saw combat at Normandy, Iwo Jima, and Okinawa. She succumbed to an atomic bomb at Bikini atoll in July 1946. (U.S. Navy)

Another view of an R-6L flying close to the water as it releases a Mark 7 torpedo during the fleet exercise of 1922. The torpedo had to be dropped from an altitude below 25 feet at an air speed of less than 65 knots. R-6Ls were first modified for dropping torpedoes in 1917. The aircraft could also be fitted with wheels for land operations. (U.S. Navy)

as "excellent."[5] No torpedoes were lost and there were no erratic runs, which was credited to the maintenance personnel of the carrier *Saratoga* and VT-3.

Only 156 Mark 13 Mod 0 torpedoes were produced. This was just enough to provide two full loads of torpedoes for each of the 18-plane torpedo squadrons assigned to the Navy's four large aircraft carriers—*Saratoga*, *Lexington* (CV 2), *Yorktown* (CV 5), and *Enterprise* (CV 6)—plus 12 in reserve. The Mark 13 was unusual in that it was the only U.S. torpedo fitted with a rail-type tail in which the propellers were placed forward of the rudders. The Torpedo Station, for reasons that remain unknown, was dissatisfied with this arrangement and began testing a new variant of the Mark 13 fitted with a non-rail-type tail. When the station received an order from the Bureau of Ordnance to manufacture 72 more Mark 13 torpedoes in the spring of 1938, Commander Johnson recommended that the non-rail tail be incorporated in the new torpedo, which would be designated as the Mark 13 Mod 1 torpedo. Produced in 1939, the first Mod 1 entered service in 1940. Unfortunately for the pilots who would later have to rely on this weapon in battle, the modified design was plagued by so many defects that one writer called it the worst piece of ordnance ever forced upon the Navy. The Mark 13 Mod 1 was so bad that four of ten torpedoes launched by VT-6 during the practice fire of July 1941 sank from sight and were never seen again. Of the remaining six, five experienced erratic runs. Only one of the ten Mark 13 Mod 1s dropped ran hot, straight, and true.

In the months that followed the Bureau of Ordnance discovered that the Mark 13-1 tended to veer left upon entry into the water. No sooner was this problem corrected than chronic depth failures were detected. The propellers were found to be too weak to stand the shock of high-speed water entry and the exploder mechanism had to be modified to keep it from arming in the air. Even when these defects were eliminated the Torpedo Station could not find a way to make it suitable for the high-speed drops that were tactically necessary.

Although Rear Admiral Blandy, the head of BuOrd, was aware of the defects in the Navy's aerial torpedoes, he was not overly concerned based on the comments he made in a letter addressed to the Commander Aircraft, Scouting Force on 1 March 1941.

> I don't think our aircraft torpedoes are so bad, compared to foreign types, as some of the Intelligence [*sic*] reports indicate. Naturally all of the belligerents tend to overemphasize the importance and value of their own weapons. Flight Commander Bruce of the Royal Air Force states that instructions to the British Service require

The first torpedo squadrons to become operational aboard U.S. aircraft carriers were VT-1 and VT-2, assigned respectively to the *Lexington* (CV 2) and *Saratoga* (CV 3) in 1928. They flew the Martin T3M-2 and its successor, the Martin T4M-1. By the following year the T3M-2 had been replaced by the T4M-1. The latter are shown on the deck of the "Sara" on 8 October 1929, armed with the Mark 7 torpedoes. (National Archives)

The Mark 7 torpedo was the standard aerial torpedo used by the U.S. Navy throughout the 1920s and most of the 1930s. The weapon suffered from poor performance and imposed severe speed and altitude restrictions on the attacking torpedo plane. (U.S. Navy)

A Douglas TBD-1 Devastator of VT-6 releases a Mark 13-1 torpedo during training operations on 10 October 1941. The TBD, which had entered service in 1937, was the first all-metal, carrier-based monoplane ordered for the U.S. Navy. Its career as a first-line aircraft lasted only until the disaster at Midway in June 1942. The Mark 13-1 was plagued by numerous defects. Four of ten torpedoes launched during an exercise conducted in November 1940 sank from sight and were never seen again; five experienced erratic runs; only one ran "hot, straight, and true." (National Archives)

> maximum launching heights of 70 feet with the hope that 120 feet will not be exceeded, and that the British expect only 50 per cent effective runs if the launching height is 140 feet. Our Torpedo Section here says that the Mark 13-1 torpedo launched from 140 feet will give 80 per cent effective runs.[6]

Vice Admiral William F. Halsey, then the Commander Aircraft, Battle Force, thought otherwise. On 21 July 1941, he dispatched a long memorandum on the subject of aircraft torpedoes addressed to the Chief of the Bureau of Ordnance. Halsey noted that the Mark 13 torpedo had many drawbacks, the most noteworthy being more range than necessary, inadequate speed, insufficient ruggedness to withstand high-speed, high-altitude drops from glide approaches, and excessive weight and size.

Admiral Halsey was not alone in his concerns. There were some in BuOrd who were also aware of the Mark 13's shortcomings. The urgent need to increase the speed of the Mark 13 to 40 knots was emphasized by representatives of the Bureau of Aeronautics who visited the Torpedo Station during the last week of September 1941. Although the Inspector of Ordnance in Charge sought authorization to initiate a design project to modify the Mark 13's design, the project would not get off the ground for two years.

The Torpedo Station was well aware of the tactical limitations of the launching requirements of the Mark 13 torpedo. It had begun investigating the problem as early as 1938, when it conducted tests of the Mark 13 torpedo fitted with "board extensions" that showed promise of improving the entrance angles that would permit torpedo launches from aircraft at greater speeds and higher altitudes. The extent of the tests conducted by the Torpedo Station is not known, but it appears they did not involve actual drops from aircraft. This was left up to the Bureau of Aeronautics.

On 7 April 1938, Captain Albert C. Read, of *NC-4* trans-Atlantic fame, requested the trial installation of a Mark 13 torpedo with board extensions using a TBD-1 Devastator from an operational torpedo squadron.[7] Read, who was acting Chief of the Bureau of Aeronautics, was reluctant to refer the matter to the forces afloat, but did so because of the urgent nature of the project and the fact that "adequate facilities are not at present available elsewhere."[8]

Nothing came of this effort, and the problem seems to have lain dormant until the spring of 1941, when the Torpedo Station—in response to a requirement to arm PBYs with the Mark 13 torpedo—began investigating torpedo launching angles on various aircraft. On 1 May 1941, the Torpedo Station issued instructions for the manufacture and use of air stabilizers for patrol plane torpedoes that would "greatly improve the launching characteristics of the torpedo and eliminate most of the difficulties that have in the past caused erratic runs."[9] Manufacture and installation of the device, which was constructed of plywood, became the responsibility of each patrol squadron.

Similar instructions for the use of stabilizers on aircraft torpedoes for TBD-1 carrier-based aircraft were not issued until 10 November 1941. It took another month for the memo to work its way through the Bureau of Aeronautics, which subsequently forwarded it to the forces afloat, requesting that they provide comments on its use. By then the U.S. Navy was engaged in a two-ocean war that found the beleaguered carriers in the Pacific focused on striking back at the enemy. The only torpedo squadron able to make such an assessment was Torpedo Squadron 4, assigned to the *Ranger* (CV 4) in the Atlantic.

The *Ranger*'s personnel had no difficulty in fabricating the stabilizers designed by the Torpedo Station, but had difficulty loading a Mark 13 Mod 1 fitted with stabilizers. The TBD-1 was so close to the deck that a torpedo fitted with the stabilizers could not be loaded unless the tail of the plane was lifted. Several men had to lift the tail about 18 inches and block it up with a packing case while the torpedo was placed in position and hoisted under the aircraft. There were other problems too. The stabilizer interfered with the Mark 35 bomb rack under the rear portion of the fuselage.

Once installed, it was found that the rear end of the stabilizer was only 3½ inches from the flight deck. But the cross-deck pendants—landing arresting wires—on the *Ranger* were 5½ inches high. In order to ensure that no damage would occur to the torpedo or the aircraft upon landing, if the torpedo had not been released, the carrier's commanding officer recommended that the lower leading corners of the stabilizer be rounded off.

THE FIRST AERIAL TORPEDO ATTACK

In the summer of 1915 the British were fighting the German-supported Turks in the eastern Mediterranean. The seaplane carrier *Ben-my-Chree*, which entered the Aegean Sea in August 1915, carried Royal Navy floatplanes that would launch the world's first aerial torpedo attack. On 28 July 1914, Navy pilot Gordon Bell made a successful torpedo drop at Calshot, flying a Short S.64 seaplane. These trials used a 14-inch diameter, 850-pound Whitehead torpedo. This led to the rapid development of a torpedo-carrying floatplane, the Short Type 184 biplane—the world's first operational torpedo aircraft. Two of these aircraft were on board the *Ben-my-Chree* when she arrived in the Aegean. On 11 August 1915, a Turkish ship was sighted on the north side of the Sea of Marmara; she would be the first victim of an airborne torpedo attack. Just before first light of 12 August the *Ben-my-Chree* arrived at the eastern end of the Gulf of Xeros. Sailors hoisted out a Short 184 seaplane with a 14-inch torpedo slung under the craft. Flight-Commander C. H. Edmonds got the heavily loaded plane into the air after a relatively short run, climbed to 800 feet, and headed north. When Edmonds sighted his target, a 5,000-ton supply ship, he maneuvered into an attack position, cutting his engine and gliding in for the kill, releasing the torpedo from an altitude of 15 feet when some 300 yards from the ship. As he restarted his engine and began to climb, Edmonds saw the torpedo strike the ship, which promptly began to settle in the water. A "first" for naval aviation? No one will ever know for certain, because the British submarine *E-14* claimed to have torpedoed the Turkish ship just before the seaplane arrived!

On 17 August another torpedo attack was made against Turkish supply ships, this time off Ak Bashi-Liman, just above the narrows of the Gallipoli Peninsula. The two planes on this raid were piloted by Edmonds and Flight-Lieutenant G. B. Dacre. Edmonds made the attack as planned, his torpedo hitting a steamer that became a flaming wreck. Dacre's plane developed engine trouble during the flight and came down at sea. Seeing an enemy tug, Dacre taxied his plane toward the craft, released his torpedo at close range, and scored a direct hit. The tug sank. As Dacre taxied away, his engine began picking up power, and, without the weight of the torpedo, he took off and flew back to the *Ben-my-Chree*. Ship-based aircraft had sunk their first ships.

While installing the torpedo it was discovered that the starting lanyards supplied for the Mark 13 would not function with the Mark 13 Mod 1. As its name implies the starting lanyard initiated the motor starting sequence within the torpedo as the torpedo fell away from the aircraft. The lanyard was physically connected to the aircraft and was pulled out as the torpedo was dropped. The *Ranger*'s crew found it necessary to modify the Mark 13 lanyards by removing the fitting for the Mark 13 torpedo and soldering in its place a small clevis. This modification was easily accomplished by the ship's crew.

A few drops were eventually made by VT-4's TBDs, but this did not occur until sometime in February 1942. It took several weeks for this information to be processed by the torpedo experts at Newport. The results, reported by the Inspector in Charge on 15 March 1942, were inconclusive:

> The torpedo drops from carrier based TBD-1 airplanes indicate that fairly good stabilization may be obtained at 100 to 110 knots, with better results obtainable at altitudes from 100 to 150 feet. The stabilization is poor at maximum plane speed, and should such speed seem desirable, further work is necessary. . . . Either project would necessitate extended assignment of a TBD-1 land plane to the Naval Torpedo Station.[10]

Further work would be done, but not with the obsolete TBD, which was being rapidly replaced by the Grumman TBF.

BATTLE OF THE CORAL SEA

Five days earlier, on 10 March, the *Lexington*'s Torpedo Squadron 2 launched the first aerial torpedoes ever fired in anger by the U.S. Navy while attacking several transports at Lae and Salamaua in New Guinea. The attacks with Mark 13 torpedoes were part of the coordinated raid conducted by *Lexington* and *Yorktown*. It was not an auspicious debut for the Mark 13: only one or two of the 13 torpedoes launched by VT-2 planes struck home, sinking the 6,000-ton transport *Yokohama Maru*. The other "fish" either ran too deep or malfunctioned. This was an extremely poor showing for an unopposed action against slow or stationary targets.

The next opportunity to test the combat effectiveness of the Mark 13 did not present itself until 4 May 1942, when 12 TBDs of VT-5 took off from the *Yorktown* to attack Japanese shipping off Tulagi. One section of three TBDs went after the transport *Azumasan Maru*, anchored off the island's northeast shore; two other TBDs set their sights on the transport *Kōei Maru* at anchor near the island of Gavutu. No hits were scored and faulty torpedoes were blamed once again. The rest of the TBDs from VT-5 made a successful anvil attack on the minelayer *Okinoshima* and the destroyers *Kikuzuki* and *Yuzuki* that were alongside. Only one of the torpedoes launched against these ships ran true, striking the *Kikuzuki* and causing enough damage for her to be beached and eventually abandoned.

After being rearmed and refueled on the *Yorktown*, the VT-5 Devastators returned to attack the minelayer *Okinoshima*, which had evaded an earlier bombing attack from *Yorktown*'s dive bombers. The 20-knot minelayer managed to avoid all of VT-5's torpedoes, leaving the squadron record for the day at one hit for 22 torpedoes expended.

The torpedo planes fared much better three days later when squadrons from both carriers took part in the coordinated attack on the Japanese light carrier *Shoho* on 7 May. Included in the 93-plane strike force that set off from the *Lexington* and *Yorktown* were 22 torpedo planes: 12 TBDs of VT-2 and 10 TBDs of VT-5. The *Lexington*'s air group was the first to reach the target. As her dive bombers began to attack, the slower Devastators of VT-2—which had been unable to keep up with the faster SBDs—approached the target from the southwest. To avoid anti-aircraft fire from the two nearest Japanese cruisers, Lieutenant Commander James H. Brett, skipper of VT-2, took the squadron around to the north where he could attack the carrier abeam through the widest gap in her screen.

As dive bombers pressed home their attacks, Brett split his squadron in half, sending the second division's six planes to attack from the starboard side, while he took the first division to the carrier's port side for the classic anvil attack from both sides. In theory, this formation, would make it impossible for the target to escape. No matter which way she maneuvered, one side or the other would always be presented to one of the attacking torpedo planes. Although easily sketched on a blackboard, the tactic was most difficult to execute, given the TBD's relatively slow speed.

The most detailed account of the VT-2 attack, reproduced here, is that written by John Lundstrum in his widely acclaimed book *The First Team*:

> At 1119 Brett released his torpedo from off the *Shoho*'s port quarter. Fanning out, the remaining TBD pilots curved toward the target, launched their fish, and sheared off to avoid antiaircraft fire. As each torpedo cut into the water, it threw out a huge splash before (hopefully) righting itself at a proper depth and heading for the target.
>
> Torpedo Two's strike, the first for an American squadron against an enemy carrier, was a masterpiece. Lieutenant (jg) Leonard W. Thronhill's fish was the first to slam home. His torpedo struck the *Shoho*'s starboard quarter, the blast half hidden by smoke already raised by [Bombing Squadron 2's] bomb hits. The explosion wrecked both electrical and back-up manual steering systems, forcing the ship to hold to a steady southeasterly heading. On the port side, Lieut. (jg) Lawrence F. Steffenhagen's torpedo struck just aft of amidships, followed in short order by that of Lieut. Robert F. Farrington smashing into the port bow. Another pilot of Brett's 1st Division sent his torpedo into her port side just forward of amidships. Gunner Harley Talkington made the fifth torpedo hit. His fish detonated aft of amidships on the starboard side, causing a huge pillar of water to mushroom far above the flight deck.[11]

The five hits (nine were claimed) tore huge holes in the *Shoho*'s hull, knocked out her boilers, and destroyed her engines.

The leading elements of *Yorktown*'s air group showed up just as the *Lexington* dive bombers were completing their attacks. By then the *Shoho* had lost headway and was dead in the water. She was a sitting duck when the *Yorktown* bombers pushed over to start their dive bombing attack. Eleven bomb hits were added to the destruction.

The last of the U.S. carrier planes to reach the scene were the TBDs of the *Yorktown*'s VT-5. They attacked at extremely close range from the starboard side and claimed ten hits. At least two of these torpedoes struck the ship's starboard bow; others probably hit farther aft or were detonated by debris in the water. It did not matter, for within minutes *Shoho* was no more, having slid beneath the waves. She was the first Japanese carrier to be sunk in the war.

"Scratch one flattop!" radioed the commanding officer of *Yorktown*'s scouting squadron.

The sinking of the *Shoho* showed what could be accomplished with the Mark 13 torpedo, provided that air opposition was light or nonexistent. The *Shoho*'s limited fighter complement of 4 type 96 fighters and 8 A6M Zeros could not contend with the 18 F4F Wildcats included in the 93-plane strike group launched against her by two large U.S. Navy carriers. Unimpeded by pesky fighters, and aided by the well-coordinated dive-bombing attack, the 12 slow Devastators of VT-2 were able to maneuver into the classic anvil position and had plenty of time to set up for the low-speed (less than 110 knots), low altitude (50 foot) runs needed to successfully drop the Mark 13 torpedo. VT-5's ten Devastators were even luckier. Not only were there no enemy fighters in the air, but their target was virtually defenseless (only two of her anti-aircraft guns were still functioning) and dead in the water. It was a perfect setup.

The tactical limitations and weakness of the TDB/Mark 13 combination would become clearly apparent during the next day's fighting. Against the *Shoho* the two U.S. carriers faced a "light-weight" opponent. On the following day, 8 May 1942, they faced two Japanese "heavyweights": the large aircraft carriers *Shokaku* and *Zuikaku*. The outcome would be much different.

The carrier battle of 8 May 1942 was an evenly matched engagement that primarily pitted the air groups of two large American carriers against the air groups of two large Japanese carriers. Although the American planes slightly outnumbered the Japanese, the latter were battle tested, had a more-reliable aerial torpedo, and had a better torpedo plane in the B5N Kate. On the downside, the Japanese lacked any inherent scouting force and its dive bomber, the D3A Val, was outclassed by the U.S. Douglas SBD Dauntless. The Japanese Zero fighter was more maneuverable, but the American F4F Wildcat was better armed and could take more punishment.

Unlike the Japanese, American air groups did not coordinate their strikes, proceeding instead independently from their individual carriers. On this occasion the *Yorktown*'s strike group was divided into two tactical elements—the first consisting of 24 SBDs and 3 F4Fs; the second 9 TBDs and 3 F4Fs. Lieutenant Commander Joseph Taylor, VT-5's commanding officer, decided to attack the trailing carrier *Shokaku* because her defensive screen appeared to be looser. The squadron descended to 50 feet and increased speed to 110 knots just as the strike's dive bombers began their attack. While their escorts contended with the Japanese fighters, the nine TBDs of VT-5 formed a single line abreast, each pilot making a run against *Shokaku*'s port beam. They released their torpedoes from 1,000 to 2,000 yards just as the dive bombers were completing their attack. None of the 30-knot torpedoes, which were fired at intermediate range, struck the *Shokaku*. Three hits were reported along with three erratics; the former where evidently near misses from 1,000-lb bombs that the torpedo pilots mistook for torpedo blasts.

The *Lexington*'s torpedo planes did not fare any better. Shortly after taking off, one of the 12 TBDs of VT-2 was forced to turn back due to engine trouble. The remaining planes continued on escorted by four Wildcats. As they approached the battle scene, the Wildcats were jumped by enemy fighters, forcing VT-4 into a high-level approach that would take them away from the fighters. Descending from 6,000 feet, the Devastators entered a

The Japanese light carrier *Shoho,* with smoke billowing from her stern, suffers the first of seven Mark 13 torpedoes that would sink the ship in minutes. The *Shoho* was the first major warship sunk by U.S. aerial torpedoes, on 7 May 1942, in the Coral Sea. The attacking aircraft were probably from VT-2 carrying the first batch of Mark 13 torpedoes to be manufactured. The evidence suggests that these were considerably more reliable than the Mark 13-1 and subsequent "mods." (National Archives)

shallow dive that increased their speed to about 180 knots. Lieutenant Commander Brett took advantage of the cloud cover to screen their approach. The *Shokaku* opened fire with her 5-inch guns as soon as the torpedo planes emerged from the overcast.

Two of Brett's planes had become separated in the overcast, breaking out about a mile ahead and far to the left from the other nine planes. These two planes lacked the speed to overtake the *Shokaku*, which had made a radical starboard swing, away from the American torpedo planes. Both planes pressed home their attacks from close range without success. The *Shokaku* completed her 180 degree turn and was now heading in the opposite direction. This maneuver set up the opportunity for a nine-plane squadron attack against her starboard side. The TDBs fanned out, each pilot choosing his own angle of attack and approaching to what he thought was an optimum range. The *Shokaku* turned away and outran the slow torpedoes. One Japanese observer thought the American flyers had done a poor job because they had released their torpedoes over a thousand yards from the target. VT-2 pilots claimed to have scored two hits.

While the American planes were attacking *Shokaku*, planes from the enemy carriers were attacking the American ships. The 71-plane enemy strike group was composed of 35 Val dive bombers and 18 Kate torpedo planes escorted by 18 Zero fighters. Four of *Zuikaku*'s Kates attacked the *Yorktown*, which easily evaded the three torpedoes that were launched in her direction. (One plane was intercepted and destroyed before it could attack.) The *Lexington* was less fortunate. She became caught in the classic anvil attack in which the torpedo planes approached from both sides simultaneously. Five Kates attacked from the starboard side and six from port. The "Lex" turned away from the nearest group, avoiding the three torpedoes that the surviving torpedo planes had been able to launch. This set her up for the port-side attackers, which had also been whittled down to three planes. All three launched torpedoes within close range and in perfect position. Fortunately for the *Lexington*, the first torpedo plunged deep into the water, failed to correct, and passed harmlessly beneath the keel, proving that Japanese torpedoes were not perfect either.

The next two torpedoes did not miss. One torpedo struck the *Lexington* forward and the other hit opposite the island. The explosions rocked the ship, jamming the two main flight deck elevators, buckling the port gasoline stowage tank, rupturing piping, and opening

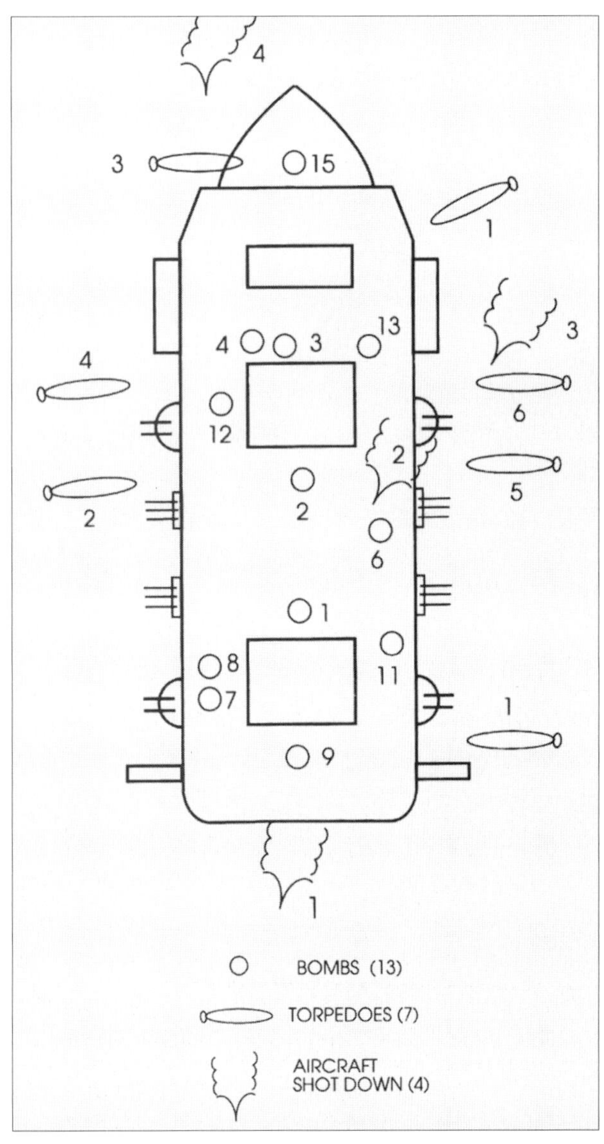

This drawing, adapted from a Japanese naval document, shows torpedo and bomb hits made on the light carrier *Shoho* in the Battle of the Coral Sea. The 11,262-ton *Shoho* was begun as a naval oiler but completed as a submarine tender in 1939. She was then converted to a carrier, being completed in that configuration in January 1942. She could embark 31 aircraft. (U.S. Navy)

several fuel bunkers to the sea. But the ship could still make 25 knots and would have survived had it not been for the secondary explosion caused by leaking gasoline fumes.

Japanese pilots were not immune from the exaggerated, overly optimistic strike assessments that seemed to plague the aviators of all nations. Aircrews returning to the Japanese carriers claimed to have achieved nine torpedo hits when in reality there had only been two. Nevertheless, this was enough to doom the *Lexington.* Although badly damaged by aerial bombs, the *Shokaku* was able to fight another day.

The Type 91 Mod 2 aerial torpedoes employed by Japanese airmen to sink the *Lexington* were superior in many respects to the Mark 13s then available to the U.S. Navy. Unlike the U.S. Navy, which had adopted a short, somewhat squat configuration for its aerial torpedo, the Type 91, with a diameter of 18 inches and a length of 17 feet, 4 inches, retained the long, slender appearance that typified most torpedoes of the era. It was slightly lighter at 1,872 pounds and carried a somewhat small explosive charge (350 pounds vs. the Mark 13's 400 pounds), but it was considerably faster, having a maximum speed of 42 knots.

It was much better from a tactical standpoint, however, as the Type 91 Mod 2 could be released from 330 feet and a maximum air speed of 162 knots. These characteristics were achieved over a ten-year period during which the Type 91 torpedo, which was first introduced in 1931, was continually modified to improve its ability to withstand drops from higher air speeds and greater altitudes. The first change came in the early 1930s, when the designers reinforced the body of the torpedo to prevent the torpedo from breaking up or deforming when it hit the water. While analyzing the performance of this torpedo, the engineers at the Yokusuka Naval Air Arsenal, who were responsible for its design, realized that the aerodynamic problems associated with air launching—entry angle, pitch, yaw, and roll—affected the course of the torpedo once it entered the water. The problem of correct entry angle was solved early on by designing a special coil-spring release mechanism fitted to the launching aircraft. It took more time to determine a solution to the problems of pitch and yaw, which was solved in the late 1930s by adding wooden extensions to the tail fins that would break away upon impact with the water. This was the same solution adopted somewhat later by the U.S. Torpedo Station at Newport, which took more time to solve the roll problem, which was not solved until just before the eve of war. The Japanese attempted to solve this problem by affixing small, gyroscopically controlled, contrarotating steel fins at the forward end of the tail cone that acted like ailerons. To increase the aerodynamic efficiency of the fins, the Japanese added wooden extensions that were also designed to break away as the torpedo entered the water. By the beginning of World War II, the Japanese had solved all of the basic aerodynamic problems associated with air launching. The U.S. Navy would not achieve this goal for at least two more years.

ON TO MIDWAY

The extremely poor performance of the Mark 13 during the Battle of the Coral Sea in May 1942, was a harbinger of worse to come. Toward the end of the month, the U.S. Navy, warned by decoded intercepts of Japanese communications, became aware of the Japanese Navy's plans to capture the strategic island of Midway, located 1,300 miles northwest of Hawaii in the Central Pacific. An all-out effort was made to defend the island, which was reinforced with U.S. Army, Navy, and Marine Corps aircraft, including torpedo-armed PBY Catalinas, B-26 Marauder bombers, and six new TBF Avengers, which arrived on 1 June.

The Avengers were just starting to enter service and would replace the obsolete TBD Devastator as the Navy's standard carrier-based torpedo bomber. They had arrived in the Pacific on board the aircraft transport *Kitty Hawk* (AKV 1), which delivered 21 TBF Avengers assigned to Torpedo Squadron 8 at Pearl Harbor on 17 May. They did not join VT-8 in the *Hornet* (CV 8) when the carrier reached Pearl Harbor because none of the pilots had ever landed an Avenger on a carrier. Instead, their combat debut would be from the sand-covered Midway atoll. On 1 June six of the Avengers under the command of Lieutenant Langdon K. Fieberling took off from Ford Island in the center of Pearl Harbor and, with two PBY navigators on board, successfully made the eight-hour, 1,300-mile flight to Midway. The flight was uneventful, and all six aircraft arrived safely.[12]

On the eve of the carrier battle, four Midway-based PBY Catalina flying boats, each lugging a single Mark 13 torpedo, attacked Japanese transports of the invasion force at 1:15 on the morning of 4 June. Three torpedoes were released in the darkness and one struck the tanker *Akebono Maru*, steaming to the rear of the transports. The explosion killed 11 sailors and wounded 13, but the tanker was only slowed momentarily and quickly rejoined the formation. Thus, the first damage inflicted on the Japanese fleet—albeit insignificant—was by a Mark 13 torpedo.

Back at Midway, Fieberling and the members of his detachment spent an uneasy night waiting for word of the approaching enemy. The stillness was broken by the whine of inertial starters and blasts of exhausts as the Catalinas assigned to that morning's search began to warm up their engines for a 4:15 takeoff. The first reports of enemy carriers' location began coming in an hour and five minutes later, when Lieutenant Howard Ady, piloting a searching Catalina, reported that he had sighted an unidentified plane. A short time later he reported 2 carriers, 3 battleships, 4 cruisers, and 6 to

THE TYPE 91 AERIAL TORPEDO

Air-dropped torpedoes tend to descend deeply in the water before rising up to their programmed depth. In the 1920s the Japanese Navy had pioneered the development of tactics that would enable torpedoes to be dropped from greater heights and higher aircraft speeds than could Western torpedo aircraft. Also, in 1939—before there was any idea of an attack on Pearl Harbor—the Japanese Navy began studying the problem of attacking ships in shallow water. When the Japanese planned the Pearl Harbor attack for December 1941, their torpedoes would normally have gotten stuck in the shallow waters—40 feet in the main channels and less in most other areas within the harbor. The solution: The torpedo was fitted with frangible wooden extensions to the fins at the rear of the torpedo's tail assembly. The wooden fins would break off upon the torpedo's entry into the water, keeping it from diving too deep and into the bottom at Pearl Harbor. These weapons, modified at the torpedo factory in Nagasaki, were rushed to the six Japanese carriers that would participate in the surprise attack on the U.S. battle fleet at Pearl Harbor. At the time there were fewer than 300 aerial torpedoes of all types available in the Japanese Navy. Of those, one-half were allocated to the Pearl Harbor carriers, with the remaining weapons shipped to Formosa and Indochina for use by land-based naval bombers. In the Pearl Harbor attack 40 Mitsubishi B5N2 attack planes carried the modified torpedoes, which inflicted grievous damage on the U.S. Pacific Fleet at anchor in the harbor.

The Consolidated PBY Catalina was an outstanding reconnaissance and attack aircraft, produced in larger numbers than any other flying boat in history. Two Mark 13 torpedoes could be carried for torpedo attack, as shown here being loaded from a utility boat and in flight. Note the removable lifting winches fitted above the wing. The absence of national and unit markings indicates that this PBY-3 was on a test flight. (National Air and Space Museum)

8 destroyers. Ady had found the Japanese strike force. Thirty minutes later, a second PBY, commanded by Lieutenant (jg) August Bartles, also sighted the Japanese carrier force.[13]

A half-hour later, at 6:00, a Marine officer drove up in a jeep, climbed on Fieberling's TBF wing, and told him of the approaching Japanese task force, now estimated to be 150 miles away on a bearing of 320 degrees. Fieberling immediately took off with his six torpedo-carrying TBFs and headed to attack the Japanese carrier striking force. The six Avengers never had a chance to release their torpedoes against Japanese carriers. (They *were* attacking the Japanese carriers when shot down.) As they approached the enemy strike force they were savagely attacked by swarms of A6M Zero fighters. Although heavily damaged, the only TBF to make it back to Midway was the one piloted by Ensign Albert K. Earnest.

By 7:30 AM, the *Enterprise* and *Hornet*, which were the two U.S. flattops nearest to the Japanese carriers, had launched separate air strikes against the Japanese force; these included 14 *Enterprise* TBDs and 15 torpedo planes from the *Hornet*. An hour later the *Yorktown*, which had been delayed as she recovered

A Grumman TBF-1 Avenger dropping a Mark 13 torpedo during a test off Quonset Point, Rhode Island, in 1942. Note that the air stabilizer added to the tail and the stream of exhaust emanating from the rear of the torpedo indicated that the water trip-valve to delay the firing of the igniter until the torpedo had entered the water had not yet been added to these torpedoes. This feature caused the torpedo to run cold during the air flight, eliminating turbine failures due to overspeeding in the air. (National Archives)

her morning scout planes, followed suit, launching 12 Devastators as part of her air strike. There was no coordination within each of the strike groups, or between the three torpedo squadrons, or within the dive bombers. The torpedo squadron from each carrier was operating independently from the other torpedo squadrons and were not in contact with each other. Torpedo Squadrons 3, 6, and 8—from the *Yorktown*, *Enterprise*, and *Hornet* respectively—each carried out its search for the Japanese carriers using navigation information provided from its own carrier. The torpedo planes planned to coordinate their attacks with their own carrier's dive-bombing squadrons, but when these groups lost contact with each other, it became necessary for the torpedo squadrons to immediately attack because of their fuel situation. Thus the torpedo squadrons abandoned any plans of coordination with their dive bombers.

The uncoordinated torpedo attacks in the face of extremely strong fighter opposition resulted in a massacre. None of the planes from Torpedo Squadrons 3 or 8 survived the attack and only four planes from Torpedo 6 returned—a 90 percent loss rate!

The performance of the American torpedo planes sent to attack the Japanese during the main day of the battle—4 June 1942—was horrendous. Of the 51 torpedo-laden aircraft launched against the Japanese carriers—6 TBFs, 4 B-26s, 41 TBDs—fewer than one dozen managed to get close enough to the target to launch a torpedo, and none of them struck an enemy ship. During the entire battle, only one Mark 13 did any damage to the Japanese forces, and that one was launched during the night by a PBY in attacking the invasion convoy.

There were many reasons for the debacle of the American torpedo planes at Midway: The attacks were all made without the benefit of fighter cover. This made the planes sitting ducks for Japanese carrier-based fighters during the slow torpedo runs dictated by the dropping restrictions on the Mark 13 torpedo. The torpedo planes were committed piecemeal and were not coordinated with other units. And they were made, for the most part, by inexperienced pilots, the majority of whom had never even dropped a practice torpedo.

Although the obsolete TBDs would soon be replaced with the much more capable TBF Avenger, it was obvious after Midway that something had to be done to improve the performance of the Mark 13 torpedo. The Torpedo Station continued to work on the Mark 13 during the summer of 1942, conducting tests of an improved stabilizer designated the Mark 2. Very little is known about this work. Bureau of Ordnance records for this period remain classified and many of the photographs of the tests were censored by having the stabilizers airbrushed out. Available evidence suggests that the Torpedo Station was unable, however, to achieve the degree of improvement needed to make the Mark 13 acceptable for use under the conditions now deemed essential for a successful torpedo attack.

In frustration, the Navy turned to the National Defense Research Committee (NDRC) for assistance. Composed of some of America's most renowned scientists, the committee, which had been established by executive order on 27 June 1940 to support scientific research on the mechanisms of war, served as the advisory body for the Office of Scientific Development and Research. In July 1943, the Navy requested the NDRC to undertake the design of an improved aerial torpedo capable of withstanding the shock of water entry when launched at 350 knots from 800 feet. These characteristics were designed to increase the flight time of the torpedo, decreasing the length of its underwater run to the target and making it much harder for the enemy to avoid by maneuvering. That fall, the NDRC issued a research contract to the California Institute of Technology (CalTech) to conduct a study of the full-scale hydromechanical phenomena associated with the high-speed water entry of a torpedo.

Once testing got under way, the scientists at CalTech quickly discovered that the drag provided by the special braces the scientists had added to reinforce the tail structures of the Mark 13 test assemblies tended to stabilize the torpedo's entry, minimizing its tendency to hook and broach. From their earlier work on "Mousetrap," a rocket-propelled anti-submarine projectile, they knew that a ring-type tail would improve the stability of the water entry trajectory of an underwater weapon. This knowledge was used to develop a low-drag shroud ring for the Mark 13 torpedo that greatly reduced the amount of hook and broach previously exhibited by the torpedo.

The success of the shroud ring led to further work on all aspects of torpedo design and to the development of an entirely new torpedo, the Mark 25. This torpedo had the same physical dimensions of the Mark 13, but it had an advanced, high-temperature turbine that gave it a 40-knot speed over a range of 2,500 yards. The development program for the Mark 25 was nearing completion when the war ended, but it was never placed in production.

The addition of the shroud ring along with two other important modifications greatly improved the reliability of the Mark 13 torpedo. The first was the addition of a water trip valve that delayed firing of the igniter until the torpedo had entered the water. This feature caused the torpedo to run cold during the air flight, eliminating turbine failures due to overspeeding in the air. The second was the addition of a drag ring, commonly called the "pickle barrel" (because of its resemblance to that object) composed of a preformed plywood cylinder made up in three segments that was slipped over the warhead. The drag ring increased the effect of the stabilizer, reduced hooks and the depths of dive during initial water entry, and acted as a shock absorber to reduce the impact of deceleration.

These and other minor improvements in the Mark 13 torpedo—Mods 2A, 3, 4, and 5—were implemented by the fall of 1944. Torpedoes so modified reached the fleet just in time for the invasion of Leyte in the Philippines and the ensuing Battle of Leyte Gulf. The four-day battle marked the eclipse of Japanese naval power; it was also the largest naval battle ever fought. It was separated into four parts, each carrying its own name: the Battle of the Sibuyan Sea, when U.S. carrier planes struck the Japanese center force and sank the super battleship *Musashi*; the Battle of Cape Engaño, where U.S. carriers destroyed the Japanese carrier force that had served as a deception; the Battle of Surigao Strait, where U.S. and Japanese warships fought the last dreadnought-versus-dreadnought engagement of all time; and lastly, the Battle of Samar, where the Japanese battle force attacked the U.S. escort carriers defending the beachhead and were soundly defeated by minuscule forces.

At the Battle of Leyte Gulf—Japan's effort to disrupt the American landings in the Philippines—*Yamato* and *Musashi* formed the core of the Japanese so-called

A Mark 13 torpedo striking a Japanese ship at anchor in Truk Harbor in the Caroline Islands during a U.S. carrier air attack on shipping conducted on 16 February 1944. The torpedoes were launched by Avengers, which served aboard carriers as level and torpedo bombers, and scouting aircraft. (National Archives)

Crewmen arm a TBM Avenger on the flight deck of the light carrier *San Jacinto* (CVL 30) during the Battle of Leyte Gulf in October 1944. The addition of the shroud ring, trip valve, and drag ring greatly improved the performance and reliability of the modified Mark 13. Note the so-called "pickle barrel" (drag ring), a plywood cylinder made up in three segments that was slipped over the warhead. (National Archives)

The death marker for the Japanese super battleship *Yamato*: She was attacked by waves of U.S. carrier planes on 7 April 1945, while en route to Okinawa in a futile effort to oppose the American landings. The dreadnought was struck by 11 torpedoes and 10 aerial bombs before she vanished in a spectacular explosion that could be seen for 100 miles. The aerial torpedoes had been the killing factor for the *Yamato* as well as for her sister ship, the *Musashi*, sunk the previous October off Leyte. (National Archives)

Center Force. At 65,000 tons full-load displacement, *Musashi* and *Yamato* were the largest battleships ever built. Their main battery of nine 18.1-inch guns were the largest guns ever to go to sea. While crossing the Sibuyan Sea in the central Philippines on 24 October 1944, the Center Force was intercepted by aircraft from the carriers of Task Force 38 (TF-38) assigned to the Third Fleet. TF-38 was composed of nine fleet carriers and eight light carriers. Although the numbers varied by one or two planes, the aircraft complement of each fast carrier was made up of one 18-plane torpedo squadron. That of the light carriers was one nine-plane squadron. The total number of torpedo planes carried by all 17 of the carriers in TF-38 was 229 Avengers (TBFs and TBMs).

The Avenger, which entered service in 1942, was a superior aircraft to the old TBD. More important, the changes made to the Mark 13 allowed the Avenger pilots to make their torpedo runs at close to the Avenger's top speed of 270 miles per hour at altitudes as high as 800 feet. This proved to be a decided advantage in avoiding the high volume of anti-aircraft fire encountered during the Battle of the Sibuyan Sea when wave after wave of U.S. torpedo planes and dive bombers repeatedly struck the *Musashi*. The ship was struck by at least 13 torpedoes and as many bombs before rolling over and sinking. She was the first major Japanese warship to be sunk primarily by aerial torpedoes since the light carrier *Shoho* went down in the Coral Sea in May 1942. The air attacks on *Musashi* and the other ships in the Central Force were so fierce that they forced the enemy to retire for the time being.

The Japanese warships returned to Leyte Gulf the following day, while the Third Fleet's carriers were off chasing the enemy's Northern Force, which was sent as a decoy to draw Halsey's carriers away from the landing beaches. Although American torpedo planes continued to participate in all phases of this historic battle, their torpedoes were not a decisive factor in any other aspect of the battle.

The Battle of Leyte Gulf marked the demise of the Imperial Japanese Navy and was the last major naval battle of World War II. It was not, however, the last major action involving U.S. torpedo planes. This occurred on 7 April 1945, when the torpedo pilots of TF-58 sank the super battleship *Yamato* en route to Okinawa, which had been invaded by U.S. forces in 1 April. The dreadnought succumbed to 11 torpedo and 10 bomb hits. The accompanying light cruiser *Yahagi* sank after being struck by six torpedoes.

During World War II, 1,287 Mark 13 torpedoes were dropped by carrier-based aircraft; another 150 were dropped by other U.S. Navy aircraft. The pilots of these aircraft claimed to have achieved 514 hits—a 40 percent success rate that seems highly suspect in light of a close analysis after the war of the damage inflicted on Japanese ships. The discrepancy between what pilots thought they saw and what actually happened is easily understood. One writer explained this phenomena:

> Calculation of damage inflicted was always the least reliable part of aircraft reports. . . . The flash of a warship's guns would be reported as a bomb hit. A pillar of water from a near-miss would become a torpedo hit. A ship sending out a smoke-screen would be reported as on fire. and with the sheer speed of air-combat, it was inevitable that several pilots, all in perfect good faith, would claim responsibility for a real success.[14]

Fewer than 10 percent of the 17,000 Mark 13 torpedoes produced during World War II were actually launched in combat against enemy warships. While troublesome in the early part of the conflict, the Mark 13 eventually achieved some notable successes, the most spectacular being the sinking of two super battleships, the *Musashi* and *Yamato*.

CHAPTER EIGHT

They Were Expendable

PT Boats at War (1941–1945)

After the Spanish-American War of 1898 the main thrust of Pacific war planning in the U.S. Navy was directed toward the defense of the Philippine Islands, which Spain had ceded to the United States after the war. While the Navy needed large numbers of destroyers to accompany the fleet, it saw little value in acquiring smaller torpedo boats, which were regarded as short-range defensive weapons only suited for the protection of coastlines and harbors—a job more suited for coastal submarines. By 1909 the last of the torpedo boats still in Navy service, most of which had been authorized in 1898, had been sent to the Charleston Navy Yard and laid up in reserve. A few continued to serve in miscellaneous functions unrelated to the purpose for which they were constructed. Several were recalled to serve as patrol boats and submarine chasers during World War I, but all were discarded by the end of 1919.

U.S. Navy interest in Motor Torpedo Boats (MTBs), as they were now classified by other navies, languished until the mid-1930s, when Rear Admiral Emery S. Land, Chief of the Bureau of Construction and Repair, took note of the progress that European navies had made in developing this type of craft. Land issued a memorandum on the subject to the Chief of Naval Operations on 5 December 1936:

> Developments since the war of the motor torpedo boat type, then known as Coastal Motor Boats, have been continuous and marked in most European Navies. . . . The results being obtained in foreign services are such to indicate that vessels of considerable military effectiveness for the defense of local areas are being built, the possibilities of which should not be allowed to go unexplored in our service. It is, of course, recognized that the general strategic situation in this country is entirely different from that in Europe, so that motor torpedo boats could not in all probability be used offensively by us. It appears very probable, however, that the type might very well be used to release offensive service ships otherwise unavoidably assigned to guard geographic points such as an advanced base.[1]

The War Plans Division of the Office of Naval Operations had recently drawn up a blueprint for crossing the Pacific—known as War Plan Orange—that involved the capture of a series of advanced bases as the fleet made its way across the Pacific from its jumping off point of Pearl Harbor to its final destination in the Philippines. From there the fleet could prepare for the counter engagement with the Japanese Fleet that was expected to be attempting to capture the American-held Philippines. The MTBs suggested by Admiral Land would be useful for defending advances bases and fleet anchorages against Japanese raids that were anticipated as the fleet made its way across the Pacific.

An Elco-type PT boat fires two Mark 8 torpedoes. This scene, repeated hundreds of times in combat, struck fear in the hearts of the captains of enemy warships, although the effectiveness of the PT boats was overrated by both American and Axis commanders. This was a test firing off the Naval Torpedo Station, Newport. Note the radome visible above the torpedoes' exhaust. (U.S. Navy)

Meanwhile, *Field Marshal* Douglas MacArthur, former U.S. Army chief of staff and since 1936 the commander of military forces of the Commonwealth of the Philippines, believed that, "A relatively small fleet of [MTBs], manned by crews thoroughly familiar with every foot of the coast line and surrounding waters, and carrying, in the torpedo, a definite threat against large ships, will have distinct effect in compelling any hostile force to approach cautiously and by small detachments."[2] MacArthur told a key adviser that "I want a Filipino navy of motor torpedo boats."[3] He envisioned some 50 MTBs as the core of his Philippine fleet. The American admirals liked the idea—torpedo boats would be cheap to build and their Filipino crews would be easy to train. (In contrast, U.S. Army generals were appalled by MacArthur's requests for hundreds of modern warplanes and for modern weapons for coastal defense and to equip a 125,000-man Filipino army.)

Meanwhile, Admiral Land's MTB recommendation for the U.S. Navy was passed to the General Board, which agreed wholeheartedly with his assessment of the potential value of torpedo boats. The Board recommended that the Navy initiate an experimental torpedo boat program, which was approved by Secretary of the Navy Claude A. Swanson on 7 May 1937. President Franklin D. Roosevelt, who was Assistant Secretary of the Navy during World War I, was convinced of the value of such small craft and personally supported the proposal.

Congress voted "the sum of $15,000,000 to be expended at the discretion of the President of the United States for the construction of experimental vessels none of which shall exceed three thousand tons displacement." The Navy quickly announced a competition for American shipyards to submit designs for advanced anti-submarine craft and motor torpedo boats.

Based on the proposals of various firms, in March 1939 the Navy awarded contracts to three yards to each build two PT—patrol torpedo—boats, while the two other boats, designed by the Navy, would be built at the Philadelphia Navy Yard. (These were the *PT 1–8*.)

But these contract-winning designs were already obsolete. While the competition was under way, the Elco Navy Division of the Electric Boat Company (Elco) began investigating Royal Navy MTB designs. The British had an extensive motor torpedo boat program and, after looking at several designs, Elco executives judged the design produced by Hubert Scott-Paine to be the best. That 70-foot MTB could carry two 21-inch diameter or four 18-inch diameter torpedoes plus several machine guns. It was powered by three Rolls-Royce gasoline engines and was fast, maneuverable, and sturdily constructed. With the personal approval of President Roosevelt, on 1 June 1939, Elco purchased a 70-foot MTB designed by Scott-Paine along with manufacturing rights for the U.S. government. This MTB—given the U.S. designation *PT 9*—would be the direct progenitor of the several hundred patrol torpedo

(PT) boats that U.S. shipyards would construct during the next six years. On 7 December 1939, the Navy awarded Elco a contract to construct 23 near-duplicates of the British-built prototype—10 MTBs (*PT 10–19*) and 12 "chasers" (*PTC 1–12*), the latter to carry depth charges in place of torpedoes to "chase" enemy submarines. The American PT boats would have three 1,200-horsepower Packard engines in place of Rolls-Royce engines. Although they proved to be good sea boats, the design was already outdated by the time the last of the craft were delivered in 1940. By then Elco and its principal competitor, Higgins Industries Inc. of New Orleans, had developed better designs.

Elco offered an improved, 77-foot PT boat to the Navy and was quickly awarded a contract for 49 boats (*PT 20–68*). These craft, which were constructed of wood, displaced just over three tons and could carry four 21-inch torpedo tubes (or two tubes plus eight depth charges). All four tubes fired forward; just prior to firing the tubes were trained outboard along a short track mounted under the front of each tube. No torpedo reloads were carried. Each boat also had two aircraft-type turrets—with Plexiglas domes—for twin .50-caliber machine guns behind the deckhouse. Two .30-caliber twin Lewis guns could be mounted on the foredeck. This was the first major production run of MTBs for the U.S. Navy, and these would be the first torpedo boats to see combat—at Pearl Harbor, in the Philippines, and at the Battle of Midway.

When the United States entered World War II on 7 December 1941, the Navy had 29 PT boats in commission; scores of additional boats were on order. Once war began the Navy standardized on three major classes of PT boats: the standard 80-foot Elco design that had superseded the company's 77-foot design, a 78-footer designed by Higgins, and a 70-footer designed by Vosper. Most of the last series were sent to Britain and the Soviet Union under Lend-Lease agreements. Wartime production for all designs totaled 774 MTBs. Another four torpedo boats were built in Canada for the U.S. Navy. And 36 were built as PTCs; these were small (63-feet long) and without sonar and were suitable for harbor patrol and little else; most were transferred to Great Britain and the Soviet Union.

By definition the principal weapon of the PT boat was the torpedo, although its guns often inflicted considerably more damage on their targets. The boats for American use initially had two or four 21-inch torpedo tubes. In Elco boats the tubes were fitted on the fore-and-aft axis; the forward tubes could be trained outboard 8 degrees, while the after tubes were trained out 12 degrees; the training gear was hand operated. On the Higgins boats the tubes were fixed, angled out 12 degrees from the centerline. Torpedoes could be fired electrically from a cockpit position or manually by a crewman striking the firing mechanism with a mallet!

PT boats initially carried the 21-inch Mark 8 torpedo, which was 21 feet, 4 inches long and weighed up to 3,176 pounds, depending upon modification, with the warhead containing either 316 or 475 pounds of TNT. (See Appendix C.) The Mark 8 torpedo had been developed before World War I and was the first 21-inch torpedo in the U.S. Navy inventory. Although the majority of the torpedoes had been produced during World War I, they had been continually improved and modified during the intervening years. Like all steam-powered, gyroscopic-controlled torpedoes, the Mark 8 required constant maintenance and adjustments. This became difficult at the MTB forward bases in the Pacific areas, which were often in jungle settings with frequent rain and high humidity, and with limited technical personnel available. For the gyro to work properly the torpedo had to be launched with the PT boat on an even keel, a difficult attitude for a high-speed craft making an attack. The Mark 8 torpedoes allocated to PT boats had a maximum speed of 29 knots and a maximum range of 15,000 yards.

As the war progressed, PT boats were armed with the smaller, 22.5-inch Mark 13 torpedo, which had been developed in the 1930s for use by aircraft. Although the range of the Mark 13 (5,500 yards) was considerably less that the Mark 8, it was a significantly

The Elco-built *PT 10* through *PT 19* were the U.S. Navy's first series-built MTBs. This November 1940 photo shows the *PT 10*; note the 18-inch torpedo tubes and aircraft-type Dewandre gun turrets with twin .50-caliber Browning machine guns. These boats were similar to the British-built *PT 9*. American PT boats were invariably dubbed the "mosquito boats" by the press. (U.S. Navy)

A Mark 8 torpedo is loaded aboard the *PT 65*, a 77-foot Elco boat, at Melville, Rhode Island, in October 1942. Melville was the U.S. Navy's PT boat training center. The Mark 8 was the Navy's first 21-inch "long" torpedo and was outdated when the war began; it proved to have limited effectiveness. (U.S. Navy)

faster (33.5 knots), and lighter (2,216 pounds) weapon, but it carried a larger explosive charge (600 pounds TNT or Torpex). It was also more reliable and could be launched from a moving torpedo boat by being released over the side from a lightweight, side-launching rack (designated Mark 1 Mod 1; it could launch either the Mark 8 or Mark 13). The release was accomplished electrically from the cockpit. Dispensing with the bulky torpedo tubes saved precious weight and space.

The performance of torpedoes launched from the Navy's PT boats was not impressive. Many failed to function properly, especially at the very shallow settings required for attacking small ships and craft in coastal waters. Equally significant was the lack of realistic training received by the PT boat crews; once deployed, there was little opportunity for them to continue their training and, of course, torpedoes were initially in short supply.

A Mark 13 torpedo is side-launched from a Mark 1 launcher aboard a PT boat. Its counter-rotating propellers are spinning and exhaust gas is escaping from the rear. These torpedoes, developed for aircraft use, and the older Mark 8 weapons could be launched from these devices. The Mark 1 launchers were lighter and took up less space than conventional torpedo tubes. (U.S. Navy)

PT BOATS AT WAR

The first PT boats to see combat were the 12 boats of MTB Squadron 1 that had been dispatched to Pearl Harbor toward the end of 1941. These boats were about to be shipped onward to the Philippines as deck cargo on the Navy oiler *Ramapo* (AO 12) when the Japanese struck on 7 December. Four PT boats were resting in cradles on the oiler's deck and two others stood by for loading in cradles on an adjacent pier. The squadron's other six PT boats were in the water, moored in a nest at the submarine base, a few hundred yards to the east. As bombs began exploding around the harbor, the PT boat crews raced to their guns and in a few seconds the .50-caliber Brownings on all six boats were firing skyward. Although credit was claimed for shooting down two Japanese planes, with so many ships firing, it was impossible to determine who was ultimately responsible for downing the attacking aircraft.

PT boat crews also raced for the guns on board the four boats already loaded on the *Ramapo*, but those boats' fuel tanks had been emptied for shipment, thus the crews could not start their gasoline engines to run the air compressors to produce the compressed air that was needed to power the hydraulic oil in the turret training mechanisms. The sailors quickly cut the hydraulic lines, permitting several men to slew each turret manually while a gunner fired at the Japanese planes. (After the 7 December attack the decision was made to keep the dozen boats of MTB Squadron 1 at Pearl Harbor to help repel possible Japanese landings on Oahu.)

A few hours after the attack on Pearl Harbor, on December 8 (local time) in the Philippines, Japanese bombers struck U.S. airfields on Luzon, destroying half of the B-17 bombers that had been sent to the islands—General MacArthur's other major defensive force for the Philippines. The Asiatic Fleet, based in the Philippines, had been partially scattered for several days in anticipation of the outbreak of war. This initial Japanese air attack concentrated on U.S. airfields. Most of the ships remaining in the Manila area escaped, including the six boats of MTB Squadron 3, which were based at the Cavite Navy Yard at the eastern end of Manila Bay under the command of Lieutenant John D. Bulkeley.[4]

The PT boats did not come under attack until Japanese bombers struck the navy yard on 10 December in a high-level bombing raid conducted from 20,000 feet. The boats easily avoided the bombs while maneuvering in the bay. For the next six weeks Bulkeley's boats were kept on a tight rein, mostly conducting uneventful patrols and routine errands.

Bulkeley's squadron was not ordered into action against major Japanese ships until the night of 18–19 January 1942, when the *PT 31* and *PT 34* were sent to investigate reports of Japanese ships off the western coast of Bataan. The *PT 34*, with Bulkeley on board, sighted a Japanese freighter and fired two torpedoes. One launched properly and moments later there was an explosion. The second torpedo stuck partially out of the tube. Every time the speeding craft hit a wave the water turned the small impeller in the nose of the torpedo. If the impeller made enough turns, the torpedo would become armed and could be detonated by the slap of a hard wave. To prevent this from happening, one of the boat's crew straddled the torpedo and stuffed toilet paper into the impeller blades to stop them from turning. (The torpedo was later jettisoned.)

The squadron's first torpedo attack against a Japanese warship occurred on the night of 1–2 February when the *PT 32*, patrolling in Subic Bay off the western coast of Bataan, sighted what appeared to be a Japanese cruiser. As the *PT 32* closed to within 5,000 yards of the enemy ship, she was illuminated by the ship's searchlights and was immediately taken under gunfire. The MTB fired two torpedoes before withdrawing under the heavy fire. Although two hits were reported, Japanese records show only that the small minelayer *Yaeyama* suffered damage in the area at that time, and that was attributed to shore fire.[5]

During this period the *PT 31* drifted aground with engine problems and, when taken under fire by Japanese shore guns, was destroyed by her crew. The *PT 33* was also lost after running aground. All of the boats suffered engine problems.

The most daring and, in some respects the most important accomplishment of Bulkeley's PT boats was the evacuation of General Douglas MacArthur, his family, and 13 U.S. military personnel from the besieged island of Corregidor in Manila Bay, American's last holdout in the northern Philippines.[6] MacArthur and his family boarded Bulkeley's *PT 41* at Corregidor's north dock on the evening of 11 March 1942, with his three surviving boats—the *PT 32, PT 34,* and *PT 45*—carrying the others. The four boats slipped out of Manila Bay led by the *PT 41*. The boats arrived at Cagayan Island off Mindinao on 13 March. From there MacArthur, his family, and his staff boarded two B-17 bombers and were flown to safety in Australia. The Belkeley formation had encountered no Japanese warships on their odyssey. (During subsequent operations in the Philippines, the *PT 34* was sunk in a Japanese strafing attack and the three other boats eventually were scuttled.) Thus ended PT boat operations in the Philippines—for some two and a half years; they had relatively little success for their efforts and the few torpedoes fired.

The next action involving PT boats occurred at the Battle of Midway in June 1942. During the battle, which was primarily a clash of carrier aircraft, six PT boats from MTB Squadron 1 that had been shipped from Pearl Harbor to Midway atoll blazed away with their .50-caliber machine guns at attacking Japanese aircraft—apparently without any success. No Japanese landing was attempted at Midway and these boats saw no further action.

ON THE OFFENSIVE

The first major PT boat actions occurred following the first U.S. offensive operation of the war. U.S. Marines landed on Japanese-held Guadalcanal in the Solomon Island chain early in August 1942. Beginning in October PT boats joined larger American warships in the lengthy contest for control of Guadalcanal and the adjacent waters. These PT boat operations were expanded when Army forces (under General MacArthur) began operations along the northern coast of New Guinea.

These torpedo-armed boats found more targets for their guns than for their torpedoes, because the Japanese used barges as well as destroyers and submarines to reinforce and supply their troops ashore. Hence, the newer PT boats were armed with a variety of guns—40-mm Bofors, 37-mm and 20-mm Oerlikon cannon, .50-caliber Browning machine guns, and even 5-inch rocket launchers. Indeed, whereas the torpedo was the principal weapon for a submarine, for the PT boat the gun could often be considered the principal weapon. And radar enhanced their night-fighting capabilities, becoming an invaluable tool for PT boats that operated mostly at night and in the coastal waters of the Solomons and New Guinea.

Thus, PT boats, operating mostly at night, proved themselves highly effective in attacking Japanese barge traffic and, at times, fighting off larger Japanese warships. In one action a PT boat launched a torpedo that went completely through the destroyer *Hatsukaze*; the Japanese ship was still able to get away at 18 knots. Another Japanese destroyer, maneuvering to avoid PT boat torpedoes, struck a mine and sank. Later a Japanese bomber put a torpedo—which did not explode—completely through the hull of the *PT 167*. (During one night operation the *PT 109* was surprised

Mark 13 torpedoes are loaded aboard a barge in a forward area in the South Pacific. Like most American torpedoes, they had low reliability and, in general, PT boat crews were poorly trained in their use. The torpedo problems for PT boats were exacerbated by the primitive maintenance and storage facilities, often in hot, humid environments. (U.S. Navy)

and sliced in half by a Japanese destroyer; the PT boat skipper, Lieutenant John F. Kennedy, and most of his crew survived.)

In discussing PT boat operations in this period, an official U.S. Navy history notes, "with relatively little damage to themselves [the PT boats] took a terrible toll of the Japanese. Along the coastline [of New Guinea] was the wreckage of hundreds of blasted barges; in former enemy encampments were bodies of thousands of soldiers who died for lack of supplies."[7]

U.S. PT boats also saw action in the Aleutians, in the Mediterranean beginning in April 1943, in the Normandy invasion of June 1944, and, finally, in the 1944–1945 recapture of the Philippines. The largest—and one of the least productive—PT boat operations occurred in October 1944 as the U.S. Navy undertook major landings on Leyte in the central Philippines. The Japanese reacted by dispatching several task forces to oppose the landings. Two separate—uncoordinated—Japanese forces were attempting to reach the invasion beaches by steaming through Surigao Strait on the night of 24–25 October. U.S. commanders were warned of the sortie by reconnaissance aircraft and a trap was prepared. The U.S. forces that would "cork" the strait consisted of 6 old battleships, most veterans of the Pearl Harbor attack, 4 heavy and 4 light cruisers, a screen of destroyers, and then 39 PT boats.

In darkness, the Japanese warships entered the 12-mile-wide channel, and history's last battleship-versus-battleship engagement began. First came a seven-ship group; in the ensuing engagement both Japanese battleships and two destroyers were sunk in the fierce torpedo-gun duel, and a heavy cruiser and another destroyer were heavily damaged. The single surviving destroyer beat a hasty retreat. Now came three Japanese cruisers and four destroyers headed toward Surigao Strait. The force commander knew that American torpedo boats had been encountered, and he was now advancing with caution. The destroyer that sped past him identified herself but said nothing of the ambush.

American PT boats fought in most theaters where the U.S. Navy operated. This is the PT boat base at Finger Bay, Adak, Alaska, in 1944. The weather was arduous, the waters treacherous, and ice forming on—and inside—the boats was a constant danger during the long Arctic winters. Four of the MTBs at right are fitted with radar—a vital "tool" for both navigation and fighting. (U.S. Navy)

Suddenly, a torpedo struck a light cruiser and exploded. The Japanese admiral realized that there was trouble ahead. This was the only torpedo hit scored by the 39 PT boats that participated in the battle.

The Japanese admiral ordered his ships to fire a spread of torpedoes at what were believed to be U.S. ships in the darkness ahead and then he directed a withdrawal. But in retreating, one of his heavy cruisers collided with the damaged cruiser from the earlier forces, after which all of the surviving ships turned away, ending the surface fight. The Japanese had lost two battleships and two destroyers and suffered damage to four other ships. U.S. casualties were one destroyer heavily damaged (primarily by friendly gunfire) and ten PT boats shot up, with one being sunk; three Americans were killed and ten wounded in the night battle. The contribution of the PT boats—which launched 34 torpedoes—was the single hit on the light cruiser *Abukuma.*

The first U.S. PT boats to reach European waters arrived at Gibraltar on 13 April 1943. They soon began operations against the German-held North African coast. Their main antagonists were the German E-boats, larger torpedo boats that prowled the coastal waters of the Mediterranean. The E-boat was a large MTB, more heavily armed than its British and U.S. contemporaries. (E-boat was an Allied term; the German designation was Schnellboot—fast boat.) There were few Axis surface warships in the Mediterranean; thus, once more it was combat against barges as well as the E-boats.

As in the Pacific, the PT boats were plagued with faulty torpedoes. For example, on the night of 19–20 October north of Leghorn, three MTBs—the PTs *208*, *211*, and *217*—fired eight torpedoes at the Italian merchant ship *Giorgio* being escorted by three German torpedo boats and a "flak" boat.[8] Six of the eight torpedoes were erratic, one of them circling the PT formation and forcing the boats to maneuver to avoid their own fish. The two other torpedoes missed the enemy freighter.

Beyond their "normal" operations—every mission was different—the U.S. PT boats were given a novel assignment in March 1944 as operations in the Mediterranean were coming to an end. Beginning on the night of 21–22 March, the PTs were ordered to fire

PT boats were short-legged and required forward bases and tenders. Here the *PT 194*, an 80-foot Elco boat, refuels from the MTB tender *Wachapreague* (AGP 8). These were part of a force of four tenders and 45 PT boats that transited from the Palau Islands to Leyte Gulf in October 1944. This Elco boat had four Mark 1 launchers; some carried a dozen depth charges or mine rails in place of the after pair of launchers. (U.S. Navy)

the remaining store of Mark 8 torpedoes into German-held harbors between Genoa and the French-Italian border. Such night attacks were made for almost a month until the last Mark 8s were expended. Several explosions resulted, damaging port facilities with many secondary explosions.

For two years three MTB squadrons with about three dozen boats had operated in the "Med." They fired 354 torpedoes and claimed to have sunk 38 ships and large craft, totaling 23,700 tons, and to have damaged 49, totaling 22,600 tons. Additional German and Italian losses were attributed to joint U.S.-British MTB operations.

Beginning in March 1944, American PT boats began operations against German forces in the main European theater, initially commanded by Lieutenant Commander John D. Bulkeley. With few enemy targets, during the invasion of Normandy in June 1944 the PT boats proved useful in the reconnaissance and rescue roles.

Of the several hundred PT boats that saw action in World War II, 69 were lost during the conflict. Most were lost through grounding and in storms—29; only 23 are confirmed to have been lost to enemy action (including 5 to shore fire and 4 to mines). Seven were probably sunk by U.S. forces, one by Australian forces, and three in PT boat collisions.

Despite their poor record of the early days in the Philippines and the failure at Surigao Strait, the Navy's PT boat efforts were highly successful, principally against Japanese and German coastal supply barges and small craft. In those actions the PT boat's principal

The Elco *PT 131* in a nest of MTBs alongside the tender *Wachapreague* in the Leyte Gulf area. The boat has twin .50-caliber machine guns and a 20-mm Oerlikon cannon forward; a twin machine gun mount is on the starboard side of the bridge (and another amidships); and two multi-rail rocket launchers. These are Mark 13 torpedoes. Some PT-boat operators wanted to dispense with torpedoes completely in favor of more rocket launchers and guns for coastal operations. (U.S. Navy)

weapon was the gun. Torpedo attacks against major warships and merchant ships were relatively few. Indeed, through the entire war in the Pacific the largest Japanese warships sunk by PT boats were three destroyers—the *Teruzuki*, *Uzuki*, and *Kiyoshimo*. The last had been crippled by two bomb hits from U.S. Army aircraft and was a "sitting duck" when she was attacked and sunk by a PT boat. In the European theater several German E-boats and other small warships were sunk by PT boats.

The PT-launched torpedoes, the Mark 8 and Mark 13, usually failed to achieve many major successes. Their "normal" failure rate coupled with the limited attack opportunities produced few big-ship kills. Perhaps the most significant kills were the two Japanese submarines sunk by PT boats, both in coastal waters. The Japanese considered the PT boats a major threat, because their submarines were impressed to carry supplies to cut-off island garrisons. One submarine captain wrote:

> American PT boats turned out to be the unconquerable enemy of Japanese submarines. They were very small, which made them hard to see, either at sea or against the shoreline. It did no good to fire torpedoes at them, as the Model 95s passed well beneath them. And they had radar. While they could hide under the smallest cover cast by an overshadowing cloud or in a cove, they could still see us at a great distance with their electronic eyes. They could dart in an attack with machine guns, torpedoes or depth charges, then race away at high speed before a submarine could do anything.[9]

A hive of mosquito boats alongside a tender at Leyte for refueling and repairs. The twin .50-caliber machine guns in the foreground have flash-suppressors in their muzzles, another manifestation of the night-fighting tactics of PT boats. The radomes—on folding masts—are for Raytheon SO-series microwave radars. These boats fought against Japanese battleships during the Battle of Surigao Strait in October 1944. (U.S. Navy)

A PT boat on nocturnal prowl off the coast of New Guinea. The greatest value of U.S. PT boats in both the Southwest Pacific and European-Mediterranean areas was in coastal operations, attacking enemy supply barges and small combat craft. (U.S. Navy)

The U.S. torpedo boat effort in World War II was well worth the cost. The boats, fabricated of wood, were relatively cheap to construct, their crews were small and required only limited training, and their maintenance and support costs were meager compared to other naval forces. But most of their success was based on their intrepid crews and guns—not torpedoes.

The Navy ordered four prototype PT boats of different designs in the late 1940s—three from private yards and one from the Philadelphia Naval Shipyard. All had aluminum hulls and were generally similar. Completed in 1950–1951, the *PT 809–812* were designed to accommodate torpedo launchers, but torpedoes were never carried. In the 1960s two became PTFs—fast patrol boats—and additional PTFs were acquired from Norway and the Norwegian design was also built in the United States. These craft were employed primarily for "special operations" in the Vietnam War, putting people ashore and picking from up in clandestine, night operations. None were used as torpedo boats.

CHAPTER NINE

Torpedoes That Didn't Work

Submarine Failures (1941–1943)

When the Japanese struck Pearl Harbor on 7 December 1941, the U.S. Navy had 51 operational submarines in the Pacific. Nine Pacific Fleet submarines were under way in the area of Hawaii–Wake Island–Midway, five were moored in Pearl Harbor, and another eight were on the West Coast. The others were assigned to the Asiatic Fleet in the Philippines, operating out of Cavite Navy Yard near Manila. Of the 29 boats based in the Philippines, 6 were older S-type submarines armed with Mark 10 torpedoes. All of the other boats were "modern fleet-type" submarines armed with the relatively new Mark 14 torpedo, which had been fitted with the top-secret magnetic influence exploder. By mid-December, 22 of the surviving submarines in the Asiatic Fleet had been sent to sea in an effort to disrupt the Japanese efforts to take over the Philippines.[1] Three of the subs based at Pearl Harbor had been dispatched toward Japan, and another four were on their way to the Marshall Islands.

In the weeks and months that followed, U.S. submariners begin to realize that there was something wrong with their torpedoes. More often than not success against Japanese ships was denied by torpedoes that ran too deep, exploded too soon, did not explode at all, or did not have enough explosive power to sink a ship when they did strike and detonate. Several factors worked against the submariners' efforts to sort out the causes behind these failures. It was often too difficult—if not impossible—for a submarine commander to accurately assess the performance of his war shots. Premature detonations were often mistaken for hits on the target, while torpedoes that ran too deep or failed to explode were simply "chalked up" as misses. Uncovering the exact nature of the flaws in the torpedoes was complicated by the fact that the torpedoes sometimes worked and sometimes they did not. Further, multiple flaws in the Mark 14 design tended to mask one another. Resolving these problems was compounded by the initial unwillingness of some senior torpedo experts to acknowledge the possibility of the defects as they were reported after each patrol.

When the Japanese began invading the Philippines on 10 December 1941, the only U.S. forces available that could realistically oppose them were the submarines and a few motor torpedo boats of the Asiatic Fleet. One of the first boats to enter the fray was the *S-38*, commanded by Lieutenant Wreford G. (Moon) Chapple. His submarine was typical of the six obsolescent coastal submarines known only by class letter and number that were assigned to the Asiatic Fleet. When Chapple's boat departed Manila Bay on 8 December 1941, she carried 12 Mark 10 Mod 3 torpedoes, with one in each of the four forward torpedo tubes—there were no after tubes—and eight reloads nested in the torpedo room racks. Although the Mark 10 was ten knots slower and had less range than the Mark 14, it was a formidable weapon that packed a 497-pound punch of TNT triggered by the Mark 3 contact exploder.

The submarine *Shark* (SS 174) shown ready for launching on 20 May 1935, was typical of the "fleet boats" built in the 1930s that served effectively in World War II. This view of the *Shark* shows the opening for her four 21-inch bow torpedo tubes; two stern tubes were also fitted and she carried 20 "fish." Later fleet boats had a standard armament of six bow and four stern tubes, and could carry 24 torpedoes. (U.S. Navy)

Initially assigned to patrol in Verde Island Passage, Chapple shifted to the west coast of Mindoro on 9 December and then moved into the Cape Calavite area. On the night of 19 and 20 December he set a course toward the Luzon coast. The *S-38* put into Camens Cove the following night to repair damage in the port engine lube oil cooler. She resumed her patrol the next day.

At 7:30 PM on the evening of 21 December 1941, Lieutenant Chapple received orders by radio to leave his assigned patrol area and to head for Lingayen Gulf, where a Japanese invasion force was about to land troops. After a 160-mile run, Chapple took the *S-38* into the gulf just before dawn and submerged. When he raised the periscope a 6:15 AM, he sighted four transports slowly moving down the Lingayen Gulf toward the enemy beachhead. Chapple set his torpedoes to run at a depth of 12 feet and fired four carefully aimed shots at the transports, but there were no explosions.

After moving off and sustaining a depth charge attack, Chapple moved in for a second try. It was difficult to understand how all four torpedoes had missed, unless, he rationalized, the enemy was using shallow draft vessels for the landings. Thus he set his next four torpedoes for nine feet and fired two torpedoes at the 5,445-ton transport *Hayo Maru* from a range of about 600 yards. Chapple watched them run straight, hot,

The Mark 14 torpedo was the primary weapon of U.S. submarines during most of World War II. While a potentially potent weapon, it had multiple flaws that were not fully remedied until the spring of 1943. Thus, for more than a year, the burden of the war in the Pacific was carried primarily by U.S. aircraft carriers and, in the Guadalcanal-Solomons campaigns, additionally by cruisers and destroyers. (U.S. Navy)

and true until a geyser of water erupted alongside the target. The blast could be heard in every compartment of the submarine.

While Chapple's decision to reset the torpedo depth of his torpedoes was correct, his rationale was wrong. Unknown to Chapple or any of the other of the S-boat commanders at the time was the fact that the Mark 10 torpedo was running four feet below its depth setting. This was caused by differences in the weight and balance between the exercise heads used during the peacetime test shots and the live warheads used once hostilities commenced. This information was not passed on to the Asiatic Fleet until 5 January 1942, when the Bureau of Ordnance sent a radio dispatch from Washington containing this information.

Even before the United States entered World War II, the Torpedo Station at Newport suspected that U.S. torpedoes were running too deep. As evidenced during the first tests of the magnetic exploder in the 1920s, maintaining proper depth control was not a new problem. In fact, the problem of depth control plagued the torpedoes of almost all navies. Many of these problems could be attributed to errors in proofing and testing that included:

- Differences between calibration shots using exercise heads and warheads,
- Failure to simulate service launching conditions during calibration,
- Changes in calibration setting over time due to deteriorating parts in the depth-engine,
- Erroneous calibration caused by a failure to check against an absolute standard (i.e., relying on the depth and roll recorder readings in the exercise head).

The Torpedo Station was guilty of all of the above. Tests of the Mark 14 torpedo conducted at Newport during October 1941 indicated that it, too, was running four feet deeper than set, but this information was not passed on to the fleet until much later.

As Lieutenant Commander Tyrell D. Jacobs, commanding the *Sargo* (SS 188), soon discovered, the depth problem was only one of several defects afflicting the Mark 14. The *Sargo*, commissioned in 1939, was one of the newer fleet boats that were bigger, faster, and had much greater endurance than the old S-boats. In addition to four torpedo tubes in the bow, these boats also had four tubes aft and could carry 20 torpedoes. The *Sargo* made her first contact with the enemy on 14 December 1941, when Jacobs spotted a 4,000-ton freighter. He set his Mark 14 torpedoes to run at 15 feet so that the Mark 6 Mod 1 magnetic exploder would detonate under the ship's keel. Eighteen seconds after firing a violent explosion shook the *Sargo*. The torpedo had exploded prematurely after running just far enough to become armed.

Jacobs, who had received two years of postgraduate training in naval ordnance at the Naval Academy and had recently completed a short course in torpedoes, suspected that there was some problem or possibly countermeasure that caused the premature detona-

The submarine *S-38* (SS 143) in the late 1930s. Several S-class submarines saw action against the Japanese in the Philippines area and off Alaska early in the war. The *S-38* began her first war patrol out of Manila on 8 December 1941, the day that war came to the Far East. The 18-year-old submarine was in almost continuous action through October 1942, scoring several successes against Japanese ships. (National Archives.)

tions. After talking the matter over with his torpedo officer, he decided to inactivate the magnetic features. This required removing the Mark 6-1 exploder from the warhead, disabling certain electric circuits, and replacing the exploder in the torpedo, making certain that it was properly seated on the rubber gasket that prevented seawater from entering the warhead. If this were done correctly, the exploder would act as an ordinary contact exploder, eliminating the possibility of a premature explosion.

The first chance that Jacobs had to test the modified exploders came on 24 December when he sighted and attacked two cargo vessels, firing two torpedoes at the lead ship and one at the trailing ship, all set for 15 feet. The targets took evasive action and there were no hits. As one of the merchantmen turned away, Jacobs got off two quick shots at the lead ship, setting the torpedoes for ten feet. Again, no hits were achieved.

He encountered a similar situation two days later when he again sighted cargo ships steaming in company. Once again he fired two more torpedoes set to run at ten feet. Like the other commanders of Asiatic Fleet submarines, Jacobs had been told to husband his torpedoes due to a shortage caused when 233 Mark 14 torpedoes were destroyed during the first Japanese air raid on the Cavite Navy Yard on 10 December. Both of these torpedoes fired by the *Sargo* missed their target.[2]

The *Sargo* had now fired eight torpedoes without achieving hits. Her skipper was angry and baffled, but he was determined to find out the cause for his misses. What transpired next was later chronicled by Captain Wilfred Jacob Holmes, a former submariner himself:

> Toward evening in the same day (December 24) *Sargo* sighted two more ships. They were making nine knots, and *Sargo* was in favorable position

> from which to make a deliberate approach. Moreover, the visibility conditions were such that almost unlimited periscope exposure could be made without serious risk of being sighted. Jacobs suspected the accuracy of the tactical data for torpedoes fired with large gyro angles. To eliminate this as a possible source of error he sought and obtained a firing position and course to give his torpedoes a zero gyro angle.[3]

Jacobs fired two torpedoes at the first ship when the range closed to within 1,200 yards. He shifted targets and fired two more at the second ship. All of the torpedoes were set to run at ten feet and were fired with zero gyro settings. All four missed.

After analyzing all the data on his attacks, Jacobs and his torpedo officer, Lieutenant Doug Rhymes, concluded that the only possible reason for their failure was that their torpedoes were running deeper than set. This, they believed, was caused by the warheads, which were heavier than the practice heads used in peacetime to proof each torpedo. Jacobs ordered Rhymes to overcome the problem by readjusting the rudder throws so that the torpedo's depth mechanism would permit shallow running.

The next target did not appear until 4 January 1942, when Jacobs spotted a slow-moving tanker. Once again Jacobs took exhaustive pains to set up an impeccable firing solution. Over a 35-minute period the submariners made 17 unhurried periscope observations until they had a perfect "setup." Jacobs fired one torpedo set at ten feet at a range of 1,200 yards. Another miss!

Jacobs, according to the Holmes account, "had now exhausted all means at his command to find and correct the trouble."[4] Jacobs broke radio silence to send a dispatch to Captain John Wilkes, the Commander of Submarines, Asiatic Fleet, to report that he had fired 13 torpedoes during six attacks made under ideal conditions yet had achieved no hits. He believed that the misses were caused by torpedoes running deeper than set, and that they had adjusted the rudder throws in an attempt to correct the problem. He also informed Wilkes that he had deactivated the magnetic portion of the Mark 6 Mod 1 exploder. Jacobs' complaints about the performance of the Mark 14 torpedo were summarily dismissed by Wilkes' staff torpedo expert, who insisted that the *Sargo*'s torpedoes ran under their targets without exploding because the exploders were flooded. In order to deactivate the Mark 6 Mod 1 exploder, the *Sargo*'s crew had been forced to withdraw the exploders from the warheads. When the exploders were replaced, according to the theory devised by Wilkes' expert, they were not properly seated on the rubber sealing gasket that was needed to keep salt water out of the mechanism

As the weeks went by, submarine skipper after skipper began reporting their suspicions about the Mark 14 running deep. These reports were initially dismissed by torpedo experts within the Bureau of Ordnance as alibis for poor torpedo shooting. Additional testing at the Torpedo Station in February–March 1942 confirmed the four-foot discrepancy referred to earlier, due to a calibration error caused by the use of lighter exercise heads. But this information was not distributed to the fleet until the end of April 1942. The depth control problem in the Mark 14 torpedo, however, was much more severe than the four feet acknowledged by the Torpedo Station.

The true magnitude of the Mark 14's depth problems were not revealed until Rear Admiral Charles Lockwood Jr. arrived in Freemantle, Australia, to relieve Wilkes (who had since been promoted to rear admiral) to command the submarine force in the Southwest Pacific. Shortly after Lockwood's arrival, the *Skipjack*, under the command of Lieutenant Commander James W. Coe, returned from a 55-day patrol off Cam Ranh Bay, Vietnam. Coe, an experienced submarine commander who had already completed three patrols and sunk five enemy ships, submitted a careful analysis of the performance of *Skipjack*'s Mark 14 torpedoes, claiming that they ran much deeper then set. After reviewing Coe's report Admiral Lockwood discussed the matter with Wilkes' former chief of staff, Commander James Fife. Fife proposed that they test fire some of Coe's remaining torpedoes using a target net to determine their actual running depth. The two-day tests, which were conducted under Fife's personal direction at Frenchman's Bay, near Freemantle, confirmed what most skippers had long suspected: the torpedoes were running deeper than set—11 feet deeper on average. The report of Fife's experiments was viewed skeptically by officers in the Bureau of Ordnance, who believed that the discrepancy was based on improper trim conditions caused by the exercise heads used during the Freemantle Bay tests.

The issue was finally resolved when a new series of scientific tests, instigated by the Chief of Naval Operations, Admiral Ernest J. King, were conducted at the Torpedo Station.[5] On 1 August 1942, the Bureau of Ordnance finally conceded that the Mark 14 normally ran ten feet deeper than set. Interim instructions for fixing the problem were quickly issued while the torpedo was redesigned. The problems with the Mark 14 depth control were attributed by the BuOrd to two factors: trim changes due to the use of warheads that were heavier than the exercise heads used to proof the torpedoes, and an improperly designed pressure sensing port. The latter introduced hydrodynamic errors in the depth reading that caused the torpedo to run deeper than indicated. This error would have been picked up if the Torpedo Station had used an absolute method of measuring the running depth of the torpedoes they tested. Unfortunately they used a depth and roll recorder that determined depth by measuring water pressure. This method was subject to the same kind of hydrodynamic error induced by the depth engine pressure sensor. Moreover, the depth and roll recorder was placed in the exercise head at a point where the error due to the hydrodynamic changes in pressure caused by the velocity of the water passing over the torpedo was the same as for the depth-sensing port located in the afterbody of the torpedo. Both the depth recorder and depth sensor were off by the same amount.

While the controversy over the running depth of the Mark 14 was at its height, another problem involving the Mark 6 Mod 1 exploder was detected—and corrected. It involved a defect in the anti-countermining safety that undoubtedly led to many of the "duds" reported by submarine commanders in the early months of the war. The anti-countermining feature had been provided in the final design to prevent the pressure wave generated from the explosion of one torpedo setting off the warhead of one that was following closely in its wake. To prevent this, the designers of the Mark 6 Mod 1 exploder had added a pressure-sensitive diaphragm that pushed a small pin into the firing mechanism train to stop the activation of the exploder. The pin, however, could also be pushed into place merely from the sea pressure within a flooded torpedo tube when a submarine was at periscope depth. When this happened, the pin could be trapped, preventing the torpedo from exploding, even when the torpedo was running at the proper depth setting. This problem was solved in the spring of 1942 by deactivating the anti-countermining feature in the Mark 6 Mod 1.

Deactivating the anti-countermining feature and correcting the depth mechanism problem in the Mark 14 torpedo failed to stop the reports of premature explosions, duds, and erratically running torpedoes that continued to plague U.S. submarines. Rear Admiral Lockwood, who had taken over as Commander, Submarines Pacific, in early February 1943, grew more concerned about the continuing complaints about torpedo performance, especially when they came from successful skippers such as Lieutenant Commander Dudley W. (Mush) Morton.

During his first two patrols in command of the submarine *Wahoo* (SS 238), Morton sank 12 Japanese ships, a record then unequalled by any other skipper. When Morton returned to Pearl Harbor after his third patrol in 21 May 1943, he was furious about the poor performance of his Mark 14 torpedoes. Although he had sunk three more ships during this patrol, faulty torpedoes had robbed him of at least three more. Morton stormed into Admiral Lockwood's office in a towering rage and proceeded to denounce the Mark 14 torpedo in a salty language diatribe that detailed the premature explosions, duds, and erratically running torpedoes that had marred his last cruise.

The submarine *Skipjack* (SS 184), shown shortly after completion in 1938, was one of the two-score fleet boats that were completed from 1935 on that carried most of the burden of the U.S. submarine campaign in the early months of World War II. These submarines were built with large conning tower structures that were cut down during the war; light guns were mounted to supplement the 3-inch/50 gun fitted aft of the conning tower. (U.S. Navy)

Two weeks earlier, on 7 May, the Bureau of Ordnance had recommended that the Mark 6 Mod 1 exploder should be rendered inoperative south of 30 degrees south latitude and should be set to run at keel depth when operated north of 30 degrees north latitude. In between these latitudes BuOrd advised that the torpedoes should be set to run at keel depth or less with the influence exploder activated. To eliminate premature detonations, the Bureau also recommended that the arming distance be increased from 450 yards to 700 yards. However, neither Admiral Lockwood nor Admiral King was willing to accept the change in arming distance.

Lockwood's patience finally ran out when the submarine *Trigger* (SS 237) failed to sink the Japanese aircraft carrier *Hiyo* in Tokyo Bay on 11 June 1943, even thought her captain, Lieutenant Commander Roy S. Benson, had fired a well-aimed spread of six torpedoes from 1,200 yards. When the *Trigger* reached Pearl Harbor on 22 June, Benson reported four hits (based on the sound of the torpedo explosions) and claimed a probable carrier sinking. But Lockwood knew from decoded radio intercepts that *Hiyo* had not gone down. Further, he knew from the same source that only one of the *Trigger*'s torpedoes had caused any damage.[6] This was the last straw. After talking to Benson, Lockwood decided to deactivate the Mark 6 Mod 1 magnetic exploder. The official order was issued by the Commander-in-Chief Pacific Fleet, Admiral Chester Nimitz, two days later, on 24 June 1943.

Rear Admiral Ralph W. Christie, the Commander Submarines, Southwest Pacific, immediately sent a dispatch to Admiral Nimitz asking the reasons for his order. Nimitz replied that it had been made "because of probable enemy counter-measures, because of the ineffectiveness of the exploder under certain conditions, and because of the impracticability of selecting the proper conditions under which to fire."[7] After much discussion with his staff, Christie decided not to deactivate the magnetic exploder. On 11 July he issued orders to his submarines directing that the magnetic feature be retained and that specified depth settings be used against various ship types. Since Christie's command, headquartered in Fremantle, Australia, was not under Nimitz's control, he was not bound by the Pacific Fleet commander's order.

The split made life complicated for the submarine skippers who were routinely shuffled from one command to another with all submarines being overhauled in Pearl Harbor. While they worked for Lockwood, they set the torpedoes one way; while working for Christie, another. A boat going from Pearl Harbor to Australia on the exchange program departed with the magnetic feature deactivated. On the way down, when the boat fell under the operational control of Fife (in command of the Brisbane Task Force) or Christie, the skippers had

The destroyer *Yamakaze* goes down after being torpedoed by the submarine *Nautilus* (SS 168) off the eastern coast of Japan on the afternoon of 25 June 1942. Earlier that month the *Nautilus* fired four torpedoes at the carrier *Kaga* at the Battle of Midway; one struck but did no damage to the ship, which was sunk along with three other carriers by U.S. Navy dive bombers. Various U.S. and Japanese records differ, but up to 39 Japanese destroyers may have been sunk by U.S. submarine torpedoes. (National Archives)

Ship killers: Lieutenant Richard H. O'Kane (left) and Lieutenant Commander Dudley (Mush) Morton, the executive officer and commanding officer, respectively, of the *Wahoo* (SS 238) in February 1943. The *Wahoo* was lost to a Japanese air attack in late 1943 with Morton and his crew of 79. O'Kane, transferred to his own command, survived the war, was awarded the Medal of Honor for his exploits, and retired as a rear admiral. (National Archives)

to reactivate the magnetic feature. Boats returning from Australia followed the opposite procedure, starting out with the magnetic feature activated, then deactivating it after falling under Lockwood's control.[8]

The last problem to be addressed was the issue of "duds"—torpedoes that did not explode when they struck a ship. The effects were similar to that of a faulty magnetic exploder: no explosion, except that in some cases the torpedo was seen to hit the target. This phenomenon was observed by Lieutenant Commander Thomas B. Klakring during his second patrol in command of the *Guardfish* (SS 217). On 25 August 1942, Klakring fired two Mark 14 torpedoes at a freighter off the northeast coast of Honshu Island in the Japanese home islands. Klakring watched the first torpedo leap vertically out of the sea and porpoise several times before passing a few yards in front of the freighter. The second torpedo struck right under the enemy ship's bridge, sending a plume of spray as high as the main deck. It did not explode. Looking through the periscope, Klakring observed the target as it slowed down and changed course. Klakring fired another torpedo, which either ran under, without exploding, or missed completely.

Klakring attempted to pursue the target, but as he surfaced, his radar detected four approaching planes, forcing him to crash dive. After the planes passed, Klakring surfaced and began steaming through the target area in pursuit of the elusive merchant ship. The *Guardfish* passed within 400 yards of one of the torpedoes she had fired. Klakring observed it floating vertically, without its warhead. The torpedo had struck the target with enough force to dislodge the warhead, yet it had not detonated: providing indisputable evidence that the exploder had failed.

The immediate reaction of the BuOrd to reports of duds was one of skepticism. As was the case with the depth control and magnetic exploder problems, the duds were blamed on the submariners, whom the Bureau accused of errors in judgment or of firing torpedoes too close to the target so that their run was not long enough to arm the torpedo. A few failures were always expected, and a mistake in installing the exploder might cause even the simplest contact mechanism to fail. Although less complex than the magnetic device, it was still an intricate mechanism.

Dramatic evidence of problems with the Mark 6 Mod 1's contact exploder would not be revealed until the submarine *Tinosa* (SS 283), under Lieutenant Commander Lawrence R. (Dan) Daspit, returned to Pearl Harbor from her second successful patrol on 4 August 1943. Daspit, who had yet to sink an enemy ship, was fuming mad. On the morning of 24 July 1943, he had attacked the *Tonan Maru No. 3*, a 19,000-ton whale factory that had been converted into a tanker, with four torpedoes. Two of them must have been hits based on the large splashes of water that Daspit observed through the periscope, but the target did not appear damaged. She turned away, leaving Daspit in poor firing position. Despite the poor setup, Daspit sent two more torpedoes her way, using a large track angle to complete the firing solution. One of these struck the stern, causing a large explosion that left the target dead in the water, but still afloat. (Unbeknownst to Daspit, it was the unfavorable track angle that led to the detonations). The *Tonan Maru* was now a sitting duck, or so Daspit thought. The *Tinosa*'s skipper recorded what happened in his patrol report:

1009 Having observed target carefully and found no evidence of a sinking, approached and fired one torpedo at starboard side. Hit, heard by sound to stop at same time I observed large splash. No apparent effect. Target had corrected list and was firing at periscope and at torpedo wakes with machine guns and one inch [gun].

1011 Fired eighth torpedo. Hit. No apparent effect.

1014 Fired ninth torpedo. Hit. No apparent effect. Target firing at periscope, when exposed, and at wake when torpedoes were running.

1039 Fired tenth torpedo. Hit. No apparent effect.

1048 Fired eleventh torpedo. Hit. No effect.

This torpedo hit well aft on the port side, made splash at the side of the ship and was then observed to have taken a right turn and to jump clear of the water about one hundred feet from the stern of the tanker. I find it hard to convince myself that I saw this.

1050 Fired twelfth torpedo. Hit. No effect.

1100 Fired thirteenth torpedo. Hit. No effect. Circled again to fire at other side.

1122 Picked up-high speed screws.

1125 Fired fourteenth torpedo. Hit. No effect.

1132-1/2 Fired fifteenth torpedo. Started deep. Destroyer range 1,000 yards. Torpedo heard to hit tanker and stop running by sound. Periscope had gone under by this time. No explosion. Had already decided to retain one torpedo for examination by base.[9]

Of the 15 torpedoes fired at the target under almost perfect conditions, 11 had been duds.

When the *Tinosa* returned to Pearl Harbor, the torpedo saved by Daspit was turned over to base's torpedo experts for inspection. Admiral Lockwood directed his torpedo officer, Lieutenant Commander Arthur H. Taylor, to find out what was wrong with the Mark 14 torpedo. He immediately proceeded to tear down and examine the weapon, but no defects were found. When bench-tested, even the detonator fired normally.

Admiral Lockwood was truly perplexed. Neither he, nor anyone in the Bureau of Ordnance could explain the reason for Daspit's 11 duds. According to the Bureau, their records showed that of the 1,811 torpedoes fired up to 30 June 1943, no more than 7 percent could be attributed to "duds, prematures, erratics and magnetic failures."[10] Although Lockwood thought the contact exploder highly suspect, there was still no hard evidence as to the cause of the defect. He expressed his concerns in a letter to Chief of the Bureau of Ordnance, Rear Admiral William H. P. Blandy, writing:

> The question as to whether this Mark 6 exploder will be fatally deformed by head-on collision is a very vital one to me. The reports of the TINOSA and WAHOO as to duds and deep running are practically unbelievable to me. On the other hand it is unbelievable that experienced submarine officers can shoot torpedoes on the simplest sort of elementary approaches with ranges near 1,000 yards at slow moving targets and still miss.[11]

At this point Captain Charles B. Momsen, the inventor of his namesake submarine escape device and hero of the *Squalus* (SS 192) rescue, came up with a practical scheme for testing the torpedo. Momsen, then at Pearl Harbor in command of Submarine Squadron 2, proposed that live torpedoes be fired at the steep cliffs of nearby Kahoolawe Island. The cliffs, which rise vertically from the sea floor 50 feet beneath the surface, provided an ideal test target. Momsen suggested that a submarine fire a series of Mark 14 war shots at the cliffs until the first dud was encountered. The errant torpedo could then be recovered and examined in an effort to determine the reason the torpedoes were not exploding as they were supposed to.

On 31 August 1943, the submarine *Muskallunge* (SS 262) started shooting Mark 14 torpedoes at the Kahoolawe cliffs. The first torpedo exploded when it hit the cliffs. The second was a dud. It was recovered and the following day inspected by Admiral Lockwood and his staff. Although the forward end of the warhead was crushed, the exploder was intact. Upon examination they discovered that the firing pins had traveled up the badly bent guidelines and had struck the fulminate caps, but not hard enough to set them off.

This shocking discovery led to additional tests of the Mark 14 warhead. These were conducted at Pearl Harbor and involved dummy warheads that were fitted with exploders and then dropped onto a steel plate from a height of 90 feet. When the warhead hit the plate at 90 degrees, as would occur during a "perfect" shot, the exploder mechanism was crushed. Only one-half of the exploders failed to work when the warhead struck the plate at 45 degrees, the equivalent to a glancing blow from a badly fired shot.

Here at long last was the explanation for the defect that had led to so many missed opportunities. The torpedo experts at Pearl Harbor immediately began work on the exploder mechanism. Three corrective measures were tried: an electric inertia switch developed by Commander Ellis A. Johnson, an ordnance expert, and two designs for modified pins that were produced independently by the torpedo shops at the Pearl Harbor submarine base and aboard the tender *Holland* (AS 3). Both of the latter solutions solved the problem by reducing the weight of the firing pin to the barest minimum, thus lessening the friction on the guide studs. Each one of the corrections worked during dry land tests, but the electric device was shelved in favor of the firing pin modifications, which were easier to implement.

The first submarine supplied with modified Mark 6 Mod 1 exploders incorporating changes to the firing pins was the *Barb* (SS 220). She left Pearl Harbor for the East China Sea on 20 September 1943, carrying 20 Mark 14 torpedoes. By mid-October enough exploders had been modified to supply all submarines leaving from Pearl Harbor with the modified exploders.

When the results of the tests conducted at Pearl Harbor reached Admiral Blandy, he lit a fire under the so-called torpedo experts at the Naval Torpedo Station at Newport, ordering them to expedite their own testing of the Mark 14 torpedo and its exploder. These later tests, conducted on the torpedo range at Newport, confirmed the results obtained at Pearl Harbor, proving without doubt that the exploder was improperly designed and had not been adequately tested. With this information in hand, Blandy ordered a complete redesign of the Mark 6 Mod 1 exploder—21 months after the start of hostilities!

Why had it taken so long to identify and correct the problems with the submarine torpedoes? Although much of the blame was attributed to "bureaucratic inertia" within the Bureau of Ordnance and the Torpedo Station, that was only part of the problem. It is certainly true that a number of senior officers within the so-called "Gun-Club," a euphemism for the BuOrd, "chose to define a 'dud' as a 'skipper's alibi to explain his miss'," yet the human element was certainly a major factor in trying to assess the Mark 14 failures.[12] After all, the torpedoes used in World War II were adjusted, aimed, and fired by officers and torpedomen's mates who were in the fleet—thousands of miles from the Navy's torpedo specialists. Faulty judgment, errors in calculation, mishandling, and poor maintenance *could* have accounted for the torpedo failures. No one knew at the time what the problem was and some, including Admiral Lockwood himself, seriously considered the possibility that the Japanese had developed a countermeasure to the influence exploder. The problem was exacerbated by the multiple defects, which tended to conceal one another. Thus, in historian Theodore Roscoe's words, "the torpedo troubles were not easily traced to their source, readily diagnosed and speedily remedied. Submarine warfare was too complex and the torpedo too complex a weapon for easy 'trouble shooting.'"[13]

There is no question that the ultimate blame for the Mark 14 debacle resided with the Torpedo Station at Newport. As the single source for all U.S. Navy torpedoes, it was solely responsible for the design, development, and testing of what was then the most complex weapon in the Navy's arsenal. But claims that the development staff was isolated from the larger U.S. technical and engineering community are unfounded. The Torpedo Station relied heavily on the General Electric Company for its advice and development of advanced electrical components used in the Mark 6 Mod 1 magnetic exploder. The Torpedo Station also worked closely with the Naval Research Laboratory in the development of the Navol Torpedo, a state-of-the-art advancement that would revolutionize torpedo performance.

The principle problem with the Mark 14 torpedo was its complexity: it was a highly intricate weapons system in an era when the realities of such systems remained unknown. The problem was aggravated by the small number of naval officers who were educated and trained as torpedo experts and the lack of money allocated for testing and evaluation. There were not enough qualified individuals, especially within the submarine service, which was the greatest user of torpe-

Another Japanese merchant ship goes down after being torpedoed by a U.S. submarine. American undersea craft are credited with sinking roughly 4.8 million tons of Japanese merchant shipping. The high scores were made after the severe torpedo problems were fixed and were possible in large part because of poor Japanese ASW strategy, tactics, and forces. The Japanese Navy had expected the war at sea to last months—not years. (National Archives)

does, to supervise the development and production of this weapon. The personnel problem was exacerbated in the years immediately preceding World War II by the rapid expansion of the submarine service, which, like every other branch of the Navy, needed every most experienced officer that it could obtain to fill the billets in the new ships that were rapidly being added to the fleet.

Finally there was a lack of money. Budgetary constraints forced on the Navy by the Great Depression severely restricted the amount of money that could be allocated for testing and evaluation, as well as development. This factor coupled with the very expensive cost of a torpedo—$10,000 in an era when an automobile cost $600—made destructive testing of a live war shot a rare event. Even when funds began to flow as war approached, the personnel at the Torpedo Station became preoccupied with the shortage of torpedoes and the need to expedite production.

THE EFFECT ON THE WAR EFFORT

The torpedo situation had profound effect on the first two years of the war in the Pacific. With the destruction of the U.S. battle force by Japanese carrier aircraft at Pearl Harbor on 7 December 1941, the only U.S. naval forces that could possibly halt or even slow the Japanese advances in the western Pacific were aircraft carriers and submarines.

As discussed previously (Chapter 7), the early U.S. carrier operations against the Japanese were limited by the problems with their Mark 13 aerial torpedoes, although at the Coral Sea in early May 1942 the TBD Devastators were able to sink a Japanese light carrier with torpedoes. A month later at Midway the TBDs were massacred, and that 41 carrier-launched torpedo planes scored no hits was the fault of tactics as well as highly restrictive flight parameters for aircraft dropping the Mark 13.

More serious was the submarine torpedo problem, because there were no SBD Dauntless dive bombers as embarked in the carriers to compensate for the submarine torpedo shortfalls. For 21 months—of a 45-month conflict—U.S. submarines were hobbled by torpedoes that rarely functioned effectively. Thus, submarines contributed relatively little to the Allied cause until the fall of 1943. For example, the first Japanese aircraft carrier to be sunk by a submarine was the small *Chuyo* in December 1943. By that time U.S. carrier-based aircraft had sunk four Japanese fleet carriers and three light carriers.

But, at least, in the fall of 1943 effective torpedoes were being loaded in U.S. submarines. The performance of U.S. submarines improved dramatically thereafter, as evidenced by the sinking of the large Japanese carriers *Shokaku* and *Taiho* in June 1944.

MAINTENANCE AND THE "TORPEDO SCANDAL"

The so-called "Great Torpedo Scandal" of 1941–1943 has created a small cottage industry of observers who have been very vocal in their criticism of the Bureau of Ordnance and the Newport Torpedo Station. Although most of the torpedo problems experienced by the Navy during this period were directly attributable to "bureaucratic inertia," there were other factors involved.

The torpedo was in many respects the most complex naval weapon of World War II. The steam torpedo was a highly sophisticated piece of ordnance that contained an intricate propulsion system, an on-board power supply, a guidance and control system, and two different warheads: one for practice runs and proofing and one for war shots. Torpedoes required careful handling and regular maintenance to achieve reliable performance. Erratic torpedo runs, "coldshots," depth-keeping problems, and premature detonations were not uncommon occurrences even with "perfect" torpedoes. The U.S. Navy instructions for torpedomen in the immediate prewar period listed 33 adjustments or maintenance issues that could lead to coldshots or other problems related to depth keeping, deflection, and speed.[14] The gyro, which served as the heart of the torpedo's guidance system "was a delicate mechanism" that required "a high degree of mechanical skill and a great amount of care and patience . . . in working with them."[15]

None of the above cautions included potential problems with warhead exploders (either impact or magnetic), which had to be tested annually and stowed in the uncocked, but armed position. For a war shot, the mechanism had to be cocked and then placed in the unarmed position. The Navy's manuals called special attention to the latter action, as failure to perform that function would result in a dud.

The experts at the Torpedo Station, knowing these issues, cast a jaundiced eye at the reports coming in from the fleet that claimed defects in torpedoes, each and every one of which had undergone proof testing prior to shipment—an action that itself contributed to the reliability problem. After a proof firing, each torpedo was disassembled, cleaned, and prepared for shipment to fleet. Each of the major components—warhead (less exploder), air flask, and afterbody—was carefully packaged in its own wooden shipping crate. The delicate gyroscope was shipped separately in a fourth crate that also contained tools and spare parts.

The crates containing the torpedo were consolidated into one large wooden box that was shipped overseas, where the torpedo was uncrated, reassembled, tested, mated with a warhead, and prepared for service use by specialized ratings (torpedomen) on board submarine tenders or in torpedo workshops at Pearl Harbor or established at advanced operating bases. Reassembling a torpedo and preparing it for sea was a complex process that required careful handling and a high degree of skill. Failure to perform these functions properly could result in a failure. (See Appendix D.)

CHAPTER TEN

Out Ranged

The Long Lance vs. the Mark 15 (1942–1944)

Few followers of naval history realize that the first surface action conducted by the U.S. Navy in World War II was a torpedo attack launched by four aging destroyers of World War I vintage. These 1,200-ton, flush-deck destroyers carried four triple torpedo tubes loaded with improved versions of the 21-inch Mark 8 torpedo that had been first introduced in 1915. Although the Mark 8 torpedo was limited to a single 29-knot speed setting, it packed a powerful punch of 475 pounds of TNT that would be detonated by the highly reliable Mark 3 contact exploder.

On 16 January 1942, Admiral Thomas C. Hart, the commander of all Allied naval forces in the Far East, was notified that a Japanese force of 16 transports escorted by a cruiser and 12 destroyers was heading for the Macassar Strait, apparently bound for Balikpapan in the Dutch East Indies. Five days earlier the Japanese had occupied the city of Tarakan in oil-rich northeastern Borneo, and the port city of Balikpapan, 350 miles to the south, was their most likely target. The only defensive force available to Hart was an American strike force, commanded by Rear Admiral William A. Glassford, consisting of the light cruisers *Boise* (CL 47) and *Marblehead* (CL 12), and eight destroyers of Destroyer Squadron 29. The latter were frequently referred to as "four stackers" because of the four smokestacks that dominated their profile.

By the time the Japanese force had moved toward Balikpapan, however, only the *Boise*, *Marblehead*, and six destroyers were within striking range of the expected invasion point. Before the force could go into action, *Boise* struck a submerged rock, which slashed a long hole in her hull. The cruiser was forced to retire to the nearest friendly port for repairs, escorted by one of the destroyers. Next, the *Marblehead* suffered a turbine casualty, which slowed her speed to 15 knots. Another destroyer was detached to protect her, leaving just four of the aging destroyers, *John D. Ford* (DD 228), *Pope* (DD 225), *Parrott* (DD 218), and *Paul Jones* (DD 230) to confront an enemy force of 12 modern destroyers, a light cruiser, and numerous auxiliaries.

With the *Marblehead* out of action, tactical command of the four destroyers passed to Commander Paul H. Talbot, the commander of Destroyer Division 29 in the *Ford*. The four destroyers steamed through heavy swells, making 25 knots, as the ships moved northward up the Makassar Strait during the afternoon of 23 January 1942. To confuse the enemy in case he was spotted from the air, Talbot ordered a deceptive course change to the east as if they were bound for Mandar Bay in the Celebes. An hour after sundown, at about 7:30 PM, Talbot swung the column into an abrupt left turn toward Balikpapan and increased speed to 27 knots.

As the destroyer crews made ready for action, Commander Talbot passed the following instructions to the other ship's captains over the TBS (Telephone-Between-Ships): "Torpedo Attack. Use own discretion attacking independently when targets are located. . . .

When all torpedoes fired, close with all guns."[1] At 2:30 AM in the early hours of 24 January 1942, Talbot's force of four destroyers was abreast of the port of Balikpapan. The port was lit up by raging fires set by the retreating Dutch in an attempt to deny the port's facilities to the Japanese invaders. As they closed on the reported position of the Japanese transports, four enemy destroyers crossed ahead of the column. One of the Japanese destroyers used a blinker light to challenge the approaching ships, which went unanswered. Talbot ordered an immediate course change, passing the enemy ships without incident. Just as suddenly the U.S. destroyers came across a large group of Japanese transports anchored some five miles off the entrance to Balikpapan Harbor.

As the U.S. column passed through the pack of anchored transports they unleashed a salvo of torpedoes. The *Parrott*, which was in the lead, fired first, sending a spread of three torpedoes toward a large transport at close range. Seconds ticked by as the *Parrott*'s crew waited for the torpedoes to complete their run to the target, anxiously waiting for the expected explosion—that never came. Two minutes later the destroyer launched five more torpedoes to starboard at another target 1,000 yards away. Again nothing happened. Simultaneously, the *Ford* fired one torpedo at another anchored transport.

The details of *Ford*'s torpedo shot were later recounted by William P. Mack, the ship's gunnery officer:

> Suddenly we found ourselves right in the midst of the Jap Transports. Down on the bridge I could hear Captain [Jacob] Cooper saying "Action Port! Action port!" and Lieutenant [John] Slaughter readying the torpedo battery. Back aft the mounts swung to follow his director. "Fire one," he said. "Fire one," repeated his telephone talker. Then came the peculiar combination of a muffled explosion, a whine, a swish, and a splash, that follows the firing of a torpedo. I watched the torpedo come to the surface once and then dive again as it steadied on its run. Astern, the *Pope*, *Paul Jones*, and *Parrott* were carefully picking targets and firing. . . . My talker was calmly counting off seconds as our first torpedo ran towards its target. "Mark!" he shouted, as the time came for it to hit. Seconds passed. Nothing happened. We knew our first [shot] had missed.[2]

The last ship in line, the *Jones*, also fired one torpedo at an enemy warship that briefly loomed out of the smoke-filled night, but it missed.

During their first pass the four American destroyers had fired ten Mark 8 torpedoes at close range. "Here," according to historian Walter G. Winslow, "were the first torpedoes fired in anger in the Pacific by highly trained destroyermen, and the scoreboard read zero hits."[3] Undaunted by their failure to inflict any damage on the enemy transports, Commander Talbot calmly turned the column back for another run. Maybe nerves had settled down, but for whatever reason this second attack proved much more successful: the *Parrott* fired three torpedoes at a transport, sinking the 3,519-ton *Somanaura Maru*. Two minutes later the *Pope*, *Parrot*, and *Paul Jones* fired torpedoes at another ship, sinking the 7,000-ton transport *Tatsukami Maru*. As the action continued the four pipers launched torpedoes at additional Japanese enemy ships, adding a large patrol boat and the 5,000-ton *Kuretake Maru* to their score. When all of their torpedoes were expended they continued to pound away with their 4-inch deck guns, wreaking havoc on the enemy transports. The destroyers escaped without damage or casualties.

The Battle of Balikpapan was the first surface action of the U.S. Navy since the Spanish-American War and the first fought in World War II. It was an incredible success fought bravely by the officers and sailors of the *John D. Ford*, *Pope*, *Parrott*, and the *Paul Jones*. These four destroyers were officially credited with sinking five Japanese ships, and they undoubtedly damaged and possibly sank several more. The Americans were greatly aided by the ineptness of the Japanese screening ships, whose lookouts mistook the U.S. destroyers for their own forces, yielding the important element of surprise to the American destroyers.

The *John D. Ford* (DD 228), shown in 1926, was one of four aging destroyers of World War I vintage that fought the first surface action and torpedo attack conducted by the U.S. Navy in World War II. These 1,200-ton, flush-deck destroyers carried four, waist-mounted, triple torpedo tubes loaded with improved versions of the 21-inch Mark 8 torpedo that were first introduced in the fleet in 1915. (U.S. Navy)

Unfortunately, the successful use of U.S. surface torpedoes would not be repeated again until the Battle of Vella Gulf on 17 August 1943. Although poor doctrine contributed greatly to the ineffectiveness of U.S. surface torpedo attacks during the early months of World War II, the torpedo that U.S. destroyermen relied upon most—the Mark 15—suffered from many of the same defects as the submariner's Mark 14.

There were two basic types of surface-launched torpedoes in the U.S. Navy at the start of World War II: single-speed torpedoes like the older, but still-effective Mark 8, and the newer multispeed torpedoes bearing the Mark 11, 12, and 15 designations. The latter were nearly two feet longer than the Mark 8 and could be set for high speed and short range, intermediate speed and moderate range, or slow speed and long range. Both types of torpedoes were 21-inch in diameter.

The torpedo tubes on all destroyers that entered service before the *Farragut* (DD 348) was completed in 1934 were designed for the Mark 8 and could not fire the multispeed torpedoes introduced in the 1930s. The *Farragut* and all subsequent U.S. destroyers were fitted with adjustable torpedo tubes that could take either type of torpedo. The torpedo most commonly supplied to these ships was the Mark 15; the Mark 11 and Mark 12 were undoubtedly furnished to some ships during the early stages of the war to make up for the shortage of Mark 15s. Like its cousin, the Mark 14 used by the submarine service, the warheads fitted to the Mark 15, as was true for the Mark 11 and Mark 12, relied upon the dubious Mark 6 magnetic exploder.

As the men on board the destroyers and scout cruisers assigned to the Pacific Fleet prepared to engage the enemy in the late summer of 1942, they had no idea of the defects in the Mark 15 torpedo.[4] It would be many months before they discovered why their torpedoes did not seem to be effective. They also lacked any knowledge of the superior enemy torpedoes they would be facing when they engaged in a series of critical night battles in the Solomon Islands from August 1942 to October 1943.

By the start of World War II the Imperial Japanese Navy had fitted all of its modern destroyers and heavy cruisers with a revolutionary new torpedo designated as the Type 93. The Type 93, which historian Samuel Eliot Morison latter dubbed the Long Lance torpedo, was the only oxygen-fueled torpedo employed by any of the combatants in World War II.[5] The development of the Type 93 was a closely guarded secret, and its existence remained unknown to the U.S. Navy until very late in the war. Although the Type 93 was a very remarkable torpedo, it was not a perfect weapon. It too

The USS *Dunlap* (DD 384) launches a Mark 15 torpedo during exercises conducted on 3 July 1942. The *Dunlap* and similar *Fanning* (DD 385) formed a two-ship destroyer class. An earlier *Fanning* (DD 37) sank the German U-boat *U-58* off Queenstown, Ireland, in 1917. Two German U-boat sinkings were credited to U.S. destroyers in World War I. (U.S. Navy)

suffered from a tendency to explode prematurely, and its accuracy, especially at extremely long range, was not good—expected error at maximum range was almost one mile. Of the 128 torpedoes launched by the Japanese during the opening phases of the Battle of the Java Sea on 27 February 1942, just one struck home, sinking the Dutch destroyer *Kortenaer*. Most were fired at ranges in excess of 15,000 yards. Nevertheless the Type 93 was a formidable weapon that would inflict severe damage on a large number of U.S. Navy warships during the war.

Although the torpedo was considered the great equalizer in surface actions, the U.S. Navy's prewar emphasis on gunfire, inadequate night fighting ability, and flawed torpedo doctrine, were major weaknesses that contributed to the heavy losses suffered by the U.S. fleet in the four major night actions fought during the campaign for Guadalcanal (Savo Island, Cape Esperance, the Naval Battle of Guadalcanal, and Tassafaronga). While considerable effort had been spent on improving torpedoes and their fire control during the interwar years, long-range gunfire was still considered to be the fleet's primary weapon. This view was reflected in the decision to remove torpedoes from all of the Navy's heavy cruisers—a step that was almost unique among the world's navies. The night actions around Guadalcanal were all marked by the U.S. Navy's reliance on cruiser gunfire to the exclusion of torpedo attacks and the detriment of the destroyers whose movements were held in strict conformity to those of the heavier ships. Most of the torpedoes unleashed by U.S. destroyers during these actions were rushed shots, fired at passing targets of opportunity.

In the first of these actions, the Battle of Savo Island that occurred in the early hours of 9 August 1942, the U.S. destroyer *Bagley* (DD 386), which was screening the Australian cruiser *Canberra*, attempted to bring her starboard torpedo batteries to bear on the Japanese cruiser column that had suddenly loomed out of the night. The opportunity passed before her harried torpedomen could insert firing primers into the black powder charges that expelled the torpedoes from their

The 22 destroyers of the similar *Craven* (DD 382) and *McCall* (DD 400) classes carried the heaviest torpedo battery of U.S. warships—16 tubes in four quadruple mounts. However, they were limited to eight tubes per broadside; some 12-tube destroyers could fire all tubes to either beam. The 16-tube *Benham* (DD 397), shown in New York in 1939, the year of her completion, exceeded 40 knots in trials. The ship's two forward 5-inch/38 guns have enclosed mounts; the two after guns are in open mounts. (U.S. Navy)

tubes. Her captain, Lieutenant Commander George A. Sinclair, did not wait for his torpedomen, but made a tight circle to the left until the port torpedoes bore. By the time the *Bagley* had fired her torpedoes, the enemy column was "disappearing so fast to the northeastward that the best-aimed torpedoes in the world could not catch up with it."[6] At least six Japanese torpedoes found their targets, sinking the heavy cruisers *Canberra* and *Vincennes* (CA 44), and severely damaging the *Chicago* (CA 29).

It would be another two months before the first Mark 15 fired in anger would actually strike a Japanese warship. This occurred six minutes before midnight on 11 October 1942, during the Battle of Cape Esperance, when a torpedo fired from the *Buchanan* (DD 484) or *Duncan* (DD 485)—both destroyers launched torpedoes at the same target—struck the Japanese heavy cruiser *Furutaka*. At least one of the torpedoes exploded in her forward fireroom, contributing further damage to the heavily shelled warship, which finally went under at 12:40 AM.

The third slugging match between the opposing surface forces contesting the waters leading to Guadalcanal took place in mid-November 1942. It was a three-day affair known as the Naval Battle of Guadalcanal. The first phase of the battle occurred during the night of 13 November, when a column of U.S. ships led by the destroyer *Cushing* (DD 376) collided head on with a Japanese force that included two Japanese battleships intent on bombarding Henderson Field on Guadalcanal. Screening the two battleships were six destroyers and a light cruiser.

The *Cushing*—the third ship to bear the name honoring the Civil War hero who had sunk the *Albemarle*—was the first to sight the enemy as two screening Japanese destroyers crossed ahead of her at 3,000 yards, moving from port to starboard.[7] A flash radio report was immediately sent to the American commander, Rear Admiral Daniel J. Callaghan, in flagship *Atlanta* (CLAA 51), warning of the enemy force. In the meantime, the *Cushing*'s captain ordered a quick turn to port in order to unmask the ship's torpedo battery. The *Cushing*'s turn caught the rest of the U.S. column by surprise, causing much confusion and a pileup in the van. Before she could begin firing, the Japanese, taking advantage of the confusion on the American side, took the initiative and opened fire. Within minutes the *Cushing* was hit by several shells, which severed all power lines and slowed her. Her captain, Lieutenant Edward N. Parker, conned her as best he could. Using

The first Mark 15 torpedo fired at an enemy warship occurred six minutes before midnight on 11 October 1942, during the Battle of Cape Esperance, when a torpedo fired from either the *Buchanan* (DD 484) or *Duncan* (DD 485)—both destroyers launched torpedoes at the same target—struck and sank the Japanese heavy cruiser *Furutaka*. The *Buchanan*, dressed in camouflage, is shown at the New York Navy Yard on 28 May 1942. (U.S. Navy)

hand-steering control and what little way remained, he swung *Cushing* right and fired six torpedoes by local control at the battleship *Hiei*, which was only 1,000 yards away. Three of the torpedoes seemed to strike home, but they had no effect on the battleship and may have exploded prematurely as the *Hiei* turned slowly away.

The *Laffey* (DD 459), directly astern of the *Cushing*, passed so close to the *Hiei* that the two torpedoes she launched did not have time to arm and bounced harmlessly off of the battleship's side. Next in line was the *Sterett* (DD 407), which launched four more torpedoes at the *Hiei*. Again no hits were recorded. The *O'Bannon* (DD 450), the forth ship in line, sent off two more carefully aimed torpedoes at the *Hiei*. The wakes from the *O'Bannon*'s torpedoes vanished into the darkness, heading straight for the enemy battleship, but they either ran too deep or failed to detonate.

Fourteen torpedoes had been fired at close range at the *Hiei*, but none had inflicted damage to the enemy battleship. As the battle progressed, 26 more torpedoes were unleashed by U.S. destroyers at other targets—*Sterett*: 2, *Monssen*: 10, *Barton*: 4, *Fletcher*: 10. Of these, only the two fired by *Sterret* near the very end of the engagement were seen to strike and sink what was then believed to be an enemy destroyer.[8]

The Type 93 torpedoes proved to be much more devastating to the American forces that night then the Mark 15 was to the Japanese. During the first 15 minutes of the engagement, Japanese destroyers launched no fewer than 44 Long Lance torpedoes at various American targets. At least six of these fish found their mark, sinking the anti-aircraft cruisers *Atlanta* and *Juneau* (CLAA 52) and the destroyers *Laffey* and *Barton* (DD 599) and damaging the heavy cruiser *Portland* (CA 33). Clearly there was something wrong with American torpedo doctrine or the torpedoes, or both.

After analyzing the action reports submitted in the aftermath of the Naval Battle of Guadalcanal, Admiral Ernest J. King, the Chief of Naval Operations, concluded that a combination of rigid procedure and the emphasis on gunnery had stripped the destroyers of their ability to conduct successful torpedo attacks. Even when the destroyers tried to operate independently at one point in the battle, Rear Admiral Daniel J. Callaghan, the officer in tactical command, had ordered them back into formation. There were many reasons why the Mark 15 torpedo failed during the battle. Some had been launched too close to the target and failed to arm, some, like those launched by the *Fletcher* (DD 445), had been fired at long range, some missed due to poor marksmanship or evasive maneuvering by their

The USS *Atlanta* was the first of 11 similar anti-aircraft cruisers; they were the last U.S. cruisers to carry 21-inch torpedo tubes. Intended to provide anti-air defense for carrier task forces, they carried 12 or 16 5-inch/38 dual-purpose guns in twin mounts plus 40-mm and 20-mm guns on a standard displacement of 6,000 tons. The *Atlanta*, shown here shortly after her completion in December 1941, and her sister ship *Juneau* (CLAA 52) were sunk by Japanese Long Lance torpedoes and gunfire off of Guadalcanal in November 1942. Note the quadruple torpedo tube mount amidships, forward of the twin 5-inch "wing" mount. (U.S. Navy)

intended targets, and a few probably failed to explode when they struck a target. The last should have raised grave concerns about U.S. torpedo performance, but it could not be easily confirmed from the scant information contained in the after-action reports. But the failings of the Mark 15 were consistent with the expectations of torpedo warfare held by most senior U.S. naval officers and did not appear to them to warrant further investigation.

Of greater concern should have been the horrible performance of the 18 Mark 15 torpedoes launched three weeks earlier during the Battle of Santa Cruz, primarily a carrier action fought on 26 October. All of the torpedoes launched that day were fired at close range at targets that were dead in the water, yet of the 18 torpedoes launched, only 9 struck their target. This was a dismal performance for a weapon fired under nearly ideal conditions—neither U.S. ship was engaged by the enemy and both had plenty of time to prepare the torpedo tubes and take careful aim. Paradoxically, the two targets were derelict U.S. warships that were being sunk to keep them from possibly falling into the hands of the enemy.

The first of the Mark 15 torpedoes launched that day were fired at the U.S. destroyer *Porter* (DD 356). The *Porter* was screening the carrier *Enterprise* (CV 6) when a damaged TBF Avenger, unable to land on the carrier, ditched some 1,500 yards ahead of the *Porter*. The pilot had been unable to release his torpedo because of damage to the aircraft and the Mark 13 torpedo was released when the plane struck the water. As the *Porter* slowed to lower a boat to rescue the TBF's crew, a lookout screamed that a torpedo was approaching. Two circling Wildcat fighters tried to detonate the porpoising torpedo by making firing passes, but could not stop the "fish." A few moments later the aircraft's torpedo slammed into the destroyer and exploded. It struck between the *Porter*'s two firerooms, knocking out all power to the ship's engines and stopping her dead in the water.

All but one of the approximately 40 Mark 15 torpedoes launched by U.S. warships during the Naval Battle of Guadalcanal missed their mark. Some, such as those launched from this quintuple torpedo mount on the USS *Fletcher* (DD 445), had been fired at too long a range, while others were launched too close to the targets and failed to arm. Note the blast shield enclosing the trainer's station atop the quintuple mount. (U.S. Navy)

The *Porter* was mortally injured, and a short time later Rear Admiral Thomas Kinkaid ordered another destroyer to remove her crew. After taking off the ship's company, the *Shaw* (DD 373) fired two torpedoes (presumably Mark 15s) at the disabled destroyer. One torpedo missed and the other passed under the target without exploding. The *Shaw* eventually sank the reluctant derelict with 17 rounds of 5-inch gunfire. The *Porter* was the only U.S. warship ever to be sunk by "friendly fire" from a Mark 13 torpedo. (At the time it was thought that a Japanese submarine, later identified as the *I-21*, had sunk the *Porter*.)

Even more ominous were results of attempts by the destroyer *Mustin* (DD 413) to send the dying *Hornet* (CV 8) under during the Battle of Santa Cruz. The carrier had taken a pounding all day, was without power and listing badly when she was abandoned at 4:00 PM on 25 October 1942. With no hope of saving the stricken carrier, the destroyer *Mustin* was ordered to sink her. The *Mustin*, a mile away off the *Hornet*'s beam, fired eight carefully aimed torpedoes in slow succession. As the historian Samuel Morison declared, the "results were not complimentary to American torpedo performance."[9] One was observed broaching astern of the target, another circled and exploded 300 yards from the *Mustin*, one exploded unseen, and nothing was seen or heard of two more. Only three of the torpedoes exploded against the *Hornet*'s hull, but she refused to sink.

The destroyer *Anderson* (DD 411) was called in to finish her off. The *Anderson* fired eight more torpedoes at the *Hornet*, of which six hit, one was premature and one missed. Unfortunately all of the torpedoes had been fired at the listing *Hornet*'s port or high side, so that the early effect of the hits was to provide counterflood-

ing. Both destroyers had to resort to shellfire, pouring 130 rounds of 5-inch ammunition into the carrier's hull before departing. By then the *Hornet* was ablaze along her entire length. The carrier stayed afloat until early the next morning, when she was struck by four Type 93 torpedoes launched from Japanese destroyers that had come across what was left of the once-proud ship that, during her brief career, had launched the Doolittle bombers that struck Tokyo and other Japanese cities on 18 April 1942.

The fiasco surrounding the attempted U.S. sinking of the *Hornet* did not go unnoticed. Admiral Nimitz forwarded a copy of the report describing the failures to Admiral King in early December. It was prepared after the battle by the commander of Destroyer Squadron 2, Captain Harold R. Holcomb, who considered the causes of these torpedo failures to be "a matter of conjecture, since no conclusive data has been uncovered definitely establishing a probable reason."[10] Holcomb attributed the two duds (those that failed to explode on contact) to the old-type firing springs, which were released to force the firing pins into the detonator. These were "recognized as a prolific source of failures on both the influence and impact firing" and were being replaced by stronger springs as the Mark 15 torpedoes underwent the overhaul processes conducted on a routine basis.[11] The magnetic influence circuits had been removed in the torpedoes fired by both destroyers, eliminating that as a possible cause of the failure. Holcomb also speculated on the reasons for the one premature, which might have been caused by an electrical fault or possibly by impact against debris in the water. As to the four erratic runs, Holcomb thought that these might have been due to malfunctions that could have been avoided if the torpedoes had been overhauled on a more frequent basis. While the torpedoes had been within the annual overhaul period, most had not been overhauled within the past six months.

According to the Bureau of Ordnance, the failures experienced in the attempt to sink the *Hornet* could be attributed to the weak firing pin springs and the failure to overhaul the torpedoes on a regular basis. The torpedo overhaul instructions issued to the Pacific Fleet in September 1942 did not contain any reference to testing of exploders. By now submarines were routinely testing the Mark 6 Mod 1 exploder after every patrol. The BuOrd now suggested that the destroyer force might prevent future exploder failures by more frequent testing of the mechanism. Although these were reasonable suppositions based on the information being received at the time, the Bureau's statement that "the failures presented no evidence of any basis in material defect," was indicative of the attitude maintained by those in charge of investigating the torpedo problems.[12] As the Commander Destroyers, Pacific Fleet pointed out in a sharply worded letter to Admiral King, this conclusion was not justified based on the known failings in the firing pin springs.

THE LONG LANCE TORPEDO

The Japanese Type 93 Long Lance was the largest torpedo and probably the fastest used by any navy in World War II. Its explosive charge of 1,080 pounds compared to 879½ pounds in the U.S. Mark 17 torpedo, the largest American torpedo of the war (although none was used in combat). The Long Lance torpedoes armed several classes of Japanese cruisers and destroyers. The secret to the Long Lance's high performance was the use of oxygen as the propellant. The French Navy had first considered using oxygen as a torpedo propellant but abandoned the concept because of the hazard of accidental explosion. The British similarly rejected the concept, but the Japanese persevered, and the Type 93 was approved for production in 1933 (year 2,593 of the Japanese calendar). It was believed that oxygen torpedoes would outrange naval guns, creating a revolution in naval tactics. By December 1941 the Japanese had produced some 1,350 Long Lance torpedoes, about one-half the total that would be manufactured. The Long Lance weighed 5,952 pounds and was 29½ feet long and 24 inches in diameter. (The standard Allied and Axis torpedoes were 21 inches in diameter.) The range varied with setting—21,900 yards at 48 to 50 knots, 35,000 yards at 40 to 42 knots, and 43,700 yards at 36 to 38 knots. A larger Type 93 Model 3 was put in production in 1945, but was not used in combat; it weighed 6,173 pounds and carried a warhead of 1,720 pounds, or twice the largest U.S. torpedo warhead. Also, oxygen-powered torpedoes did not leave a surface wake as did steam-driven torpedoes. With their great speed and range, and large warhead, they were highly effective against Allied warships in the early surface battles of the Pacific War.

The USS *Porter* (DD 356) was one of 13 large "destroyer leaders" built during the 1930s, intended to serve as destroyer squadron flagships. They mounted eight 5-inch/38 guns in twin turrets and eight (as *Porter*) or 12 21-inch torpedo tubes in quadruple mounts. The *Porter*, shown in 1938, two years after being commissioned, was accidentally sunk by a Mark 13 torpedo from a U.S. Navy TBF Avenger torpedo aircraft off the Santa Cruz Islands on 26 October 1942. (U.S. Navy)

The USS *Mustin* (DD 413), shown here at Pearl Harbor in June 1942, was ordered to sink the dying *Hornet* (CV 8) after the carrier was abandoned during the Battle of Santa Cruz in October 1942. The *Mustin* fired eight torpedoes in slow succession; the results were not complimentary to American torpedo performance: One broached astern of the carrier, another circled and exploded 300 yards from the *Mustin*, one exploded unseen, and nothing was seen or heard of two more; only three of the torpedoes exploded against the *Hornet*. (National Archives)

The last of the night actions in the waters near Guadalcanal—the Battle of Tassafaronga—was another debacle for the U.S. Navy. Rear Admiral Carleton H. Wright, the task force commander, again made the mistake of tying his destroyers to the cruiser column. Although his destroyers beat the Japanese to the punch, none of the 20 Mark 15s launched by the three destroyer's in the van of Wright's column struck enemy ships. All were launched too far from their targets, which were steaming away at high speed, for them to be overtaken. Despite their initial handicaps of surprise, cluttered decks, and U.S. gunfire, the disciplined crews on the Japanese destroyers managed to counterattack, unleashing 20 Type 93 torpedoes, which were faster, longer range, and harder hitting than their American counterparts. Within minutes they had crippled the cruisers *Minneapolis* (CA 36), hit by two torpedoes, and the *New Orleans* (CA 32) and *Pensacola* (CA 24).

A few minutes later, the Japanese destroyer *Oyashio* trained her torpedo tubes on the *Northampton* (CA 26), adjusted her fire-control settings, and launched eight torpedoes in the direction of the American ships. Although the *Northampton* sighted the wakes coming toward her on the port bow, the torpedoes could not be avoided. She was struck twice in the after part of the ship. Although the crew fought valiantly to save the cruiser, the damage was too great and she went down slowly with little loss of life.

Tassafaronga—fought during the night of 30 November–1 December 1942—"was a sharp defeat inflicted on an alert and superior cruiser force by a partially surprised and inferior destroyer force," wrote Morison.[13] In exchange for one destroyer (sunk by U.S. gunfire) the Japanese sank one U.S. cruiser and put three others out of action for more than a year.

A smaller U.S.-Japanese surface action occurred in a remote area of the Pacific on 26 March 1943 that again illustrated the difficulties with surface torpedo attacks—for both sides. Earlier the Japanese had captured two remote islands in the Aleutian chain, which stretches from southern Alaska, across the International Date Line, and almost to the coast of Soviet Siberia. The action occurred when a Japanese force was escorting and screening three supply ships bound for the Japanese-held islands. A U.S. task force was encountered in what became known as the Battle of the Komandorski Islands, the westernmost of the Aleutians. The battle lasted for three and one-half hours and was at times shrouded in fog. This condition gave the American ships an advantage. All were fitted with radar; no Japanese ships had the device.[14]

The Japanese, with two heavy, two light cruisers, and four destroyers significantly outgunned the Americans. All carried Long Lance torpedoes. (Two of these destroyers did not participate in the battle.)

The Americans had a single heavy cruiser, the *Salt Lake City* (CA 25), the outdated scout cruiser *Richmond* (CL 9), and four destroyers. The heavy cruiser did not have torpedoes; she had been built with tubes, but they were removed shortly after her completion in 1929. The five other U.S. ships carried Mark 15 torpedoes.

The battle was primarily a gunnery duel, with several ships on both sides being damaged. At one point the U.S. force commander, Rear Admiral Charles H. McMorris, ordered his destroyers to make a torpedo attack. They closed to within 9,500 yards of the enemy cruisers, but only the *Bailey* (DD 492) was able to gain a firing angle and unleashed five torpedoes. None hit.

During the battle six of the Japanese warships—including the two heavy cruisers—fired 42 or 43 torpedoes at the U.S. ships, all at ranges far beyond the comprehension of American commanders. One commander, seeing torpedo tracks and knowing the distance to the Japanese ships, declared that they could only be fish. Then-Lieutenant (jg) Jerry Miller in the *Richmond* recalled seeing a torpedo that passed under his ship, "It struck terror in [us] and I am satisfied that if that torpedo had been a few feet shallower, [the] part [of the story] about the *Richmond* would end here." He later told one of the authors of this book, "it was a close call—some would say a miracle—that not one of those long-range Japanese torpedoes struck our ships."[15]

Historian John Lorelli wrote of the Komandorski battle:

> [Japanese] destroyers also performed poorly, showing nothing of the usual élan of Japanese destroyermen. A measure of the American's good luck and [Japanese] lack of luck is found in the fact that forty-two torpedoes were fired by the Japanese and none struck home. The outcome of the battle would surely have been different had even one of those devastating weapons hit.[16]

The Japanese ships, suffering damage from U.S. gunfire, withdrew. Several U.S. ships had been damaged by Japanese gunfire.

But no further surface actions of consequence occurred until July 1943, when U.S. ships again took on Japanese forces coming down the so-called Slot—the broad sea-lane formed by the Solomon Islands—at the Battle of Kula Gulf (5–6 July 1943) and the Battle of Komombangara (13 July 1943). Japanese torpedoes sank another U.S. cruiser, the *Helena* (CL 50), and

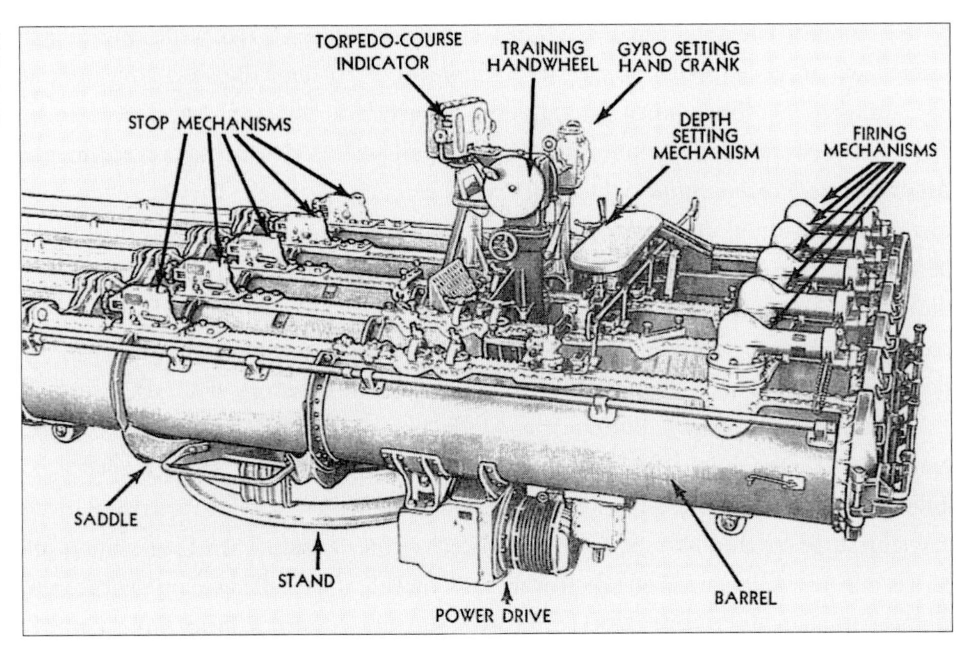

All U.S. destroyers built during the World War II era had ten Mark 14 21-inch tubes in quintuple mounts of the type illustrated here. (U.S. Navy)

damaged three others, the *St. Louis* (CL 49), *Honolulu* (CL 48), and the New Zealand *Leander*. The U.S. destroyer *Gwin* (DD 433) was also sunk. The only loss to the Japanese was the light cruiser *Jintsu* and a destroyer, both sunk by American torpedoes. Several other Japanese destroyers were damaged or sunk by American gunfire.

In early August 1943, the U.S. Navy received intelligence of another attempt by the Japanese to reinforce their army garrisons in the Central Solomons. (They had evacuated Guadalcanal in February 1943 and had been reinforcing their forces around Munda and Kolumbangara.) With few U.S. cruisers left in the South Pacific—20 Allied cruisers had been sunk or put out of action since August 1942—the only U.S. forces immediately available were six destroyers of Commander Frederick Moosbrugger's task group.

On 5 August 1943, Moosbrugger was ordered to intercept the "Tokyo Express" in Vella Gulf. He was given complete command of the operation and allowed to adopt whatever tactics he chose. His force consisted of the three destroyers of Destroyer Division (DesDiv) 15 and the three destroyers of his own DesDiv 12. The latter had trained extensively in night torpedo attacks under radar direction. Moosbrugger divided the force into two divisions: A-1) *Dunlap* (DD 384), *Craven* (DD 382), and *Maury* (DD 401); and A-2) *Lang* (DD 399), *Sterett* (DD 407), and *Stack* (DD 406). They would steam up the gulf in separate columns about two miles apart. If they encountered enemy destroyers, A-1, Moosbrugger's division, would strike first with torpedoes. If barges were encountered, A-2, which had lost half of its original torpedo tubes (four tubes per ship) in favor of a quadruple mount of 40mm anti-aircraft guns, would attack with gunfire.

This arrangement was an adoption of the basic destroyer doctrine established by Moosbrugger's predecessor, Commander Arleigh A. Burke. Burke had believed that the best destroyer tactic to be used in the "Slot" was to divide his destroyers into two groups. The group closer to the enemy when contact was made would attack with torpedoes. The second group would maneuver to cross the enemy force's "T" and strike it with gunfire once the torpedoes had struck. The groups would thus catch the enemy in a cross fire. In order to function correctly, this doctrine required a complete understanding of what each division would do in any expected circumstance.

To make certain that the A-2 group knew what to do, Moosbrugger invited Commander Roger Simpson

Commander Frederick Moosbrugger (U.S. Navy)

to breakfast on the morning of 6 August 1943 to review plans for the coming operation. As the commander of DesRon 15, Simpson would be in charge of the A-2 group. After breakfast with Simpson, Moosbrugger called **all of** the destroyer captains to a conference in which he handed out the operational order for the engagement.

The order distributed included specific instructions concerning the firing of torpedoes. According to Russell S. Crenshaw Jr., who served in the *Maury* at the time, "Moosbrugger's Op-Order instructed all ships to be prepared to fire their torpedoes at intermediate speed with depth set alternately at five and nine feet in the spread."[17] This was five feet less than the settings specified by the destroyer doctrine issued to the Pacific Fleet in February 1943. It was a distinct improvement according to Crenshaw, but he wondered if it was enough of a change.

Crenshaw was fairly certain that the Mark 15s were running too deep. He had seen three of the torpedoes fired at close range by the *Ralph Talbot* (DD 390) pass under the derelict hull of the *Gwin* two weeks earlier, when she had been hit and disabled by a Type 93 torpedo during the Battle of Kolombangara. The depth settings of the first two torpedoes fired by the *Talbot* were six feet. The next two were set at 12 feet. Only the fourth torpedo, which hit farther aft, where her decks were awash and her keel deeper in the water, exploded. As Crenshaw later explained:

> We had seen the same thing happen when *Shaw* tried to sink the *Porter* at Santa Cruz the previous October. Those torpedoes should certainly have been set shallow enough to hit the hull, but they didn't. They just continued hot, straight, and normal out the other side. The only possibility was that the torpedoes were running deeper than set. Much deeper! . . . If Moosbrugger's setting of nine feet proved to be too deep, we'd be throwing away half of our torpedoes. One—possibly two—of the torpedoes fired by *Ralph Talbot* that passed harmlessly under *Gwin*'s keel had been actually set for six feet. Why not set them all at the absolute minimum and give them the best chance to hit all of them?[18]

The magnetic exploders had been deactivated in response to Admiral Nimitz's order of 24 June. If the torpedoes were to have any chance of success, it was imperative that they strike the target. As Crenshaw explained to Commander Gelzer S. Sims, the *Maury*'s skipper, a shallow hit (which would cause less damage) was better than to miss completely under a target's keel. With Sims's approval, Crenshaw set all 16 of the *Maury*'s torpedoes to the shallowest depth setting possible—five feet.

The *Maury*, in company with the other five destroyers of Moosbrugger's command, steamed out of Tulagi Harbor at 11:30 AM on 6 August. By sunset the ships were steaming at 25 knots, headed for the Gizo Strait. As the ships entered Vella Gulf, Moosbrugger ordered them into battle formation, i.e., two columns. After probing the waters off Blackett Strait, they headed due north to scan the coast of Komombangara, with Moosbrugger's division leading to seaward. At 11:18

The USS *Ralph Talbot* (DD 390) plying placid waters off Hawaii in early 1943, a 16-tube destroyer in "war paint." She, too, experienced malfunctioning torpedoes, both in action against the Japanese and in trying to scuttle the U.S. destroyer *Gwin* (DD 433) after she was shattered by a Japanese Long-Lance torpedo off Kolombangara in the Solomons in July 1943. Unlike many U.S. destroyers that were built between the wars, the *Talbot* retained her full 5-inch gun and torpedo battery while still being fitted with additional anti-aircraft weapons. (U.S. Navy)

PM the *Dunlap*, his leading ship, made the first radar contact, which quickly multiplied into four distinct "pips"—one for each of the troop-laden Japanese destroyers that were headed toward them.

As the range quickly closed, Commander Moosbrugger passed the word over the TBS to his A-1 division to "Stand by to fire torpedoes." When the range to the target measured 6,500 yards, he issued the command "Eight William Two," ordering each ship of the division to fire eight torpedoes to port. The *Dunlap*, *Craven*, and *Maury* each responded with a spread of eight torpedoes. Crenshaw is convinced that five of the *Maury*'s torpedoes found their mark, "Five times out of eight, when Medler [the ship's torpedo officer] had called a shot, Bridge reported a big explosion on the bearing."[19]

The torpedo attack conducted by the destroyers of Moosbrugger's A-1 division was the most successful U.S. torpedo strike of the war. Seven of the 24 torpedoes resulted in effective hits—four torpedoes slammed into the destroyer *Arashi*, one struck *Kawakaze* in her magazine, and two hit *Hagikaze* in her engine room. An eighth hit and holed the destroyer *Shigure*'s rudder without exploding. Two more passed within 20 yards of her as she swung to starboard.

In less than half an hour, Moosbrugger's force had sunk three Japanese destroyers and damaged a fourth. The Japanese never touched the U.S. ships, and the U.S. force only suffered one casualty, which occurred when one of the 5-inch loader's hands was accidentally crushed. In the space of a few minutes, in the night Battle of Vella Gulf, U.S. destroyers had proven themselves in a successful torpedo attack for the first time since the night actions against Japanese transports off Balikpappan in January 1942. Captain Tameichi Hara, a leading Japanese torpedo expert who had served in the *Shigure* during the night action of 6 August, later wrote:

> Thus ended the Battle of Vella Gulf in a perfect American Victory. Three Japanese destroyers were sunk. Of their 700 crewman and 820 troops, only

Captain Arleigh A. Burke, Commander Destroyer Squadron 23, reading a letter on the bridge of his flagship, the *Charles Ausburne* (DD 570), during operations in the Solomon Islands in 1943–1944. Burke and Commander Moosbrugger were innovative and effective destroyer tacticians, Note "scoreboard" painted on the side of ship's Mark 37 gun director with a canvas-covered 20-mm Oerlikon gun forward of it; the DesRon 23 "Little Beaver" insignia is painted on the side of the bridge (forward of the life ring). (U.S. Navy)

> 310 survived. . . . The American score for this action was very high by any standard, and it was shockingly so to Japanese experts who thought the enemy weak in torpedo effectiveness.[20]

More battles would be fought as the U.S. forces moved up the Solomon Islands, but at least the warships now had a workable torpedo. The magnetic exploder had been disarmed, the original firing pin springs had been replaced with stronger ones, and the torpedoes were now set to run at shallower depths. The last became official policy on 25 October 1943, when the Commander Destroyers Pacific Fleet ordered a depth setting of six feet for all targets except those definitely identified as battleships; previously the standard depth setting had been ten feet. Although the Mark 15 used the same depth control system as the Mark 14, it was somewhat slower at the high-speed setting and was more likely to be fired at the 36-knot medium speed, which had less effect on the hydrodynamic error created by the improperly placed depth sensor.[21] This problem and the defective firing spring guides discussed in the preceding chapter were both corrected in the last half of 1943. Now, thanks to Commanders Arleigh Burke and Frederick Moosbrugger, the U.S. Navy had destroyer tactics that were highly effective and, thanks to many individuals and commanders—primarily in the fleet—the Navy had torpedoes that worked. And, there was a third important factor in the new American successes: radar.

The impact of radar on U.S. naval operations against the Japanese Navy—in daylight operations as well as night—cannot be overestimated. British naval historian S. W. Roskill, in writing of the high state of Japanese submarines, carrier aviation, and night tactics at the start of World War II, observed, "Only in Radar development did the Japanese lag seriously behind Britain and the USA; and that deficiency was in the long run to offset the advantages gained by their other developments."[22] The British and American Navies had begun fitting radar to their warships in 1940; the Japanese Navy would not install radar in warships until May 1941.[23] By the time of the destroyer actions in the Solomons all U.S. destroyers and cruisers had radar, providing an overwhelming advantage in night actions—if properly exploited. But until the Vella Gulf engagement, most U.S. cruiser-destroyer force commanders failed to comprehend the proper use of radar and how it could enhance their ships' effectiveness.

By 1943 U.S. warships were provided with multiple radars for air search, surface search/navigation, and gunfire direction. Even U.S. motor torpedo boats had radar installed by 1943. (See Chapter 8.) Meanwhile, the Japanese struggled to provide their major warships and submarines with relatively simple, less-reliable search radars.[24]

By the fall of 1943 the U.S. submarine torpedo problems had been effectively solved and the problems experienced in destroyer torpedo attacks had similarly been rectified. From the American perspective, unfortunately, many good sailors and many fine ships had been lost because of the poor U.S. naval doctrine and, especially, torpedo performance. Now the Japanese Navy, even fighting with excellent night tactics and good torpedoes, could be confronted and defeated by superior U.S. tactics, warships, and radar, as well as by torpedoes that worked.

HUMAN TORPEDOES

In the latter half of 1942, with the Japanese Navy moving to the defensive, two naval officers—Ensign Sekio Nishina and Lieutenant (jg) Hiroshi Kuroki—conceived of the kaiten, or "human torpedo," based on the large Long-Lance torpedo (see separate sidebar in this chapter). Permission was given for them to develop the weapon only if a means were provided for the pilot to escape at the last moment. In February 1944 the Naval General Staff gave permission to produce the kaiten; 100 torpedoes were converted in 1944 and 230 in 1945. Late in the war several destroyers were modified to launch two or four kaiten over the stern and a light cruiser was provided with rails to launch eight kaiten. Also, several large submarines were modified to carry four kaitens, with the pilots able to move from the submarines into their torpedoes through a hatch connection while the mother submarine was submerged. Submarine-launched kaitens attacked several U.S. ships, but their only successes were sinking a tanker (with 50 of her crew killed) and a destroyer escort (with the death of 113 Americans). Several Japanese submarines and numerous kaitens were lost in scoring those two successes. The one-man kaitens weighed 18½ tons submerged, were 48⅓ feet long, and had oxygen-kerosene engines that could drive them up to 30 knots, and for 85,000 yards at slower speeds. The warhead contained 3,400 pounds of high explosives.

CHAPTER ELEVEN

Victory Assured

Torpedoes That Did Work (1943–1945)

The introduction of electrically driven torpedoes in the U.S. Navy had its genesis in a strange and unusual event—the capture of a submarine by an aircraft. The German *U-570* was on her maiden voyage in August 1941, with a captain on his first war patrol, as were a number of his crew. Operating south of Iceland, the U-boat was periodically sighted and attacked by British aircraft. On 27 August a Royal Air Force (RAF) Hudson bomber came down on the *U-570* as the submarine was surfacing. Straddled by a stick of four depth charges, the submarine was badly damaged, with poisonous chlorine gas escaping from her batteries.

The German sailors climbed onto her deck and, with one waving a white dress shirt, surrendered to the aircraft. The Hudson remained over her captive until relieved by a RAF Catalina flying boat. Soon surface ships and more aircraft were on the scene and the *U-570* was taken in tow. Although the U-boat's crew had jettisoned the Enigma enciphering machine and all codebooks before she was boarded, she still had her loadout of G7e torpedoes.[1]

The G7e was an electrically driven torpedo.[2] Such a weapon was easier to manufacture than the traditional steam torpedo and offered certain tactical advantages desired by submariners—no wake and single speed. The British had earlier indications that U-boats were employing electric torpedoes, because fragments of the four electric torpedoes that sank the battleship *Royal Oak* in harbor in September 1939 had been recovered, as were fragments from one of the torpedoes that damaged the Dutch passenger ship *Volendam* in August 1940.[3] The *U-570* provided the first intact G7e torpedoes. (The British also obtained a German T-5 acoustic torpedo from the Soviet Navy after the *U-250* was sunk in the Baltic in 1944 and salvaged. The torpedo was given to Britain on Josef Stalin's orders, although he later imprisoned four admirals for giving away the weapon, including a former commander-in-chief of the Navy, N. G. Kuznetsov.)

In January 1942, with the United States in the conflict against Nazi Germany, the British transferred one of the G7e torpedoes to the U.S. Navy. Additional examples of these torpedoes were obtained after they washed up on American beaches as German U-boats attacked U.S. shipping along the U.S. East Coast in early 1942. Although the U.S. Navy's Bureau of Ordnance was in the process of developing its own electrically driven torpedo, it decided to copy the German design to expedite production.

The U.S. Navy's interest in electric torpedoes dates from 1915 when a development contract for a torpedo of this type was issued to the Sperry Gyroscope Company of Brooklyn, New York. The 6-foot long, 7¼-inch diameter torpedo was to have a speed of 25 knots and a range of 3,800 yards. The propulsion motor of the proposed torpedo was to act as a gyroscope to stabilize the torpedo in azimuth, as in the ear-

THE GERMAN NAVAL ACOUSTIC TORPEDO

Known by the Allies as GNAT, the German Naval Acoustic Torpedo was the world's first operational acoustic homing torpedo. It was used by U-boats against Allied anti-submarine ships in World War II. The acoustic-homing torpedo, called Zaunkönig (wren) and designated T5 by the German Navy, had passive acoustic guidance to search out the cavitation noises made by a surface ship's propellers. The first use of the GNAT was on 20 September 1943, when U-boats attacking convoy ON.202 damaged a frigate with an acoustic torpedo, requiring the ship to be taken in tow. In three days the GNATs were used to sink a destroyer, frigate, and corvette and to cause major damage to two other escort ships. The three-day action involved two adjacent convoys, with a total of 69 merchant ships, which were attacked by 21 U-boats; six merchant ships and three U-boats were sunk in the actions. The British had learned of GNAT prior to its use through prisoner of war interrogations and codebreaking. Forewarned, the British quickly developed the Foxer devices to decoy the torpedoes away from warships. These were put into service only 18 days after the convoy action described above. The Foxers totally defeated the acoustic torpedoes. Several hundred GNATs were used by German submarines.

lier Howell torpedo. The project was terminated in 1918 before any torpedoes were produced.

The successful development of an electric torpedo by Germany in World War I prompted the Navy to continue the in-house development of its own electric torpedo at the Navy Experimental Station at New London, Connecticut. On 13 June 1918, the Bureau of Ordnance authorized the expenditure of $50,000 for the manufacture of six electric-drive torpedoes. The new torpedo, designated as the Type EL, was to have the following design characteristics:

Diameter	17.7 inches
Length	16.25 feet
Speed	25 knots
Range	1,000 yards
Explosive	125 pounds, minimum

After the Navy Experimental Station was closed in 1919 as an economy measure, the project was transferred to the Naval Torpedo Station at Newport, where it was designated as the Mark 1 Electric Torpedo. Work on the project continued for the next 12 years as contracts were let to several manufacturers to develop a suitable motor and battery. The best efforts were a motor developed by the General Electric (GE) Company and a thin-plate storage battery developed by the Electric Storage Battery Company (Exide) of Philadelphia, Pennsylvania. But peacetime experimentation was expensive. Batteries were costly and torpedoes were frequently lost. The lack of a wake—the most valuable military feature of the electric torpedo—made them difficult to follow on the range. Because of the low power factors involved, the speed and range of electric torpedoes was "decidedly under what could be obtained with steam torpedoes."[4] Another means of obtaining a wakeless torpedo—the oxygen-powered torpedo—was more promising with respect to performance. This, coupled with the severe lack of funds imposed by the Great Depression, led to the demise of the Mark 1 project, which was terminated in 1931.

Work on an electric torpedo was renewed in July 1941, when the Bureau of Ordnance set aside $500,000 to develop the Mark 2 electric torpedo. The project was conducted by a small team of design engineers assigned to the research division of the Naval Torpedo Station at Newport. Again, working with General Electric and Exide, the team drew up plans to convert ten submarine torpedoes (five Mark 14s and five Mark 10s) to electric propulsion. The new torpedo was to have a speed of 33 knots and a range of 3,500 yards. The project was moving ahead slowly until the discovery of the German G7e torpedoes focused renewed attention on the urgent need to provide a similar weapon for U.S. submarines. At Admiral Ernest J. King's urging, the Bureau of Ordnance began taking steps to expedite the introduction of the Mark 2 torpedo, which was designed around a compound-wound, high-speed motor developed by GE. The motor was more efficient than the series-wound used in the G7e and promised to eliminate the unbalanced torque, which caused rolling in torpedoes that did not have balanced turbines, as in all conventional U.S. torpedoes.

The Torpedo Station, which was working around the clock to produce badly needed steam torpedoes, lacked the resources necessary to manufacture electric torpedoes, so production was shifted to GE. Seeking to obtain a second source, BuOrd contacted the Westinghouse Electric and Manufacturing Company, asking for its assistance. The company had a large, modern transformer manufacturing plant at Sharon, Pennsylvania, which had space for government work. Westinghouse, which had no experience with torpedoes, did not particularly want an electric torpedo contract. Nevertheless, representatives of the company met with members of the Bureau's staff on 10 March 1942. A week later, company representatives visited Newport to collect data and design material on the electric torpedo. After reviewing this data, which presumably included an inspection of a captured G7e, company officials decided that the quickest way to produce an electric torpedo was to copy the German models as closely as possible.

The Bureau of Ordnance concurred and awarded the company a development contract to produce five experimental models of an electric torpedo, designated Mark 18, for test at the Naval Torpedo Station by June 1942. The captured German torpedo was turned over to Westinghouse for study and the battery given to Exide, which had been selected on the advice of the Bureau as the best-available subcontractor for the storage battery. "Westinghouse engineers went to work with a speed and fervor that was dazzling," wrote submarine historian Clay Blair Jr.[5] Westinghouse rushed through the preliminary design work and was able to submit engineering drawings by mid-April.

On 2 May 1942, before any of the torpedoes of the development contract had been delivered, the Bureau of Ordnance issued a letter of intent for the manufacture of 2,000 Mark 18 torpedoes with 2,020 warheads and 543 exercise heads, plus spare parts and workshop equipment. This was done so that the company could begin setting up production concurrent with the development program. Rear Admiral William H. Blandy, the chief of the BuOrd, adopted this policy to save time, because more torpedoes were urgently needed by the forces at sea. Blandy took a calculated risk in rushing the undeveloped torpedo into production, but at the time it seemed that it would be a simple task to copy the German torpedo and a great deal of time could be saved if development and production proceeded simultaneously. Unfortunately, this would not be the case.

When the Bureau of Ordnance embarked on the Mark 18 project the concept was to make a "Chinese copy" of the torpedo for use in U.S. submarines. However, the torpedo tubes, ejection systems, tube shutters, and firing systems in the U.S. submarines were different from those used in Germans U-boats. There were no instruction manuals, and American engineers were frequently at a loss as to the purpose of certain features in the German design. Lastly, variations in the manufacturing methods and philosophies used in the two countries made it impossible to directly copy the G7e torpedo.

Thus, the production model of the U.S. Mark 18 varied considerably from its German counterpart. Its warhead, which was similar to other U.S. Navy warheads, was initially designed around the Mark 4-0 contact exploder (with subsequent Mark 18s using both the Mark 8 and 9 exploders). The magnetic influence exploder was not employed in the early Mark 18 models because of a shielding problem (later mods used the Mark 9, a combined contact/influence exploder). The battery compartment, which takes the place of the air flask in a steam torpedo, was adapted from the German torpedo using a new, thin-plate battery developed by Exide. The afterbody contained the motor and the control mechanism. The former was a series-wound, six-pole motor developing 90 horsepower that was similar to the German model. The German control system, which also relied on pneumatics, was replaced by the standard control mechanism that was already being mass-produced for U.S. torpedoes. This required the addition of a small air flask to provide pneumatic power. Only the tail section was taken in its entirety from the German design. With the Bureau of Ordnance's approval, Westinghouse copied the tail of the G7e, including propellers, vanes, rudders, and guide shoes.

The Navy took delivery of the first Mark 18 on 24 June 1942, only 15 weeks after Westinghouse had first met with the Bureau of Ordnance officials on the project. This was a remarkable achievement for a company that had never before designed or manufactured a torpedo. Unfortunately the Torpedo Station at Newport was unable to test the new torpedo, as it had no way of tracking it. The station spent months developing an effective means of following the course of the wakeless torpedo while it was running. It was November 1942 before night firing of electrically lighted torpedoes and range rafts were developed and put into operation. By that time the best part of the firing season with its good weather was gone. Further development and proofing had to be carried out at night during the winter season on Narragansett Bay.[6]

Delivery of the first production models, which one optimistic officer in Washington had promised to occur within a few weeks after delivery of the first prototype, stretched to more that a year as one design deficiency after another was discovered during the Mark 18 tests. Some of the design flaws, such as the problems with the commutator connections in the torpedo's motor and the failed tail cone bearings, were the type of "teething" problems one would expect in any new complex torpedo. Another flaw resulted from the Navy's rigid insistence on accurate speed control. The German G7e had no speed governor, relying instead on the characteristics of the storage battery to control the average speed. As the battery voltage slowly dropped, so did the speed. At the insistence of the Bureau of Ordnance, Westinghouse designed a clever, but highly complicated, brush-shifting arrangement that controlled the torpedo's speed to an accuracy of 1 percent. It proved to be unreliable and had to be removed. As the Navy's inspector general concluded, "It was a case of one more gadget, over refinement, and poor understanding of the problem."[7]

Other problems resulted from a lack of coordination. The Torpedo Station was in the process of developing its own Mark 2 electric torpedo and the personnel assigned to the Mark 18 were electrical engineers who had little practical torpedo experience. Instead of cooperating, the two groups acted as competitors, leading to unforeseen difficulties in mating the torpedo's guide stud and tail structure with U.S. torpedo tubes. The guide stud in the German torpedo served two purposes: it held the torpedo in its proper place in the tube and it kept the torpedo in the upright position while the motor was building up speed, preventing roll or heel. When this was redesigned in the Mark 18 to fit U.S. tubes, the engineers failed to take into account the second requirement. The torpedo heeled in the tube and was damaged as it exited the tube. A second guide had to be added so that both requirements could be met.

Allowing Westinghouse to duplicate the tail led to further problems, because the tail structure of the German torpedoes was weaker then that used in American torpedoes. It was too weak to withstand firing from the tubes of U.S. submarines. The tail vane shoes also had sharp edges, which cut into the tube shutters. In certain submarines the tail's vane shoes caught on the shutters and were carried away or damaged.

There were also problems with the battery cells supplied by Exide, which were produced while still in the experimental stage. Although several design conferences were held with Westinghouse engineers, the battery requirements were not made clear. Later models of the Mark 18 torpedo contained an improved Mark 2 Exide battery.

Months passed before all of the defects could be identified and corrected. It was April 1943 before the first production Mark 18s emerged from the Westinghouse factory. More months were required for proofing and shipment overseas. It was late summer before the first Mark 18s arrived at Pearl Harbor. Although it had not been battle tested, the submariner's discontent with the Mark 14 was so great that Rear Admiral Charles Lockwood, Commander Submarines, Pacific Fleet, thought it prudent to employ the electric torpedo as soon as possible.

On 10 September, the submarines *Sawfish* (SS 276) and *Wahoo* (SS 238) got under way from Pearl Harbor with a mixed load of Mark 14 and Mark 18 torpedoes—the first of the latter to be taken into combat. Unfortunately the performance of the Mark 18s proved

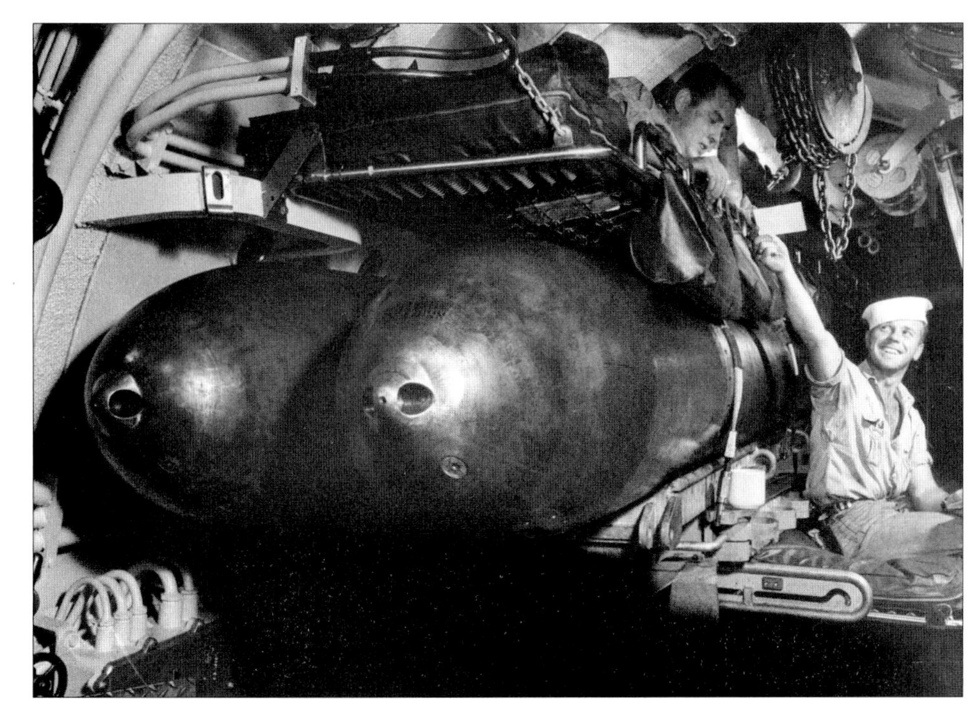

Crewmen relax on folding bunks in the forward torpedo room of the USS *Cero* (SS 225) in this "PR" photo taken in August 1943. At the time the boat was moored in New London, Connecticut. These submarines were manned by about 80 officers and enlisted men, with accommodations and mess facilities unequaled by any other undersea craft. They normally carried out patrols of up to 60 days. (U.S. Navy)

disappointing. Of the 12 fired by the *Sawfish*, three fishtailed while leaving the tubes, struck the shutters, and were never heard to run thereafter; one plunged straight to the bottom; and the remaining seven ran astern of their intended targets. *Wahoo* never returned from patrol, so nothing is known about the performance of her Mark 18s.[8]

It was also discovered that the Mark 18 batteries leaked hydrogen gas and had to be removed periodically from torpedo tubes for ventilation. Several instances of hydrogen fires occurred, including one in the submarine *Flying Fish* (SS 229) that heated the Torpex in the warhead until it melted and ran out.[9] Other problems included failure of the guide studs during depth charging, which could cause the torpedo to run "hot" in the tube, and warping of the thin shell of the battery compartment, which resulted in binding in the tube. Cold water operations lowered the battery temperature, causing the Mark 18 to run slower, ruining the fire control solution. All of these problems were quickly tracked down and eliminated by the end of the year.

Initially many of the commanding officers were skeptical of the Mark 18 because of its low speed. But the popularity of the electric torpedo grew steadily once its reliability was firmly established. Thirty percent of the U.S. submarine torpedoes fired in 1944 were electric; in 1945 the percentage went up to 65. Because of its slow speed, the Mark 18 was more likely to be used against a merchant ship. There were few naval targets left by 1945, hence most of the fish were fired against "Marus."

THE TORPEDO PROBLEM SOLVED

By the end of 1943 the U.S. Navy's torpedo situation had began to turn around as improved versions of both the Mark 14 and Mark 18 began to reach the fleet. A critical test of the improved Mark 14 occurred during the night of December 3–4, when the *Sailfish* (SS 192), guided by an Ultra radio intercept, made contact with a Japanese carrier group that had stopped zigzagging due to the foul weather. The *Sailfish*'s captain, Lieutenant Commander Robert E. M. Ward, caught up with the enemy ships in the midst of a raging typhoon. The seas were huge and a 50-knot wind was blowing. The Mark 14s he was about to fire would have to withstand the pounding of the waves without exploding, maintain

The *Ronquil* (SS 396) was typical of the just over 200 submarines of the *Gato* (SS 212), *Balao* (SS 285), and *Tench* (SS 417) classes completed during World War II. Their designs were similar—all had six bow and four stern tubes and carried a total of 24 torpedoes. The *Ronquil* has two 5-inch/25 guns and two 40-mm guns, plus lighter weapons for attacking small surface ships. She could carry mines in place of torpedoes. (U.S. Navy)

accurate depth control under arduous conditions, and function properly on impact if he were to have any chance of sinking the enemy carriers.

Ward abandoned any idea of a methodical approach and submerged after the carrier group had been detected on radar at 9,000 yards. Just after midnight—at 12 minutes past the hour—he fired four "fish" when 2,100 yards from the largest target on *Sailfish*'s radar scope. He heard two solid hits before diving deep to avoid a pursuing escort that dropped 21 depth charges.

At about two o'clock in the morning, the *Sailfish*, with her tubes reloaded, surfaced. Her radar showed pips all around, including one moving very slowly, which he tracked until he was in position to fire again. With morning light coming on, he hurried to shoot again, firing three more torpedoes from 3,200 yards. Two more hits were heard as the submarine submerged once more. Ward finally got a glimpse of the target through the periscope at 7:58 AM, when he sighted an aircraft carrier dead in the water, down by the stern, with a slight list to port. He wasted no time in putting three more Mark 14s into the carrier, which broke up and disappeared from sight. The *Sailfish*, under Commander Ward's leadership, had become the first U.S. submarine to sink a Japanese aircraft carrier in World War II, his victim being the escort carrier *Chuyo*. The *Chuyo* had been attacked twice before by U.S. submarines in 1943, but had suffered no damage. She was taken in two after being torpedoed by the *Sailfish*, but she sank with heavy loss of life—1,250 crewmen and passengers, plus 20 American prisoners who were survivors of the sunken submarine *Sculpin* (SS 191).

After 25 months of war, the U.S. submarine service finally had a steam torpedo that it could rely on. Despite the problems with U.S. torpedoes and the Mark 6 exploder, U.S. submarines managed to sink 477 Japanese ships—almost all of their merchant vessels and naval auxiliaries. In the process they fired 5,686 torpedoes, 70 percent of them Mark 14s. As Clay Blair Jr. explained, it was "a time of testing, of weeding out, of fixing defects in weapons, strategy, and tactics, of waiting for sufficient numbers of submarines and workable torpedoes."[10]

In the last 20 months of World War II, U.S. submarines, guided by Ultra intercepts, aided by improved radar and other equipment, and armed with better torpedoes sank another 847 ships. By the time the war ended in August 1945, U.S. submarines were credited with sinking 1 battleship, 8 aircraft carriers, 11 cruisers, 40 destroyers, and 18 submarines. During the entire war U.S. submarines fired 14,748 torpedoes, sinking a total of 1,314 ships—11.2 torpedoes for every ship

"Let pass safely" was the wording in messages sent to U.S. submarines in the Pacific when the Japanese notified the International Red Cross of the transit of a hospital ship, such as this one. Early on 1 April 1945, in the Taiwan Strait, the submarine *Queenfish* (SS 393) sank the merchant ship *Awa Maru*, which had been given safe passage because she was supposed to be carrying wounded Japanese. But the one survivor—of 2,003 men on board—revealed that she had contraband material on board. Still, the U.S. Navy court-martialed the submarine's commanding officer for the sinking. (U.S. Navy)

The transport *Teiko Maru*, at 15,100 tons, was one of Japan's largest merchant ships. The submarine *Puffer* (SS 268) sent the ex-French liner to the bottom off Borneo on 22 February 1944. Many U.S. submarine successes were made possible by Allied codebreaking, with the Japanese Army's water transportation codes revealing many ship sailings and their routes to the far-flung Japanese empire. (U.S. Navy)

The 18,115-ton light carrier *Chuyo* was the first Japanese carrier to be sunk by a submarine, on 4 December 1943. After being attacked twice by U.S. submarines in October without success, the USS *Sailfish* (SS 192)—in three separate attacks—obtained four or five torpedo hits on the *Chuyo*. Although taken in tow, the carrier sank, taking with her 1,250 crewmen and passengers and 20 American prisoners of war who were on board. (Courtesy Shizuo Fukui)

The largest warship ever sunk by a submarine was the Japanese aircraft carrier *Shinano*, converted during construction from a super battleship of the *Yamato* class. Not yet operational, the *Shinano* was sunk in Japanese coastal waters on 29 November 1944, by four torpedoes fired by the USS *Archerfish* (SS 311). Poor damage-control procedures by a largely untrained crew led to her loss with more than 1,400 men. This photo of the *Archerfish* was taken in June 1945. (U.S. Navy)

sunk. Almost all of these were launched by the submarine forces in the Southwest Pacific and the Pacific theaters (only 47 torpedoes were fired in other theaters). The high rate of torpedoes expended was due in some part to the poor quality of U.S. torpedoes at the start of the war. Had all these torpedoes run straight at the proper depth and detonated as designed, the number of ships sunk would have been higher.

How much higher is hard to determine. According to one study, the depth problem alone cost U.S submariners at least 30 ships. Erratic runs and exploder problems led to other missed opportunities. While these failures affected sinkings, the primary torpedo problem faced by the submarine force during the first 13 months of World War II was the shortage of torpedoes. The torpedo shortage was a nightmare. Torpedoes were rationed, and commanding officers were encouraged to fire one or two torpedoes per salvo, even at important targets. But, the single torpedo salvo nearly always resulted in a miss and therefore a wasted torpedo.

Sailors rig victory flags on the *Batfish* (SS 310) showing her credited "kills" of Japanese warships and merchant vessels. Not all would be confirmed by post-war analysis. The *Batfish* sank three enemy submarines during the war, a record unequaled by any other undersea craft of any nation. SV air search and SJ surface search radars are mounted above her periscope shears (housings). (U.S. Navy)

After the war Navy historians concluded that:

> The shortage of torpedoes undoubtedly delayed the formulation of a correct doctrine in regard to the use of spreads. In the early days of the war, 1 or 2 torpedoes would be fired in situations which in the latter part of the war were considered to merit a spread of 4 or 6 torpedoes. When a spread was fired, probability of getting 1 or 2 hits was greatly increased, thus increasing the likelihood of sinking the target. While this required more torpedoes it did not affect the ratio of torpedoes fired per ship sunk.[11]

Had the submarine force had an unlimited supply of effective torpedoes, it is likely that the number of Japanese ships sunk between 7 December 1941 and December 1942 would have increased by about 100 ships. One source estimates that this would have shortened the war by perhaps a month or two. Of course, this analysis does not take into account the effect that workable torpedoes might have had on interrupting the Japanese invasion of the Philippines, nor the amount of Japanese resources that might have been diverted to anti-submarine efforts had the U.S. submarine threat been taken more seriously by the Japanese Navy during the critical naval battles of 1942.

TABLE 11-1

TORPEDOES FIRED BY U.S. SUBMARINES IN WORLD WAR II

TYPE	FIRED	HITS*	PERCENTAGE
Mk 9, 10 11	1,722	502	29
Mk 14	6,852	2,437	36
Mk 18	3,536	979	28
Mk 23	2,058	822	40
All others	139	39	33
Total	14,881	4,889	33

**Note: The number of hits appears to have been based on the patrol reports filed by U.S. commanders after their return to port and are notoriously suspect. According to the data presented in these reports the percentage of hits obtained from all torpedo firings between 1941 and 1943 was almost identical to that reported between 1944 and 1945 despite the fact that U.S. torpedoes and torpedo-firing techniques had been markedly improved.*

CHAPTER TWELVE

Smart Torpedoes

The Fido and Cutie (1941–1945)

Even before the United States was at war, the government undertook several actions to ensure that, when the country entered the conflict, the nations' top scientific personnel could be exploited to support the war effort. This would be done in large part through the National Academy of Sciences, which dated to 1863, and the National Defense Research Committee (NDRC), which was established on 27 June 1940. The NDRC was created "to correlate and support scientific research on the mechanisms and devices of warfare."[1] Dr. Vannevar Bush, the director of the Carnegie Institution, became chairman of the NDRC.

Shortly after the establishment of the NDRC, Secretary of the Navy Frank Knox asked that the committee form a panel to advise him and the director of the Naval Research Laboratory, Rear Admiral Harold G. Bowen, on innovative ways to defeat enemy submarines. Working directly with Bowen, a division would pursue the most complete investigation possible of all factors and phenomena involved in the accurate detection of submerged or partially submerged submarines and in anti-submarine devices.[2] The panel—which would shortly evolve into NDRC's Division 6—would be responsible for the scientific studies, i.e., organizing the research, finding personnel, and issuing research contracts to industrial and academic institutions. Other aspects of the project would be handled directly by the Navy.

Selected to chair the division was Dr. John T. Tate, a physicist from the University of Minnesota. In November 1941, the leaders of Tate's division had their first joint meeting with representatives from the various Navy commands interested in developing weapons for Anti-Submarine Warfare (ASW). It appears likely that one of those present was Commander Louis W. McKeehan, then assigned to the Bureau of Ordnance. A Naval Reserve officer, McKeehan had been director of the physics laboratories at Yale University before being called to active duty. On 24 November, McKeehan wrote a memo asking the NDRC if it was "feasible to devise acoustic equipment for homing control of a self-propelled, torpedo-like body." And, if so, how would they go about doing it?[3]

An acoustic-guided ASW weapon was highly desirable because the depth charge—the principle air ASW weapon—was "dumb"; it was released when an aircraft sighted the submarine or, later, with the help of expendable sonobouys or magnetic anomaly detection. A diving submarine could quickly change depth and direction, evading depth charges, which were set to detonate at a predetermined depth. Once submerged the submarine was difficult, if not impossible, to locate precisely enough to effectively employ depth charges. The solution would be an ASW weapon that would seek out the enemy submarine moving in three dimensions in the depths.

Before November 1941 was over, McKeehan had met with Dr. Frederick V. Hunt, the director of Harvard University's Cruft Laboratory, to discuss the idea for a self-propelled depth charge or acoustic homing torpe-

Prior to the advent of the Mark 24 Fido acoustic homing torpedo, aircraft used depth bombs, rockets, and guns to attack enemy submarines. Here TBF-1 Avengers from the escort carrier *Card* (CVE 11) attack the submarines *U-66* (left) and *U-117* on 7 August 1943. The *U-117*, a submarine tanker, was sunk in this action by a Fido torpedo after being damaged by depth bombs. The *U-66* survived until 6 May 1944, when she was sunk by depth charges, ramming, and gunfire from the carrier *Block Island* (CVE 21) task group. Three Fidos dropped on the *U-66* failed to find their target. (National Archives)

do. The laboratory, which would later become known as the Harvard Underwater Sound Laboratory (HUSL), had been started when the NDRC awarded a small grant to Harvard in June to outfit a laboratory and to cover the initial expenses involved in undertaking research on underwater acoustics.

Hunt wrote to Tate on 9 December 1941 to express his confidence in the laboratory's ability to handle the project with a very small expansion in personnel. By then Hunt had already discussed an acoustic homing torpedo with Dr. A. B. Focke, a civilian scientist in the Bureau of Ordnance, who agreed to supply Hunt's group with the external torpedo casings together with the propulsion and rudder mechanisms for use in an acoustic guided torpedo. Hunt felt that his lab could complete the research within two to six months.

The following day the Subsurface Warfare Section of the NDRC met at New London, Connecticut, and the members and the Navy liaison officers—including Commander McKeehan—agreed that a project to develop a homing torpedo was of the upmost importance and should be given the highest priority possible. On 12 December 1941, Tate asked William V. Houston, a physics professor at Cal Tech, to assume leadership for the yet-to-be-named project. A week later, Tate wrote to the Secretary of the Navy asking if Houston "might be put in touch with someone in torpedo design with whom he [Houston] might talk freely."[4] This occurred on the 22nd, when Houston met with Dr. Focke and other Navy technical representatives at the David Taylor Model Basin (DTMB) in Carderock, Maryland, a few miles west of Washington, D.C. The Navy, wishing to keep the concept secret, had already dubbed it "Project Fido."[5] Details were discussed at Carderock, where preliminary specifications were drawn up for what would eventually become the Mark 24 Torpedo, initially designated the Mark 24 *Mine*. According to Harvey Brooks, a physicist who worked on the project, it was initially called a mine because McKeehan believed that this was the only way to get the project off dead center. Reclassifying the project as a mine removed it from the purview of the torpedo specialists, who had too many other projects and problems to worry about and were, therefore, reluctant to undertake the development of yet another torpedo. Other sources claim that it was called a mine for security reasons.

The initial specifications described an acoustically controlled, air-launched homing torpedo:

> The mine (torpedo) was to be an antisubmarine weapon that would be dropped by aircraft from heights of some 200 to 300 feet at air speeds of about 120 knots. Its dimensions and weight were to be such that it would fit into the 1000-pound bomb rack. It was to be used against submerged submarines only, and should not endanger surface vessels. Its speed should be of the order of 12 knots, thus affording a margin of 4 knots over the maximum submerged speed of a submarine and should be capable of operating for some 15 minutes. In the absence of a signal, or for weak signals, the mine should travel under hydrostatic pressure control at a fixed depth with provision for disabling the control when the signal levels reached a predetermined small value.[6]

These specifications were presented to researchers at the Bell Telephone Laboratories on Christmas Eve at a conference that took place in the director's office in New York. The Bell Telephone Laboratories, founded in 1925, was one of the nation's leading scientific and engineering establishments. During World War II some 2,000 separate projects were pursued for the Army, Navy, and NDRC, including major radar, communications, fire control, and torpedo projects. Within two weeks the lab had worked out the principles that would be used to provide sound control of the weapon.

In the meantime, Tate had written to William D. Coolidge, the Director of Research for the General Electric Company, advising him of the developments relating to the homing torpedo. Coolidge responded by telling Tate that GE was already developing an electric motor for the Navy's torpedoes offering his company's assistance in developing a sound-guided homing torpedo.

Thus, all of the essential "players" were in place when the Committee on Sub-Surface Warfare met on 13 January 1942 to review the proposals for Project Fido. HUSL was working measuring submarine noise levels, Bell Labs on guidance and control, GE on propulsion, and the David Taylor Model Basin on shell design. During the next nine months one of the most complex naval weapons developed up to that time was produced by this team. A most unusual—and important—aspect of the acoustic torpedo was the propeller. An acoustic torpedo required self-noise suppression for its sonar to effectively detect the target. The propeller was a major noise source.

> Neither Bell Labs nor Harvard had any background in low-noise propeller design, so a Bell Labs mathematician and an engineer went to the David Taylor Model Basin to see their expert, Karl Schoenherr. Hoping to get some advice, references, and perhaps even some help, the Bell Labs visitors discussed the problem for about 20 minutes while Schoenherr sketched freehand on a large piece of paper. He then handed over the paper with the comment, "Here is your prop." When they protested that this did not seem useful, he called a draftsman to scale off his sketch and make a dimensional drawing. After lunch, Bell Labs had its design for a propeller. . . . This procedure was most unsatisfying from a scientific viewpoint, and work to improve on the design was undertaken at Harvard and later at Penn State, *but it was 1950 before a better propeller for the purpose was made.*[7] [emphasis added]

By October 1942, the design had been completed and tested and a production contract for 5,200 Mark 24 mines was issued by the Bureau of Ordnance to the Western Electric Company. The entire design process from initial concept to production contract was accomplished in the remarkably short time of ten months.

The Mark 24 torpedo that emerged from this process was 19 inches in diameter, 7 feet long, and weighed 680 pounds, with a 92-pound HBX warhead.[8] It was

The Mark 24 torpedo—nicknamed Fido—was the first Allied acoustic homing torpedo. Launched from aircraft, it proved highly effective against German and Japanese submarines. Few photos of the Mark 24 were taken because of wartime secrecy. This is a Mark 24 being installed on a Navy airship in 1949. Note the protective "collar" that dropped away upon striking the water. The torpedo remained in U.S. Navy service until about 1949. (U.S. Navy courtesy William F. Althoff)

propelled by a General Electric 5½ horsepower motor using an Exide lead acid storage battery for power. It was the world's first truly lightweight torpedo.

Target detection was accomplished by four hydrophones symmetrically arranged around the circumference of the torpedo midsection in the left, right, up, and down positions. The four hydrophones could provide essentially all-direction coverage. The concept was to compare the signals from the left and right hydrophones to move the rudder to steer toward the target's signal. The controls produced rudder angles that were proportional to the difference in strength between the acoustic signals. The use of proportional control was a radical departure from the standard "bang-bang" (rudder hard left or hard right) control scheme that had been utilized since gyroscopic control had first been introduced to torpedoes.

When the horizontal hydrophones received an equal target signal, the torpedo was steered toward the target submarine in azimuth only, keeping at the running depth of 125 feet. As the signal picked up by the vertical hydrophones increased, a relay bypassed the depth and pendulum controls and the torpedo's depth was set by the acoustic signals. A ceiling switch, activated at 30 feet, kept the torpedo from targeting surface ships.

The Mark 24 was to be dropped into the sea by aircraft, initially diving to a depth of 125 feet. The torpedo initially used passive hydrophones to search for the target submarine, with its effective detection range being approximately 1,500 yards. If no propeller sounds were detected upon entering the water, the torpedo would begin a circular search, which it could maintain for 10 to 15 minutes. At least one submarine, the *U-926*, was sunk by a Fido after a run of 13 minutes. On another occasion, against the *U-1107*, a Mark 24 entered the water only 80 yards from the submarine but ran for three minutes before detonating. Apparently, it had not made initial contact and had gone into a circular pattern before acquiring its target.

Production units of the Mark 24 began to enter service in early May 1943. The first submarine kill was recorded by an RAF Liberator of No. 86 Squadron, which dropped a Mark 24 on 12 May 1943, seriously damaging the *U-456* and driving it to the surface. Approaching destroyers forced the U-boat to submerge, which then sank due to the damage inflicted by the Mark 24. Two days later, on 14 May, the U.S. Navy scored its first Mark 24 kill when a PBY Catalina from Patrol Squadron 84 sank the *U-640*.

The Mark 24 Fido was a major improvement over the "dumb" depth charges that had been the principle ASW weapon used by Allied aircraft. In 1943–1945 some 340 Fido torpedoes were used by Allied aircraft—142 by U.S. aircraft and 204 by British—against U-boats and Japanese submarines. Most were dropped by Liberators, Catalinas, and Avengers. These torpedoes are credited with sinking 68 submarines—including five Japanese submarines—and damaging another 33, a very high success rate for an anti-submarine weapon. Allied aircraft using depth charges against U-boats achieved a 9.5 percent kill rate compared to 22 percent for Fido torpedoes.

The successful development of this air-launched acoustic homing torpedo fueled the U.S. Navy's interest in adapting the Mark 24 control system for a submarine-launched torpedo that could be used as an anti-escort weapon. This was the role of the German G7e acoustic torpedo (see page 131). The idea for such a U.S. weapon was first broached at a conference in Tate's office on 22 November 1943, in which the NDRC was asked "to give most urgent consideration to means for modifying the Mark 24 to permit its use by submarines against small surface craft."[9] It was initially thought that this torpedo could be ejected from a torpedo tube using compressed air, but the Navy quickly decided that it would be preferable, from a tactical standpoint—so as not to reveal the submarine's position—to have the torpedo "swim" out of the tube under its own power. In less than a month Bell Labs had converted a Mark 24 torpedo into a prototype, which was successfully launched under its own power from a torpedo tube at a test barge at Solomons Island, Maryland.

The new torpedo, code-named "Cutie," was designated the Mark 27 torpedo. The original Mark 27 Mod 0 was a modified Mark 24 torpedo that had a slightly longer body and a set of wood rails to enable it to fit a standard 21-inch torpedo tube. It had a floor switch (instead of a ceiling switch) to keep it safe from the launching submarine and had various arming, warm-up, and started controls needed to fulfill its swim-out launch mode. At 12 knots, the Mark 27 had a maxi-

BRITISH CHARIOTS

The Royal Navy used "manned torpedoes"—called Chariots—during World War II. In 1909 British Commander Godfrey Herbert had obtained a patent for a manned torpedo, but it was rejected by the War Office as impracticable and unsafe. The Royal Navy subsequently developed the Chariot in 1941–1942 based on an Italian Maiale submersible captured at Gibraltar. Resembling a large torpedo, Chariots, manned by two swimmers wearing wet suits and breathing gear, were carried into the target area by a surface ship or submarine and then released to attack the enemy ship, attaching limpet mines or anchoring their torpedo's warhead under the enemy ship. The swimmers were then to return to their host ship or submarine on the after section of the Chariot (although this happened in only one operation). The basic Chariot was 25 feet, 21 inches in diameter, and carried a warhead of 600 pounds of explosives. The operator seats and controls were mounted on top. It weighed 1½ tons and could travel up to four hours at four knots. The larger Mark II, also carrying two men, had an enclosed cockpit and a warhead of 1,100 pounds of Torpex. Chariots were used in several wartime operations, but accomplished little. The film "Above Us the Waves" (1955), starring John Mills, told about Operation Title, an audacious and failed attempt to attack the German battleship *Tirpitz* in harbor. Chariots did sink the unfinished Italian cruiser *Ulpio Traiano* and the cruiser *Bolzano* in the Mediterranean, and two ex-Italian liners in a Japanese port in Thailand. And, in preparation for the Normandy landings on D-day (6 June 1944), British-manned Chariots surveyed the seabed of the landing areas. Twenty-eight Chariots were lost during the war, with almost all of their crews killed or captured.

mum range of 5,000 yards. Eleven hundred Mark 27 "Cuties" were built by the Western Electric Company and delivered between June 1944 and April 1945.

The submarine *Sea Owl* (SS 405) is credited with the first Cutie success, early on 11 December 1944, sinking a 135-ton Japanese patrol craft in the Yellow Sea with a single Mark 27 torpedo. Only 106 Mark 27s were fired against enemy escort ships, of which 33 struck home, sinking 24 ships and damaging 9 others—a 31 percent success rate. This was similar to the success rate for conventional U.S. torpedoes at the end of the war. Later versions of the Mark 27 were longer and heavier. The Mod 3, which was slightly over ten feet long and was faster, had a 200-pound warhead and a gyro for straight runout before beginning its acoustic search. Only six were completed before the project was terminated at the end of the war.

While the NDRC and various Navy and civilian facilities were working on Fido, scientists associated with the program suggested that additional experiments be conducted with echo-ranging systems and adding acoustical control to a standard Navy torpedo. Dr. Tate suggested that latter in a letter dated 24 December 1942 to Rear Admiral Julius A. Furer, the Navy's Coordinator of Research and Development. But the schedule at the torpedo ranges was so full that undertaking further experimental work, according to Furer's response, required "careful consideration, with the assignment of a definite priority with respect to other work under way."[10] The Bureau of Ordnance put the matter of acoustically directed torpedoes on hold until it was discovered that the Germans were about to introduce an acoustically controlled torpedo—the Type T "Falke." An improved version of this torpedo, code-named GNAT (German Naval Acoustic Torpedo) by the Allies, was introduced in September 1943.[11]

In April 1943, the Navy hastily shipped two Mark 18 electric torpedoes to the Harvard Underwater Sound Laboratory for use in developing an acoustic torpedo. The first problems tackled by the HUSL group was to install hydrophones, amplifiers, and recording equipment in the torpedo's nose (in place of the warheads) so that the self-noise of the torpedo under operating conditions could be determined. In addition to revealing the high noise levels of the Mark 18, the program of noise measurement led to an improved understanding of the role played by cavitation from propellers. There were enough other sources of self-noise to justify the long-held skepticism in the feasibility of acoustical control for high-speed torpedoes.

Nevertheless, the scientists at HUSL found that they could obtain favorable results by isolating the gears on the propeller shafts and by adding a noise-reducing gasket between the nose section and the body to avoid metal-to-metal contact. After an intensive effort to improve the performance of the hydrophones, they were able to design a direction unit that would discriminate between the sounds arriving from ahead along the axis of the hydrophone, and sounds arriving from the rear along the axis of the torpedo. By late 1944, the com-

THE BRITISH MARK VIII TORPEDO

The Mark VIII 21-inch-diameter weapon was used more than any other British torpedo in World War II and also scored the first "kill" of an enemy warship by a nuclear-propelled submarine. The Mark VIII entered service in 1927 and by World War II was used to arm most British submarines as well as destroyers and motor torpedo boats. The Mark VIII** (with stars indicating modification) was the principal version, being 21 feet, 7 inches long, weighing 3,450 pounds, and having a warhead of 805 pounds of Torpex. At 45.5 knots it could travel 5,000 yards, and at 41 knots its range was 7,000 yards. Mark VIII variants were kept in service after the war and were eventually carried in British nuclear-propelled submarines. In the Falklands conflict of 1982—55 years after the torpedo was developed—the nuclear-propelled submarine *Conqueror* was tracking the Argentine cruiser *General Belgrano*, which was accompanied by two missile-armed destroyers. The *Conqueror* was carrying Mark VIII Mod 4 torpedoes and the more advanced, wire-guided Tigerfish torpedoes. But there had been difficulties with the latter, and the commanding officer decided to use the older, straight-running Mark VIIIs. On the evening of 2 May 1982, the *Conqueror* launched three torpedoes at the Argentine cruiser; two torpedoes struck the 12,500-ton, ex-U.S. Navy warship. She sank, with the loss of 321 men, as the destroyers, after releasing a few depth charges, fled the scene. Of the 897 men who escaped the sinking in rafts and small boats, many did not survive the freezing night at sea. The Mark VIII remained in British submarine service until the early 1990s.

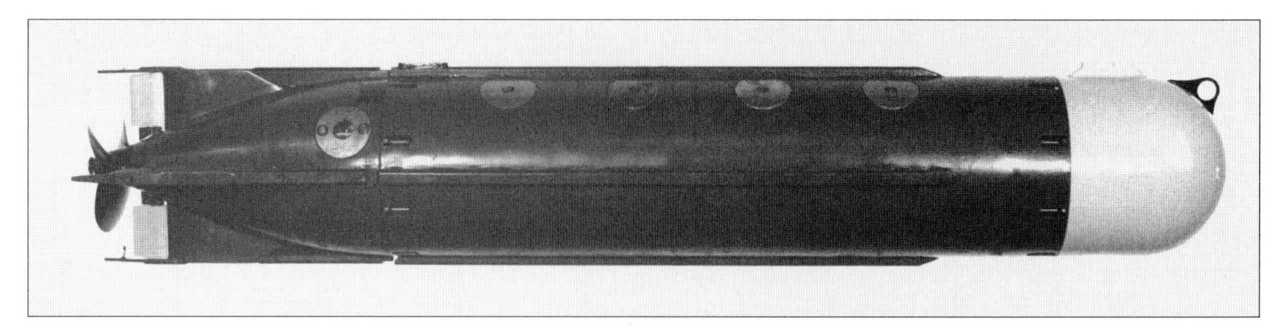

Mark 27 Cutie torpedo. (Naval Undersea Museum, Keyport, Washington)

The Mark 27 torpedo—called Cutie—was a submarine-launched variant of the Mark 24 Fido. However, while the Fido was an ASW weapon, the Cutie was a submarine-launched anti-escort weapon. That was the same role as the early German acoustic torpedoes. This unarmed Cutie is in the forward torpedo room of the submarine *Drum* (SS 228), moored as a memorial in Mobile, Alabama. The folded bunk was typical of World War II and postwar U.S. submarines, because sailors shared space with torpedoes. (USS *Alabama* Battleship Memorial Park)

bination of these modifications together with the development of a new, four-channel comparison amplifier made it possible to design an acoustic torpedo that could operate at a speed of 28 knots.

In the interim, studies conducted at both HUSL and Bell Labs convinced the engineers that it would be feasible to undertake the production of an acoustic torpedo that would operate at 20 knots with a single propeller, thus eliminating gear noise. The Mark 28 torpedo design was frozen for production at the prevailing stage of development, and Westinghouse Electric Company was assigned to manufacture the torpedo. Several hundred Mark 28s were manufactured and delivered to the Navy before the end of World War II, but arrived too late to be used in combat.

As related research continued, the engineers at Bell Labs and Western Electric worked out the design for a 25-knot successor to the Mark 28 designated Mark 29. Among the new features of the Mark 29 was a counter-rotating electric motor that could be coupled with two propellers to eliminate the noisy gears. Although this arrangement reduced noise, it failed to achieve either the weight reduction or the advantage of automatically balancing torque, and only three experimental versions of the Mark 29 Mod were produced. An improved version, the Mark 29 Mod 1, was also developed by Westinghouse, but was cancelled after 35 development models were completed.

While the Mark 28 was being produced, versions of the Mark 18 were upgraded by Westinghouse to become the acoustically controlled Mark 31 torpedoes by providing noise reduction and installing the directional hydrophone system that had been developed. An experimental lot of these torpedoes was manufactured for comparison against the Mark 28. These studies eventually aided in the design of the Mark 32, which was introduced after the end of World War II.

THE ACTIVE ACOUSTIC TORPEDO

While work on the Mark 24 was progressing in late spring of 1942, the General Electric Company proposed that investigations be made of the practicability of steering the Mark 24 by means of an echo-ranging system rather than by homing on the noise produced by the target. Such a system would overcome the problem of background noise—including that generated by the launching submarine—by projecting enough sound energy so that the return echo would stand out.

GE was given a contract to develop an echo-ranging anti-submarine torpedo, and HUSL was given a contract to make a systematic study of echo-ranging methods that might be employed. GE's engineers came up with operational methodology for the new torpedo that would protect the launching submarine or surface ships while leading to an almost certain hit on the target. After the torpedo was launched, it would slide downward at a small, fixed angle while turning in fixed circles, "pinging" as it went. With a reasonable detection range, the speed of the torpedo was such that the target would be intercepted before it escaped the vicinity of the launch point. The proposed weapon was attractive for use from both surface ships and aircraft and would retain its effectiveness in the face of noisemakers that could decoy a passive torpedo, such as the Mark 24.

The detection system devised by GE used a magnetostrive transducer, four elements wide and eight elements high, that was split into an upper and a lower half. Homing signals in the vertical plane were derived by comparing the phase of the signals from the two halves of the transducer to provide proportional control in the vertical plane where the target—the submarine's hull—was only some 25 feet wide. A simpler on-off control mechanism was used in the horizontal plane where the target was much larger. When no echo was received, the torpedo's rudder was automatically thrown hard to port, causing the torpedo to circle to the left. When an echo was received, the rudder was shifted to the other side (starboard) until approximately one second after the last echo was received. The system then repeated the evolution, causing the torpedo to again circle to the left. Under this scenario the torpedo would normally have headed alternately for the bow or stern of the target, but the dynamics of the torpedo and time constraints built into the control circuits kept the torpedo aimed at the midpoint of the target's hull once it was detected.

The first successful sound-controlled, three-dimensional run of a Mark 32 torpedo was made in February 1944, 22 months after the project was formally initiated and numerous successful hits on submarine targets were made. Because GE's manufacturing facilities were too busy, arrangements were made by the Navy with Leeds and Northrop to be the supplier. Only ten preproduction Mark 32 torpedoes were completed before the end of the war when the program terminated. It was resurrected after the war as the Mark 32 Mod 2, and saw limited use in destroyers with over-the-side launchers beginning in 1950 until it was replaced by the Mark 43 torpedo.

Thus, by the end of the war the U.S. Navy had a successful air-launched acoustic ASW torpedo and had developed a submarine-launched counterpart for use against surface ships. (The German Navy had a successful submarine-launched acoustic anti-surface ship weapon and the British were engaged in development efforts.) Significantly, there was no U.S. submarine weapon suitable for attacking a submerged submarine.

Beyond having no suitable submarine-launched ASW torpedo, at the end of the war the Navy's highly successful Mark 24 Fido and its immediate successors faced the German Type XXI U-boat. This craft had an underwater speed of 17 knots, a submerged endurance measured in days, and a snorkel breathing tube that was to enable it to remain submerged for an entire patrol.

In particular, the Type XXI's speed could make it invulnerable to existing acoustic torpedoes. By the end of the war German shipyards had produced more than one hundred of these advanced submarines, and one was on patrol when the conflict ended in early May 1945.

At war's end the German and Japanese fleets had been vanquished and policies were in hand to ensure that those states would not again become military powers in the 20th Century that could threaten Allied interests. Rather, the only potential "enemy" on the horizon was the Soviet Union, a key ally during the war. At the end of the war Soviet intransigence in Eastern Europe and vanquished Germany, the occupation of Manchuria, and control of the northern portion of the Korean Peninsula, led perceptive Western military leaders to look at the massive Soviet military establishment as a threat to Western Europe and possibly even Japan. But the Soviets had no viable surface fleet and few ocean-going merchant ships, hardly a target for U.S. naval forces. The Soviet Navy, however, had possessed the world's largest submarine fleet at the beginning of World War II. Now there was the specter of Red shipyards producing flotillas of Type XXI submarines to fight a third Battle of the Atlantic.

CHAPTER THIRTEEN

Cold War Torpedoes

Submarines (1946–1991)

During the later 1940s and the 1950s NATO war planners were most concerned about a Soviet submarine force based on derivatives of the German Type XXI U-boat. Acoustic homing torpedoes would be the most efficient means to kill those undersea craft. The Type XXI was the most advanced submarine of World War II. According to the U-boat Command's manual for employing the submarine,

> Type XXI is a boat with strongly pronounced under-water fighting qualities which are capable of largely eliminating the superiority of the enemy's A/S [Anti-Submarine] operations, resulting from his command of air and surface radar. With this boat and [other new] types, it will be possible to start a new successful U-boat war.[1]

This submarine was faster than any other combat submarine and was to undertake 60-day patrols, remaining completely submerged. A large submarine, the Type XXI displaced 1,621 tons on the surface, more than twice the displacement of the Type VII U-boat, the type that had almost won the Battle of the Atlantic in the spring of 1943. The Type XXI had a streamlined hull, devoid of protuberances such as chocks, cleats, or gun mounts. A "sail," or "fairwater," was provided instead of a traditional conning tower, and to further reduce drag, no major deck guns were fitted. Submerged, the Type XXI could make up to 17 knots, the fastest of any undersea craft and, at lesser speeds and using a snorkel "breathing tube," it could remain submerged for an entire patrol. Beyond standard diesel-electric propulsion, the Type XXI submarine had two *schleich* ("crawling") motors for quiet operation. The Type XXI carried 20 torpedoes with six bow torpedo tubes and a high-speed reloading system.[2]

As Allied armies overran German shipyards after the war, submarine blueprints, components, and other material were scooped up by the victors, as were some German submarine engineers and technicians. And, in accord with the Potsdam Agreement of July 1945, Great Britain, the Soviet Union, and the United States each took possession of ten completed U-boats. Among those 30 were 11 Type XXI submarines. Subsequently, both the U.S. and Soviet Navies introduced Type XXI features into their submarines.

U.S. Navy intelligence in 1948 predicted that by the 1960s it was possible for the Soviets to have 1,200 or more submarines at sea. One U.S. admiral made two assumptions in discussing these numbers:

> First, an assumption that the Russians will maintain their numbers of submarines in approximately the same amount that they have now but improve their types and replace older types with new ones, and the second assumption, that by 1960 or within ten years, 1958, that the Russians could have two thousand up-to-date submarines.

The German Type XXI U-boat was the most advanced submarine of World War II in terms of performance. The submarine's torpedoes consisted of the Lüt, a pattern-running torpedo, and the T11, a passive acoustic homing weapon. Data collected by the Type XXI's advanced sonar was automatically set in the Lüt torpedoes, which were then fired in spreads of six. After launching, the torpedoes fanned out until their spread covered the extent of the convoy, when they began running loops across its mean course. In theory these torpedoes were certain of hitting six ships of from 197 to 328 feet in length with the theoretical success rate of 95 to 99 percent. In firing trials, such high scores were in fact achieved! (Royal Navy)

> I have chosen that [latter] figure because I believe it is within their industrial capability of producing that number and I believe if they really intend to employ the submarines as a means of preventing the United States or her Allies from operating overseas that two thousand would be the number they would require for their forces.[3]

A Soviet admiral, also in 1948, reportedly alluded to the possibility of a Red undersea force of 1,200 submarines.[4] U.S. naval intelligence saw the major limitation on the Soviet submarine buildup as a shortage of dock space and fuel. These factors could, it was believed, probably limit the force to 400 submarines.[5] If Soviet shipyards could produce about 16 submarines per month, the 1,200 number could be attained in less than a decade. (Germany in World War II had reached a maximum average of some 25 submarines per month, but U.S. intelligence officials felt that a number of factors would prevent the Soviets from attaining that production rate.)

Allied naval planners saw the Type XXI as the principal future naval threat.

SUBMARINE VS. SUBMARINE

As early as 1946 the U.S. Navy's Operational Evaluation Group proposed the use of submarines as a primary platform for Anti-Submarine Warfare (ASW) operations against advanced Soviet submarines. That September the chairman of the planning group for the Submarine Officers Conference noted that, "with the further development and construction in effective numbers of new submarines by any foreign power the employment of our submarines in anti-submarine work may well become imperative."[6] The emphasis on the ASW role for U.S. submarines as well as for surface ships and naval aircraft was based on the belief that, employing captured German technology, designs, and even engineers, the Soviets could mass produce submarines analogous to the German Type XXI.

The U.S. Navy's decision to employ submarines primarily as an ASW weapon was engaging because, of 178 German U-boats sunk in World War I, only about 10 percent were torpedoed by other submarines, and of 911 German, Italian, and Japanese submarines sunk in World War II, only 8 percent were torpedoed by submarines.[7] Of the approximately estimated 90 enemy submarines sunk by Allied undersea craft in the two conflicts, *only one*, the *U-864*, sunk off Norway by the British *Venturer* on 9 February 1945, was detected, tracked, attacked, and sunk while submerged. Having sighted the U-boat's periscope, Lieutenant J. S. (Jimmy) Launders, commanding the *Venturer*, planned to attack the U-boat when she surfaced. The captain of the *U-864*, realizing that there was another submarine in the area, began a zigzag course. After a three-hour chase, as the submarines were nearing the German-held port of Bergen, Launders realized that time was running out and, anticipating the U-boat's course based on passive sonar information, he fired a spread of four torpedoes set for varying depths. The fourth torpedo struck the *U-864*.[8]

In 1946 the U.S. Navy had no submarine- or surface-launched torpedoes suitable for use against submerged submarines. Straight-running torpedoes could not be used effectively against submarine targets that could rapidly change depth. The available submarine-launched acoustic torpedoes were suitable only for engaging surface ships. Still, that year the Navy's ASW Conference proposed equal priority for a specialized, small ASW submarine as well as a new, large "attack" submarine based on the German Type XXI design. The construction of the large *Tang*-class was begun in 1947; these 1,670-ton, 278-foot submarines incorporated many Type XXI features to enhance their underwater performance. At the same time, the Navy initiated the conversion of 52 war-built fleet submarines to the GUPPY configuration that would incorporate many Type XXI features.[9]

A series of Navy ASW conferences and exercises begun in 1947 confirmed the potential role of small, specialized hunter-killer submarines (SSK) that would wait in ambush to torpedo enemy submarines off Soviet ports and in channels and straits where those submarines would transit—on the surface or snorkeling—en route to and from the Atlantic shipping routes.[10]

The American SSK was envisioned as a relatively small, simply constructed submarine capable of being mass-produced by shipyards not previously engaged in building submarines. The SSK plan of 1948 called for 964 hunter-killer boats to counter the potential Soviet submarine force. This number included SSKs that would be at sea, in transit to and from patrol areas, undergoing overhaul, and in port being rearmed.[11] Construction of the *K-1* (SSK 1) class began in 1948. The submarines displaced 765 tons on the surface, with a length of 196 feet, and were manned by a crew of 37. The submarines had four torpedo tubes and carried eight torpedoes. The central component of the SSK design was long-range, passive sonar coupled with effective torpedoes that "would destroy any submarine which passed within detection range" with a very high degree of probability.[12]

Passive sonar was necessary because the best active sonars at the time had a practical range of under 2,000 yards at best. Passive array sonar, copied from German GHG sets, could detect submarines at greater distances, especially if the hunting submarine could be quieted to reduce self-noise interference.[13]

Providing "effective torpedoes" for U.S. submarines was another problem. The decision to employ submarines in the ASW role was severely handicapped by the lack of submarine torpedoes that could effectively engage a submerged submarine. At the end of the war the U.S. submarine force had only the straight-running Marks 14, 18, and 23 anti-ship torpedoes plus the Mark 27 Cutie and Mark 28 acoustic homing torpedoes that were only intended for use against surface ships. Several other acoustic torpedo programs were under way when the war ended. The Mark 33—fitted with acoustic homing guidance—was a 21-inch, 1,795-pound, submarine-launched weapon for the anti-ship and ASW roles. It was the only submarine torpedo that could engage a submerged submarine to reach even the prototype stage by the end of the war. Only 30 torpedoes had been produced when the Mark 33 was cancelled in 1945 in favor of the more-versatile Mark 35 program.

The Mark 35—known as the "universal torpedo"—was a 21-inch, 1,770-pound weapon compatible with submarine launch tubes as well as the fixed tubes in surface ships, making it the first surface-launched ASW torpedo. Initially the Mark 35 was also to be carried by anti-submarine aircraft, but that requirement was dropped after weight-performance compromises necessary for the universal launching design proved to be intractable. The torpedo was an offshoot of the innovative Mark 24, but was larger and more capable. The Mark 35 was 13 feet, 5 inches in length, and had contra-rotating propellers, synchronous settings in depth and

Only once in the history of submarine warfare has a submerged submarine sunk another submerged submarine: the British *Venturer* sank the German *U-864* off of the coast of Norway on 9 February 1945. The Type IXD2 submarine began her first patrol in late 1944; she sank no Allied ships. All 73 crewmen were lost when she was torpedoed by HMS *Venturer*. Earlier on the same patrol under Lt. J. S. (Jimmy) Launders, the *Venturer* had sunk the surfaced *U-771*. (Royal Navy)

The USS *K-1* (hull number SSK 1) was the first small "hunter-killer" submarine. Completed in 1951, she has to have been followed by several hundred similar SSKs intended to ambush Soviet submarines in narrow straits and passages. The massive bow dome housed an AN/BQR-4 sonar, which was complemented by the smaller AN/BQR-2—a copy of the German GHG sonar—fitted in a keel dome. These SSKs had four torpedo tubes and carried eight torpedoes. Only three SSKs were built. (U.S. Navy)

azimuth, and simplified electronics for increased reliability. It could run for 15,000 yards at 27 knots and carried a warhead of 270 pounds of HBX. It was the first fleet-issue torpedo with seawater batteries.

An unguided, straight-running, anti-ship version of the torpedo for submarine use was built by General Electric during the Mark 35 development program. This was the Mark 36, a 21-inch weapon that was 21 feet, 6 inches long and carried an 800-pound HBX warhead. The Mark 36 had a range of 7,000 yards at the high speed of 47 knots.

The Navy procured about 400 Mark 35s, which were in the fleet from 1949 to 1960. By the latter date most large (21-inch) torpedo tubes had been discarded from surface ships, replaced by the ubiquitous Mark 32 tubes that launched short, lightweight ASW torpedoes.

In response to the ominous intelligence reports of Soviet submarine developments, in 1948 the Navy directed Pennsylvania State University's Ordnance Research Laboratory to accelerate development of the Mark 27 Mod 4, which would become the first U.S. submarine-launched torpedo capable of attacking a submerged opponent. The basic Mark 27 Cutie had entered the submarine force in 1944 as an anti-escort torpedo (developed from the air-launched Mark 24).

The Mod 4 was a 19-inch diameter weapon that weighed 1,175 pounds with a 128-pound HBX warhead. It entered service in 1949 and some 3,000 were produced. Although its 16-knot speed did provide the advantage of relatively quiet operation, it was considered vulnerable to submarine-launched noisemakers and decoys as well as being limited to targets 14 knots or slower.[14]

"A MAJOR MILESTONE"

To overcome the anticipated limitations of the Mark 27 Mod 4, in 1946 Penn State and Westinghouse teamed to develop the Mark 37—the first postwar ASW torpedo to enter U.S. submarine service. It was "clearly a major milestone in torpedo development," according to torpedo historian Frederick Milford.[15] Wire guidance and other features marked a major departure—and capability—in anti-submarine torpedoes. The first Mark 37 test torpedoes were produced in 1955, followed by the production of more than 3,300 weapons that entered service beginning in 1960.

A 19-inch-diameter weapon, the Mark 37 Mod 0 weighed 1,430 pounds, including a 330-pound HBX-3 warhead. The wire-guidance feature lengthened the torpedo by 26 inches (to 13⅓ feet) and added 260 pounds to its weight.[16] If the 26-knot Mark 37 were launched at a range of several thousand yards against an "alerted," 16-knot or faster submarine, the torpedo could be in a stern-chase situation and could run out of power before it could close with the target. Still, it was superior to the Mark 27 Mod 4 and became one of the longest-serving torpedoes in U.S. history. The Mark 37 Mod 1, which entered service in 1960, was a larger, slower version, but it had improved target acquisition and wire guidance. Earlier versions of the Mark 37 were converted to later mods. Regardless of these limitations, Mark 37 Mod 4 torpedoes armed U.S. submarines for much of the Cold War, remaining in service until 1982. (Subsequently, the upgraded and reengineered NT37 variant was provided to several allied navies.)

The Mark 37 was launched from a submarine on a preset, straight, gyro-controlled run on an estimated intercept course; after travelling a preset distance, the torpedo began a passive acoustic search using a "snake," or circular, search pattern. When a target was acquired by its passive sonar, the Mark 37 homed on the target until the echo strength in the guidance became sufficient for active acoustic homing on the target. The U.S. Navy listed an acquisition range of about 700 yards, although such a distance appeared optimistic. The homing system initially used miniature vacuum tubes that were subsequently replaced by solid-state semiconductors.

The Mark 37 Mod 0 suffered from the initial, straight run, taking up to 15 minutes at 17 knots (for a theoretical 23,000 yards) or less time at 26 knots (10,000 yards). During this period the target submarine—probably having heard the Mark 37 launch—could speed up, take evasive action, or deploy jammers and decoys. The Mark 37 Mods 1 and 2 had wire guidance that enabled the launching submarine to control the torpedo's initial maneuvers through the wire unreeling from both the submarine and the torpedo. This feature—also based in part on earlier German technology—permitted initial guidance by the more-capable submarine sonar rather than that of the torpedo.

There were, however, several disadvantages to wire guidance: the submarine's maneuverability was restricted because of the wire, only one torpedo could be launched at a time (the two wires could become entangled if two torpedoes were in the water at the same time), and the empty torpedo tube could not be reloaded until the guiding wire was cut. The latter occurred when the torpedo's internal guidance system had locked on to the enemy submarine.[17]

Still, the fleet liked the Mark 37. One former submarine officer recalled in the mid-1960s that, "compared to the Mark 14 and Mark 16 torpedoes that we carried, the Mark 37 was great. It was easy to maintain and it was reliable."[18]

One other wire-guided torpedo entered the fleet in this period—the Mark 39, which was a Mark 27 Mod 4 modified by the addition of a wire dispenser and improved propulsion. The modifications were developed by Penn State and the Vitro Corporation. The Philco Corporation converted 120 torpedoes from 1956, mainly for use in fleet familiarization and evaluation

The Mark 37 torpedo was the standard ASW weapon of all U.S. and many Allied submarines for more than two decades. It was the first specialized ASW torpedo to enter U.S. service after World War II and had several innovative features, including wire guidance. Its successor, the Mark 48 was intended for surface ship and submarine use but, like the Mark 37, ultimately armed only submarines. Note the "lugs" to enable the 19-inch weapon to be launched from 21-inch tubes. (U.S. Navy)

with wire-guided torpedoes by the seven fleet submarines converted in the early 1950s to the SSK role.

NUCLEAR TORPEDOES

Another torpedo of this era was the Mark 45—fitted with a nuclear warhead. Nuclear-propelled submarines, which first appeared in the U.S. Fleet in 1955 with the USS *Nautilus* (SSN 571) and with the Project 627/November in the Soviet Navy in 1958, again changed the ASW equation. While early U.S. nuclear submarines were relatively noisy (especially compared to early Soviet submarines), their unlimited underwater endurance and relatively high speed made them virtually invulnerable to existing ASW systems. Writing in 1960, the commanding officer of the early nuclear submarine *Seadragon* (SSN 584), declared:

> In fact, surface and air ASW forces today normally detect a nuclear submarine only when she attacks—and often the detection consists [in exercises] of sighting the submarine's flare firing signal, or of hearing his announcement by sonar.
>
> Such circumstances demand a deadly weapon for the ASW forces of air and surface with which to club the submarine quickly. We do not have it in usable form.
>
> Conventional depth charges and ahead-thrown weapons are totally inadequate against such a high-speed, deep diving enemy. Service torpedoes are not sophisticated enough. The ability to localize the nuclear submarine is not good enough. Nuclear depth-bombs would do the trick, but the submarine seems always to be too near friendly ships to use one, or at large in an area of uncertainty too great to bomb. And one must be just a little uneasy about pinning everything on the nuclear blast.[19]

Still, nuclear weapons—especially torpedoes—offered the means of countering a Soviet nuclear submarine that could run fast and dive deep to avoid available ASW weapons. A nuclear torpedo had been proposed for the U.S. Navy as early as 1943: Captain William S. Parsons, head of the ordnance division of the Manhattan (atomic bomb) project, recommended providing a "gun," or uranium-type nuclear warhead, in the Mark 13 aircraft-launched torpedo. But no action was taken on his proposal, primarily because of all available resources being allocated to the (two) atomic bomb projects.

By the early 1950s several factors converged that led to the development of nuclear torpedoes by the U.S. Navy, among them (1) the dramatic reduction in the size of nuclear warheads and (2) the development of nuclear-propelled submarines by the Soviet Union. The higher speeds and greater operating depths of advanced Soviet submarines could enable them to evade conventional ASW weapons. This situation would be exacerbated if Soviet submarines employed decoys or jammers to counter U.S. acoustic homing torpedoes.

The USS *Nautilus* (SSN 571)—the world's first nuclear-propelled submarine—shown loading a torpedo. The *Nautilus* went to sea in January 1955 to lead a revolution in undersea warfare. Nuclear submarines can operate at high speeds for essentially unlimited periods, never having to surface or project up a snorkel tube to breath in air for their engines. The bulbous dome on the starboard side is an under-ice sonar; the light patch at right covers the tethered rescue buoy, released if the submarine is disabled in relatively shallow water (U.S. Navy)

Work began in 1956 on the Mark 45 submarine-launched torpedo. Development of the weapon—later given the name ASTOR for Anti-Submarine Torpedo—was directed by the University of Washington's Applied Physics Laboratory. The first developmental model of the Mark 45 was ready in December 1958, and the production prototype was delivered for service use in July 1959, a relatively rapid peacetime development effort.

The Mark 45 was a 19-inch diameter torpedo with a length of 18 feet, 9 inches (18 feet, 11 inches in the Mod 1 and 2 variants); the Mod 0 weighed 2,330 pounds. The weapon was unique among modern torpedoes in having no on-board guidance system or self-detonating device. As the Mark 45 was launched, a wire payed out, connecting it to the submarine. But whereas in previous wire-guided torpedoes the wire provided control from the submarine until the torpedo's guidance could acquire the target submarine, at which point the wire was severed, in the Mark 45 it was deemed necessary for wire control for the torpedo's entire run to the target to ensure "positive control" of the nuclear warhead, which could only be detonated by a command signal through the wire.

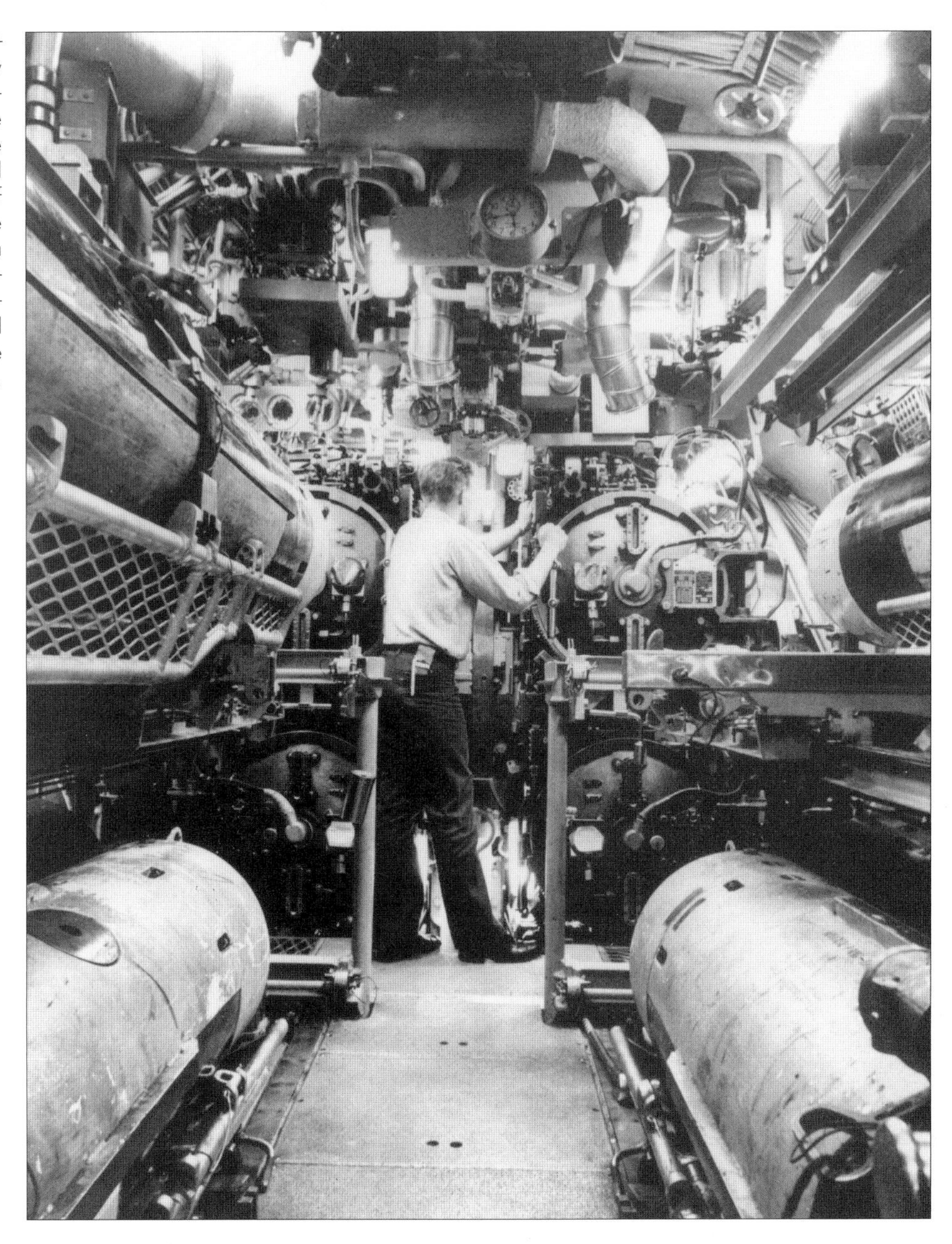

The torpedo room of the *Nautilus*, which had six 21-inch bow torpedo tubes with a total load-out of 20 torpedoes. Note the complexity of equipment in the compartment, which also served as a berthing space for some of the crew, which numbered more than 100. The *Nautilus* was a one-of-a-kind "experimental" submarine but with full combat capabilities. She had a conventional submarine hull derived from the German Type XXI design. (U.S. Navy)

The torpedo's warhead was the W34 nuclear device, which was also used in the Lulu nuclear depth bomb and the Hotpoint aerial bomb. The warhead, 17 inches wide and 32 inches long, produced a yield of 11 kilotons (compared to 15 kilotons for the atomic bombs dropped on Hiroshima and Nagasaki in 1945).

Details of the Mark 45's service remain classified. However, most or possibly all U.S. nuclear-propelled attack submarines and ballistic missile submarines carried two of the torpedoes on patrol. When the attack submarine *Scorpion* (SSN 589) was lost in 1968 she carried two Mark 45s. In 1966 an Atlantic Fleet submarine fired a Mark 45 exercise (unarmed) unit on what was then the longest torpedo run in U.S. Navy history. The torpedo traveled more than 15,000 yards under wire guidance.

Although the nuclear-armed SUBROC was also carried in some U.S. nuclear-propelled attack submarines, the Mark 45 imposed major handling, security, and administrative burdens on submarine crews. And, because of the lethal radius of its warhead, submariners often cited the Mark 45 as having a Pk (probability of

The dubious honor of being the first ship to be sunk by a nuclear-propelled submarine went to the discarded landing ship *Chittendon County* (LST 561), torpedoed in Hawaiian waters by the USS *Sargo* (SSN 583) on 21 October 1958. The only ship to be sunk in combat by a nuclear submarine was the Argentine cruiser *General Belgrano*, torpedoed by the British *Conqueror* on 2 May 1982, during the Falklands conflict. The *Conqueror*, under Commander Christopher Wreford-Brown, used outdated Mark VIII torpedoes in the attack although more modern torpedoes were on board. (U.S. Navy)

kill) of 2.0—both the target submarine *and the firing submarine*!

The Mark 45 was retired—the last being taken out of service in 1977—because of the estimated effectiveness of the Mark 48 carrying a conventional warhead. It was believed that the later torpedo's range, speed, and guidance could outperform the Mark 45's lethal capabilities.

When a proposal was made in 1983 to reintroduce nuclear torpedoes to the submarine force, the nuclear submarine community was adamant in its opposition. The subject was raised during the 1983 study "Torpedoes and ASW Weapons" by the Secretary of the Navy's principal advisory group, the Naval Research Advisory Committee. A subsequent minority report by Dr. Donald Kerr, at the time director of the Los Alamos National Laboratory, and naval analyst and author Norman Polmar raised the question of providing nuclear inserts for U.S. submarine torpedoes.[20] Their concern was that U.S. Mark 44 and Mark 46 lightweight ASW torpedoes, with relatively small warheads, and possibly the new, heavy-weight Mark 48 could not cope with modern Soviet submarines, which had significant standoff distances between their inner (pressure) and outer hulls and multiple compartments as well as enhanced performance.[21] And Soviet submarine tactics emphasized the extensive use of decoys and jammers to counter acoustic homing torpedoes.

There was also the specter of massive U.S. torpedo failures—as had occurred in World War II—caused either by technical problems or Soviet countermeasures. Indeed, even the much-vaunted Mark 48 would experience significant "teething problems." Thus, Kerr and Polmar proposed a nuclear insert for heavy-weight ASW torpedoes, enabling them to be used in the conventional or nuclear mode. In the nuclear mode the weapons would have a low-kiloton or even sub-kiloton yield.

The only torpedo developed in the West with a nuclear warhead was the Mark 45 ASTOR—Anti-Submarine Torpedo. It was unique among U.S. torpedoes in that it did not have any contact or proximity detonation mechanism; its W34 nuclear warhead could only be detonated by the launching submarine sending a signal through the control cable. Although unpopular in the fleet, the ASTOR reflected the widespread deployment of nuclear weapons during the Cold War. (U.S. Navy)

An unclassified exposition of their views concluded that torpedoes with nuclear warheads address specificcally the following issues:

- Increased-lethality heavy torpedo for use against Typhoon SSBNs [Project 941], Oscar SSGNs [Project 949], and their successors;
- Lightweight torpedo homing warhead for delivery by aircraft or missiles against time-urgent, hard targets;
- Response to sudden deployment of effective Soviet countermeasures;
- Response to ineffective conventional torpedoes against.[22]

The United States, they concluded, "must maintain a capability to conduct tactical nuclear warfare at sea for two reasons: first, as a deterrent; second, as a hedge against the catastrophic failure of essential weapons—in this case, conventional torpedoes—as happened to the U.S. and German submarine torpedoes in the early phases of World War II.[23]

The article thus questioned the effectiveness of U.S. torpedoes against modern Soviet submarines and consequently implied that U.S. submarines were highly vulnerable to Soviet torpedoes. It unleashed a firestorm of comment and criticism by the U.S. nuclear submarine community. A few days after the *Proceedings* article was published, on 8 August 1986, Vice Admiral Bruce DeMars, the Deputy Chief of Naval Operations for Submarine Warfare, held up a copy of the article at his staff meeting and declared, "We need to squash dissent like this."[24] Without bothering to determine if the article had been cleared for publication—which it had by the Department of Energy—DeMars initiated an investigation by the FBI and the Naval Investigative Service into the authors' possible violation of their security clearances by publishing the article.[25] Next, DeMars caused a series of comments by naval officers and Navy contractors to attack the authors in the Naval Institute *Proceedings*. In the event, after hundreds of hours of investigative time, the charges of security violations were simply dropped.

But the nuclear issue continued to be raised. The appearance of the Soviet Project 941/Typhoon and Project 949/Oscar classes of missile submarines, the largest submarines ever constructed, had double hulls with significant "stand-off" distances between them that could absorb explosive forces, major compartmentation, and other features to enhance survivability, led to proposals for a so-called Sub-Kiloton Insertable Nuclear Component (SKINC) that could provide a yield as small as 1⁄10th of a kiloton for a Mark 48 heavy torpedo, and possibly for a lightweight ASW torpedo. A 1⁄10th kiloton warhead has the equivalent of 8,800 pounds of high explosive force. One proponent of this concept, John J. Englehardt, a naval architect responsible for Navy weapons effects and submarine vulnerability studies, wrote,

> Because SKINC torpedoes pack very small nuclear yields, it is unlikely that full nuclear release authority is needed. Rules of engagement (ROE) could be developed to give submarine or carrier battle group commanders engagement options that include using a SKINC torpedo in the nuclear mode without specific release by the National command Authority [i.e., president and Secretary of Defense]. For example, SKINC torpedoes could be restricted for use in the nuclear mode against highly survivable Soviet submarines . . . and deep-diving/high-strength Alfa and Mike attack subs.[26]

But the nuclear submarine community remained resolute against even an open discussion of the nuclear torpedo issue. Such a discussion, it was feared, would highlight the disparity between U.S. and Soviet submarine performance and survivability. No further efforts were undertaken by the United States to develop nuclear torpedoes.

(The Soviet Navy, which had first provided nuclear-armed torpedoes in submarines in 1958, continued to deploy those weapons, at least until the end of the Cold War in 1991.)

SUBMARINE ROCKETS

The new, passive-array AN/BQQ-2 sonar being developed in the mid-1950s by the U.S. Navy gave the promise of passive acoustic detection of target submarines at greater ranges than could be effectively reached by Mark 37 and Mark 45 torpedoes. Thus, despite the disdain for tactical nuclear weapons by many senior U.S. naval officers, the Mark 45 ASTOR was followed on board U.S. torpedo-attack submarines by the nuclear-armed SUBROC (UUM-44A), which entered the fleet in 1965.[27] Although not a torpedo, it was launched from submarine torpedo tubes and displaced torpedoes in torpedo room reload racks. To some degree an underwater analogy to the surface-launched ASROC anti-submarine weapon, the SUBROC was a rocket-propelled nuclear depth bomb. Unlike the ASROC on board surface ships, it could not be fitted with a conventional homing torpedo.

After being launched from the submarine, the SUBROC streaked to the surface, left the water, and travelled on a ballistic trajectory for a predetermined time to a distance out to 30 nautical miles. In flight the booster fell away and the bomb reentered the water at supersonic speed, detonating at a preset depth. The SUBROC carried the W55 nuclear warhead with a "dialable" yield of 1 to 5 kilotons.

THE T-5 NUCLEAR TORPEDO

The world's first nuclear-armed torpedo to become operational was the Soviet T-5, launched from submarines against surface ships. Nuclear warheads in anti-ship torpedoes would improve their "kill" radius, meaning that a direct hit on an enemy ship would not be required. A nuclear warhead could thus compensate for poor acoustic homing by the torpedo or for last-minute maneuvering by the target, and later for overcoming countermeasures or decoys that could confuse a torpedo's guidance. With the RDS-9 nuclear warhead, the T-5 was test launched from a submarine in 1957; the nuclear explosion had a yield of ten kilotons at a distance of just over six miles from the launching submarine. (That warhead size was more than half the size of the Hiroshima and Nagasaki atomic bombs.) The T-5 was the first nuclear weapon to enter service in Soviet submarines, becoming operational in 1958 as the Type 53-58. It was a 21-inch diameter weapon.

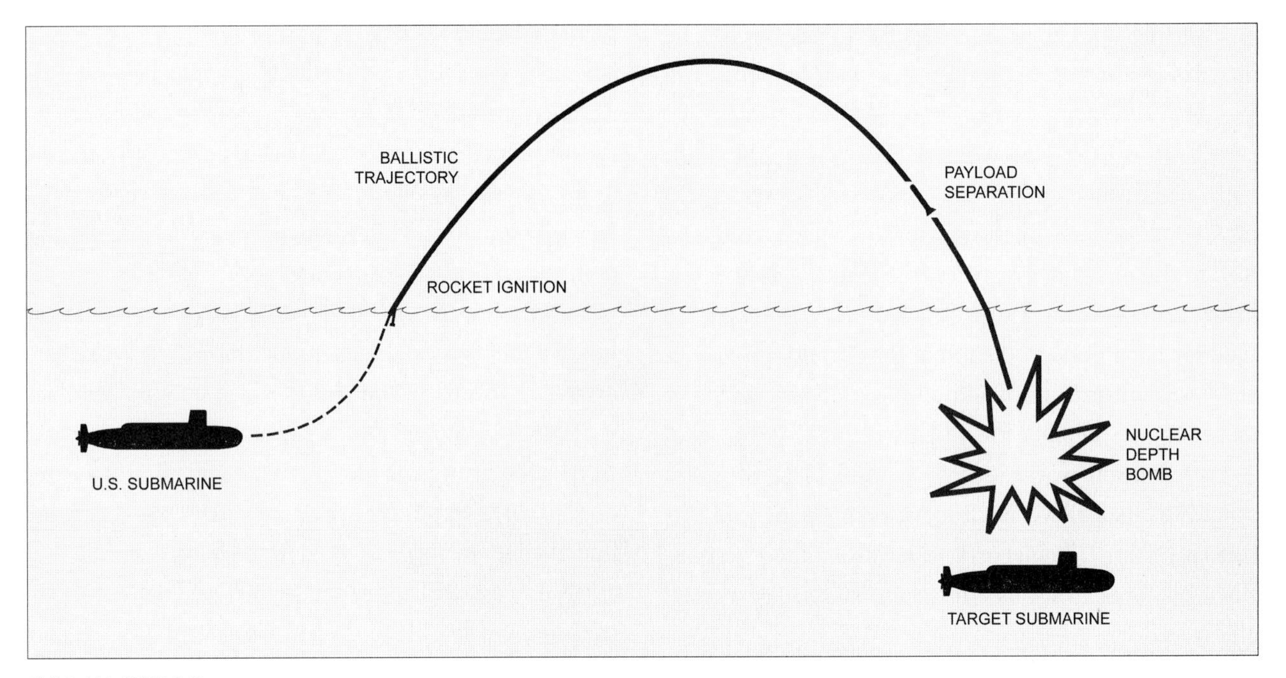

UUM-44A SUBROC operation.

Launching SUBROC required the submarine to produce an active acoustic "ping" to determine the range to the target submarine (which was normally detected by passive sonar). The use of active sonar to produce even a single emission could have alerted the enemy submarine to the launching submarine's presence in the area, hence SUBROC was even less popular with U.S. submariners than was the Mark 45 ASTOR.

The SUBROC was analog and thus not compatible with later U.S. attack submarines having the Mark 117 digital fire-control system. Only about 25 submarines of the *Permit* (SSN 594) and *Sturgeon* (SSN 637) classes that were fitted with the Mark 113 analog fire-control system carried the weapon. The missile, which became operational in 1965, outlasted the Mark 45 ASTOR aboard submarines, albeit in a relatively small number of boats. The SUBROC was in service until 1989, a relatively long service life—more than two and a half decades. In the late 1950s, with the cancellation of the Regulus II land-attack cruise missile, the SUBROC was also considered for use against shore targets with an air-burst warhead. The missile weighed some 4,000 pounds, was 21 inches in diameter and 21 feet long, and was powered by a two-stage, solid-propellant rocket.[28]

An advanced weapon known as the Sea Lance (UUM-125B) was to have succeeded SUBROC in the fleet. By the 1980s SUBROC was obsolete, with safety considerations as well as warhead viability being major factors in seeking a replacement stand-off ASW weapon. In 1979 Vice Admiral David Emerson, the head of Navy research and development, told Congress that "The urgency of the requirement is severe."

At times during its development Sea Lance was to have been launched from both surface ships and submarines, and was to be able to carry either a nuclear (W89) or conventional (Mark 50 torpedo) payload. The Sea Lance was designed for attacks out to the third sonar Convergence Zone (CZ), i.e, approximately 90 to 100 nautical miles; however, when fitted with the Mark 50 torpedo the effective range would probably be only the first CZ, some 30 nautical miles.

After a very convoluted evolution, the Sea Lance was cancelled in January 1990—less than a year after the last SUBROC was removed from inventory. Within the next two years, with the demise of the Soviet Union

A torpedo tube-launched SUBROC is loaded aboard the attack submarine *Permit* (SSN 594). In submarines of the *Thresher* (SSN 593) class the torpedo loading chute was fitted aft of the diminutive sail structure. The large AN/BQQ-2 bow sonar dome—some 15 feet in diameter—required that the torpedo tubes be placed aft of the dome, angled out at about 15 degrees, two per side. All subsequent U.S. nuclear submarines had four torpedo tubes except for the three submarines of the *Seawolf* (SSN 21) class, which had eight 26½-inch tubes. (U.S. Navy)

A SUBROC missile breaks through the surface during a submerged-launch test. Carrying a W55 nuclear warhead, SUBROC had a range of up to 30 nautical miles, using inertial guidance; what it lacked in accuracy compared to an acoustic homing weapon was compensated for by the nuclear blast area. And, unlike homing torpedoes, SUBROC was not susceptible to jammers or decoys launched by the target submarine, and its transit through the water could not be detected. (U.S. Navy)

UUM-44 SUBROC

A submarine-launched ASW ballistic missile carrying a nuclear warhead, the SUBROC was launched from standard 21-inch torpedo tubes in *Permit* (SSN 594) and *Sturgeon* (SSN 637) submarines fitted with analog fire control systems. After leaving the tube the SUBROC streaked to the surface, its rocket engine ignited, and it flew a ballistic trajectory to the target where the warhead—and nuclear depth bomb—was released.

Length	21 ft
Diameter	21 in
Weight	4,000 lb
Warhead	W55 (200-kiloton) thermonuclear
Rocket motor	Thiokol TE-260G solid-propellant
Speed	supersonic
Range	30 nautical miles
Guidance:	Kearfott SD-510 inertial

in December 1991, all nuclear weapons were removed from U.S. warships except for the Trident submarine-launched ballistic (strategic) missiles.

The threat of the Soviet submarine force guided U.S. torpedo development during the Cold War. Indeed, heavy-weight torpedoes were developed only for submarines during the Cold War and not for surface ships. (See Chapter 14.) Further, in an era of nuclear weapon proliferation, U.S. surface ships and submarines were armed with nuclear ASW weapons—ASROC in surface ships and the Mark 45 ASTOR torpedo and SUBROC in submarines. Similarly, the Sea Lance missile as conceived was to have had a nuclear warhead.

The planned UUM-125B Sea Lance ASW missile was originally envisioned as a torpedo tube-launched weapon to succeed the SUBROC aboard attack submarines. During its convoluted development, the Sea Lance was also planned for surface ships (to replace ASROC), and to carry a Mark 50 lightweight torpedo or a nuclear warhead to ranges out to almost 100 nautical miles. A combination of technical difficulties, program delays, and continual changing requirements led to the cancellation of the weapon on 3 January 1990. (Boeing Aerospace)

But eventually, a single, advanced, conventional torpedo—the Mark 48—would be the sole armament of U.S. attack submarines, and the "tactical" weapon of strategic missile submarines.

CHAPTER FOURTEEN

Cold War Torpedoes

Surface Ships and Aircraft (1946–2010)

As World War II in the Pacific was reaching its climax, the U.S. Navy was developing designs for the next generation of warships, including destroyers to succeed or at least to complement the highly successful *Allen M. Sumner* (DD 692) and *Gearing* (DD 710) classes. Those ships, at 2,200 and 2,425 tons standard, respectively, carried six 5-inch guns in twin gun mounts, numerous 40-mm and 20-mm anti-aircraft guns, and ten 21-inch torpedo tubes in two mounts. As the Navy began planning post-war destroyers, there was initial interest in providing heavy torpedo batteries, with proposals to fit new ships with 5-tube, 10-tube, and even 15-tube mounts. The last would weigh about the same as a twin 5-inch gun mount—just over 60 tons.[1]

But the destroyer-type ships constructed after World War II would not have large torpedo batteries for fleet actions. Rather, the ships were intended primarily to provide air defense for carrier groups with a secondary anti-submarine role. The air threat was postulated on the expectation that the Soviets would soon acquire turbojet engine technology and guided bomb designs from the vanquished Germans.[2] And, if the principal roles for destroyers would be anti-air and Anti-Submarine Warfare (ASW), the principal use of their torpedoes would be for anti-submarine rather than anti-ship operations.

There were a large number of U.S. war-built, torpedo-armed destroyers (and submarines) available in the late 1940s. The Navy retained in commission some 140 war-built destroyers, most of which had a five-tube 21-inch torpedo mount; about 200 more torpedo-armed destroyers were "mothballed" in the reserve fleet. The second five-tube mount had been removed in active ships to provide for the installation of additional 40-mm Bofors and, subsequently, 3-inch/50-caliber anti-aircraft guns. With 150 fleet and modernized (GUPPY) submarines still in service, and new submarines under construction, there were "plenty" of 21-inch torpedo tubes available in the active fleet.

Available to the fleet were thousands of straight-running, electric and Navol torpedoes available for these ships: Marks 14, 18, and 23 for submarines, and Marks 15 and 17 for surface ships. Mark 13 heavy, anti-ship torpedoes were also available for aircraft. With the huge inventory and variety of war-built heavy torpedoes, there was no need for new-production heavy torpedoes to fill destroyer (or submarine) tubes for the anti-ship role. Indeed, during the 45 years of the Cold War, no heavy torpedo was developed specifically for use by surface ships.[3]

CONCENTRATION ON ASW

In 1948 two different destroyer-type classes were begun. The first, completed in 1953, was the *Norfolk* (DL 1), a 5,600-ton ship begun as an ASW cruiser.[4] However, the *Norfolk* was too large and too expensive for series production. Simultaneously, the Navy began the *Mitscher* (DD 927)-class destroyers of 3,500 tons; these ships were also commissioned as frigates (DLs), the first in 1953.

In keeping with her primary anti-air role—despite her initial ASW designation—the *Norfolk* carried eight of the new 3-inch/70-caliber guns. For ASW she had four of the new Weapon A launchers that fired a 12.75-inch anti-submarine projectile to a maximum of 760 yards. Similarly, the *Mitschers*—four were built—carried two of the new 5-inch/54-caliber guns and four 3-inch/70-caliber weapons for the anti-air role, and two Weapon A launchers.[5]

The *Norfolk* had eight fixed Mark 24 tubes for heavy ASW torpedoes (30 carried) and the *Mitschers* had four fixed Mark 24 tubes for ASW torpedoes (10 carried). But the Weapon A—also known as Weapon Able and, subsequently, Weapon Alfa—was considered the ships' principal ASW weapon.[6]

THE SHKVAL HIGH-SPEED TORPEDO

During World War II there was some rocket-propelled torpedo development in Germany and Italy. After the war interest was shown in this type of weapon in the Soviet Union and the United States, but only the Soviet Union developed rocket-propelled torpedoes. The aircraft-launched RAT-52 torpedo of 1952, for use against surface ships, had an underwater speed of almost 70 knots; the air-launched anti-submarine torpedoes APR-1 and APR-2 used solid-rocket propulsion to attain speeds just over 60 knots. In 1960 development of a submarine-launched rocket torpedo began. A solid-propellant rocket motor was coupled with moving the weapon through the water surrounded by a gaseous envelope, or "bubble" that is created by the torpedo's shape and sustained by the rocket's exhaust. This technique is known as "artificial cavitation." The preliminary design of the Shkval (squall) was completed as early as 1963, and a submarine began test launches in May 1966. The tests were halted in 1972 because of problems, but after seven launches from the submarine *S-65* in June–December 1976 the torpedo—given the designation VA-111—was declared ready for operational use in November 1977—17 years after the project was initiated. The 21-inch-diameter torpedo is 27 feet long and weighs 5,950 pounds. Estimated range is more than 10,000 yards at approximately 200 knots. Initially fitted with a nuclear warhead, later variants are reported to have terminal guidance and a conventional, high-explosive warhead of 460 pounds. There were later reports that a 300-knot version was under development. Even the 200-knot weapon was beyond the capabilities of U.S. torpedo countermeasures.

Weapon A was one of several efforts to increase the range of ship-launched ASW weapons. The wartime-developed Hedgehog and Mousetrap (spigot mortars) fired projectiles a few hundred yards ahead of the ship, but their purpose was to enable the use of shipboard sonar without the acoustic disruption that depth charges caused as they exploded. Hedgehog and Mousetrap projectiles detonated only on contact with a submarine's hull.

The advances in shipboard sonar after the war required longer-range ASW weapons. And, because a surface ship had little possibility of delivering a surprise attack at short range, the surface ship always sought to attack at maximum range to reduce the danger of a counterattack from the target submarine.[7] Weapon A and the Norwegian-developed Terne III were the first major efforts to provide that capability in U.S. warships. Their maximum ranges were 800 yards and 920 yards, respectively. But greater weapon ranges were needed.

Seeking methods to extend torpedo ranges, as early as 1941 both the Army and Navy began looking into the feasibility of torpedo-carrying missiles; although the Navy had initially taken over that project, its efforts soon were concentrated to the Bat, an air-launched, anti-ship glide bomb carrying a 1,000-pound warhead.[8] A variation was the Pelican glide bomb with radar homing intended to attack surfaced submarines with depth charges; its payload was subsequently changed to a 2,000-pound bomb. Significantly, in August 1944 the Bureau of Ordnance initiated a derivative of the Bat to carry an anti-ship torpedo. Several variations of the weapon—named Kingfisher—were proposed.[9]

Most variants were not pursued, but the Kingfisher C became the Petrel (AUM-N-2), an air-launched, turbojet-powered weapon carrying the Mark 21 homing torpedo.[10] The 2,130-pound torpedo was a passive homing variant of the venerable Mark 13; the ASW variant of this torpedo—the Mark 21 Mod 2—had been in development since 1943. It had a range of 6,000 yards at a speed of 33.5 knots. The Naval Ordnance Plant in Forest Park produced 312 of these torpedoes between 1946 and 1955.

The USS *Norfolk* was begun as a hunter-killer cruiser (CLK 1), redesignated a destroyer leader (DL 1), but completed as a frigate (DL 1), with primarily an anti-aircraft armament of eight 3-inch/70 rapid-fire guns. She spent most of her active service as an experimental ship, testing the Weapon A and ASROC anti-submarine weapons. In this 1969 photo she has an eight-cell ASROC launcher abaft her second funnel; the four Weapon A launchers previously carried have been removed. (U.S. Navy)

The Petrel had wooden wings and fins and was propelled by a Fairchild J44 turbojet engine. The missile was released some 200 feet above the water and cruised at some 375 mph toward the target. Its range was almost 20 nautical miles, with the Mark 21 having a range of 6,000 yards at a speed of 33.5 knots. When 4,600 feet from the target, the engine shut down and the wings and fins were jettisoned. The torpedo then fell to the water and initiated an acoustic search for the target—a surface ship or surfaced submarine. The weapon was carried primarily by P2V Neptune maritime patrol aircraft. Petrel tests began in 1951, with development transferred to Fairchild Engine Co. The missile became operational in 1956, but it had only a brief service life.

The Kingfisher E was another torpedo-carrying variant—for launch from surface ships to counter deep-diving submarines. The Chief of Naval Operations, Fleet Admiral Chester Nimitz, gave this project the highest priority in March 1946. The "E" evolved into the Grebe missile, designated the SUM-N-2 (Surface-to-Underwater Missile, No. 2 in the Navy series). It was intended to carry the Mark 35 heavy-weight torpedo, but was reoriented to take the lighter Mark 41 or, possibly, the Mark 34. All were ASW homing weapons. The Grebe effort looked at two configurations: with a rocket booster to be used for short distances (5,000 to 8,000 yards), and with the carrying vehicle longer distances (out to 40,000 yards). Although prototype flights were

The destroyer *Waller* (DD 466) fires a Weapon A rocket during 1959 ASW exercises in the Eastern Atlantic. The Weapon A launcher replaced the ship's No. 2 5-inch/38 gun mount. A hedgehog ASW mortar can be seen at right, immediately below the bridge. Although Weapon A was mounted in numerous U.S. destroyers and escort ships, its limited range and effectiveness led to its rapid replacement by the torpedo-carrying ASROC. (U.S. Navy)

The AUM-N-2 Petrel air-launched missile provided a turbojet booster for the Mark 21 Mod 2 acoustic homing torpedo. This was one of several air-launched guided weapons developed by the U.S. Navy and Army Air Forces during World War II. This Petrel is mounted under the wing of a Lockheed P2V-6M Neptune, a maritime patrol aircraft flown in large numbers by the U.S. Navy and by several other navies and air forces. (U.S. Navy).

conducted, the project was cancelled in 1950 because no existing sonar could provide accurate submarine detections out to some 20 nautical miles.

In May 1948 General Electric (GE) began development of an air-launched version of the Mark 35 torpedo to be known as the Mark 41. The contract called for the development of a turbine-drive in competition with a reciprocating-engine development by another company. Gerry Hoyt, the head of GE's torpedo program, realizing both the drives would require extremely long development times, initiated the design of a new electric drive that proved to be superior and easier to develop. The new drive was used on GE production units. The Mark 41 had many features in common with the Mark 35 (see p. 151), but was only ten feet long and was reinforced to withstand the impact of water entry at 200 knots.

Early in 1950, Gerry Hoyt initiated preparation of an unsolicited proposal for a new, smaller, antisubmarine torpedo for use by aircraft accompanying convoys. A week before submission, GE and four other companies were formally asked to submit competitive proposals on just such a torpedo, the Mark 43. Although somewhat doubtful that the Navy would award another torpedo contract to GE at the same time they were working on the Mark 35 and Mark 41, Hoyt obtained $30,000 of company money to develop a prototype unit, and Harry Burkart was appointed project engineer. This unit was in the water and running before the signing of the Mk 43 development contract, which was awarded to GE. The Mark 43 used a single propeller. It was only 12¾ inches in diameter by about 90 inches long and ran at 15 knots.

In another effort to extend torpedo range—and one compatible with available sonars—several destroyers were fitted with RAT (Rocket-Assisted Torpedo). This twin-arm device could be fitted to standard 5-inch/38-caliber gun mounts. The RAT launched a rocket-assisted lightweight torpedo to a distance of 5,000 yards. The RAT test fired a Mark 24 torpedo in 1952 and the larger Mark 43 in 1954. The RAT was looked upon as a replacement for the ahead-firing Hedgehog and Weapon A weapons. But RAT, which entered the fleet in the mid-1950s as the Navy's first "long-range" ASW torpedo system, suffered from unsatisfactory ballistic characteristics and it failed to achieve the predicted accuracy. It was soon overtaken by a more capable torpedo-delivery weapon—the ASROC (Anti-Submarine Rocket).

The ASROC was a ballistic weapon with a solid-propellant booster that could be employed with two warhead options—a lightweight Mark 44 or (after 1965) Mark 46 homing torpedo, or a W44 nuclear depth bomb. With a range of about seven nautical miles, the ASROC (RUR-5A) torpedo was launched on a ballistic trajectory, with its range determined by a timer (set before launch) that cut off the solid-rocket motor. The torpedo was lowered to the water by parachute, after which it began its normal underwater search pattern. An ASROC was tested with a nuclear warhead,

THE WORLD'S LARGEST TORPEDO

The largest torpedo known to have been proposed was the Soviet T-15 nuclear torpedo for use against U.S. and British naval bases. In 1949–1950 the development of a submarine-launched nuclear torpedo was begun in the Soviet Union. The torpedo was to carry a thermonuclear (hydrogen) warhead a distance of some 16 nautical miles. It would have a diameter of just over 5 feet and a length of approximately 77 feet. The 40-ton underwater weapon would be propelled to its target by a battery-powered electric motor, providing a speed of about 30 knots. Obviously, a new submarine would have to be designed to carry the torpedo—one torpedo per submarine. The submarine carrying a T-15 would surface immediately before launching the torpedo to determine its precise location by stellar navigation and using radar to identify coastal landmarks. The long submerged distances that the submarine would have to transit to reach its targets demanded that it have nuclear propulsion. This was the beginning of Project 627, the Soviet Navy's first nuclear submarine (NATO code-name November). The project was initiated without knowledge—or approval—of senior naval officers. In the fall of 1952 Soviet dictator Josef Stalin formally approved the project. Stalin died in early 1953 and in 1955—in response to the Navy's objections and recommendations—the requirement for Project 627 was revised for attacks against enemy shipping, to be armed with conventional (high-explosive) torpedoes. The forward section of the submarine was redesigned for eight 21-inch torpedo tubes, with 12 reloads provided. (The submarine carried a total of 20 weapons.) Later nuclear-warhead 21-inch torpedoes were added to the loadout of conventional torpedoes.

The RAT—Rocket Assisted Torpedo—was a means of launching ASW torpedoes from surface ships with rocket boosters to increase their range. This photo shows the twin-arm RAT launcher installed on the No. 3 twin 5-inch/38 gun mount of the destroyer *De Haven* (DD 727). At the end of its trajectory the lightweight torpedo would be lowered to the water by parachute after the rocket booster fell away. RAT also gave way to ASROC. (U.S. Navy)

being fired from the destroyer *Agerholm* (DD 826) on 11 May 1962, in a test known as Swordfish that was part of the Dominic test series. Operating about 370 nautical miles southwest of San Diego, the *Agerholm* launched the ASROC at a target raft 4,000 yards away with the one-kiloton warhead detonating underwater.

Most ASROC ships were provided with the Mark 116 "pepper box" launcher that held eight rockets. The number of reloads—in an adjacent deckhouse—varied, with converted Fleet Rehabilitation and Modernization (FRAM) destroyers having 10, ASW frigates 8, and the nuclear cruiser *Long Beach* (CGN 9) 12, but most other ships with the Mark 116 launcher having no reloads. In some of the later frigates (see endnote) and subsequent ships the ASROC was launched from the Mark 26 surface-to-air missile launchers, and up to 20 ASROCs were carried in those ships.[11] This latter arrangement required less deck space, although it reduced the number of surface-to-air missiles that could be carried in those ships.

The ASROC became operational in 1961 and was fitted in 251 U.S. warships:

35 cruisers/frigates
151 destroyers
65 escorts/frigates[12]

The Honeywell Corporation manufactured 12,000 ASROC rockets from 1960 through 1970. Although all ASROC with nuclear warheads were removed from the fleet in 1989, ASROCs with the conventional warheads remained in U.S. warships until the early 1990s.

NEW ROLES FOR OLD SHIPS

In the early stages of the Cold War new roles were being developed for the large force of war-built destroyers: *Sumner* and *Gearing* classes were converted to radar pickets and specialized anti-submarine ships. These modifications included the removal of both quintuple 21-inch torpedo mounts from these ships to compensate for the weight and space needed for new weapons and electronic installations. In all, 36 destroyers were converted to radar pickets to provide early warning of air attacks on the fleet, and 15 ships were modified to the ASW role. Plans were considered for modifying most of the earlier *Fletcher* (DD 445)-class ships to anti-submarine ships, but only 19 were converted, and those units retained five torpedo tubes.

Next came a very ambitious destroyer modernization effort and with more impact on torpedo armament: the Fleet Rehabilitation and Modernization program. FRAM encompassed 131 war-built destroyers. Seventy-nine of these were *Gearing*-class ships that were up-

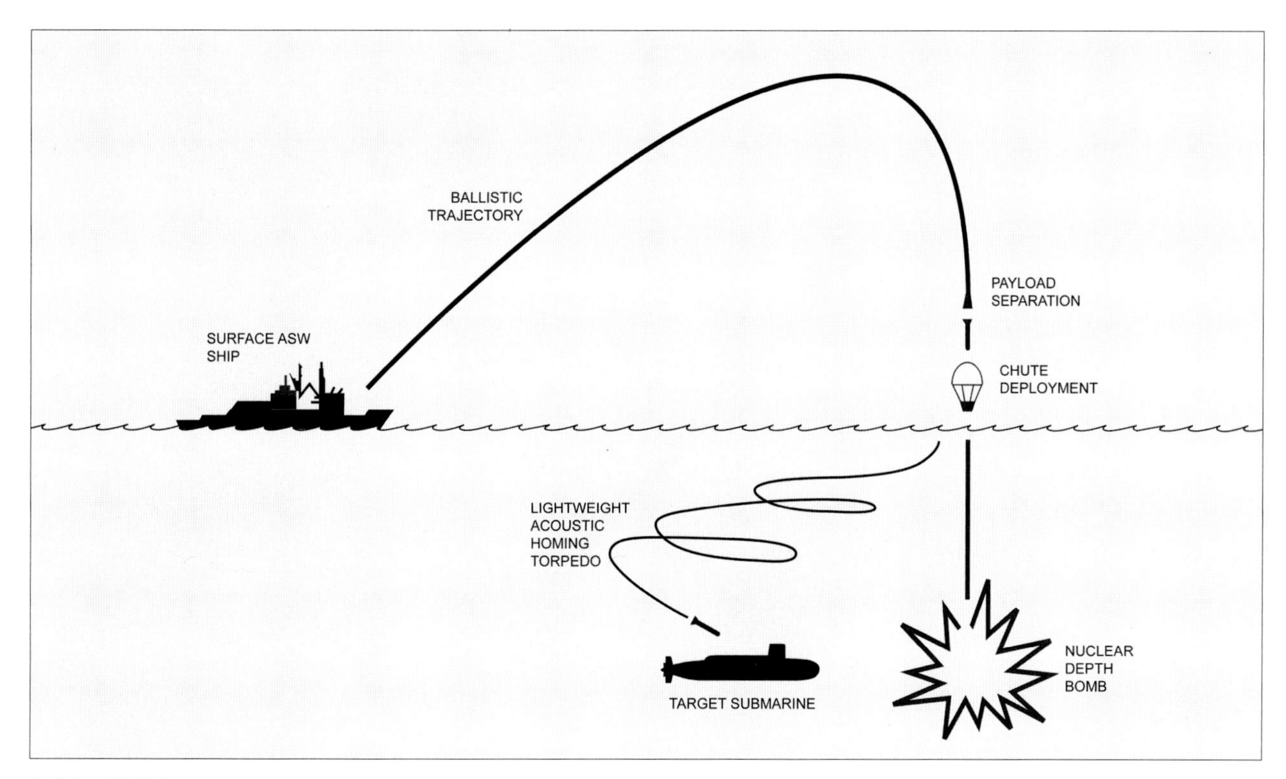

RUR-5A ASROC operation.

graded to the FRAM I configuration with installation of a Mark 116 "pepper box" ASROC launcher as well as torpedo tubes and facilities for an unmanned ASW helicopter.[13] The first FRAM I ship was the USS *Perry* (DD 844), converted at the Boston Naval Shipyard from April 1959 to May 1960. The *Perry* and subsequent FRAM I ships initially were fitted with a pair of fixed, 21-inch Mark 25 torpedo tubes for launching Mark 16 anti-ship and Mark 35 anti-submarine "heavy" torpedoes, and two triple Mark 32 torpedo tubes that could launch 12¾-inch diameter Mark 44 and Mark 46 lightweight torpedoes.

The remaining 52 *Sumner* and *Gearing* conversions were FRAM II ships. These ships were generally similar to the FRAM I units, but without the ASROC launcher and reloads, and a few ships did not have DASH facilities. All FRAM destroyers were eventually stripped of their large Mark 25 torpedo tubes, retaining only the "short" Mark 32 tubes for lightweight torpedoes.

The unmanned, radio-controlled helicopters embarked in most FRAM ships were known as Drone Anti-Submarine Helicopters (DASH). These helicopters could carry two Mark 44 or one Mark 46 lightweight torpedoes. The DASH had a range of about 30 nautical miles from the ship, limited by the range of radar/radio control. The drones carried no sensors, but relied on shipboard sensors and weapon release commands. The DASH helicopters could also deliver the B57 nuclear depth bomb, but those weapons were not embarked in surface combatant ships for their helicopters.[14]

The DASH helicopters were employed as torpedo carriers on board U.S. warships from 1962 to 1971. Of 746 produced, however, more than 400 were lost, most because of operational and training issues.[15] No ship-based replacement for DASH was readily apparent. The Royal Navy and Canadian Navy were flying the large SH-3 Sea King from destroyers, but the facilities in U.S. warships—sized for the diminutive DASH—could not handle so large a helicopter. In the late 1960s the Navy began a program to fit the small Kaman HU2K-1 Seasprite utility helicopter for shipboard ASW operations. The term LAMPS I—for Light

The destroyer *Brinkley Bass* (DD 887) launches a torpedo-carrying ASROC. An ASROC could carry a Mark 44 or Mark 46 lightweight torpedo or a W44 nuclear depth bomb to a range out to seven miles. Most ships were fitted with an eight-cell Mark 16 ASROC launcher, with some ships carrying reload missiles. Larger missile-armed ships could launch ASROCs from their Mark 26 surface-to-air missile launchers. These weapons have all been discarded from U.S. ships. (U.S. Navy)

An ASROC rocket booster leaves an exhaust trail after launch from the escort ship *Bowen* (DE 1079) during a 1972 exercise in the Caribbean Sea. Most ASROC-armed ships also operated DASH helicopters and, subsequently, manned ASW helicopters that could carry lightweight torpedoes to significantly greater ranges than could ASROC. (U.S. Navy)

The FRAM—Fleet Rehabilitation And Modernization—program updated World War II destroyers for Cold War missions. Among the more significant upgrades were (in most ships) provision of Mark 32 triple 12.75-inch torpedo tubes for lightweight ASW torpedoes, an ASROC launcher, provisions for two unmanned helicopters, and more-capable sonar. The *Myles C. Fox* (DD 829) was a FRAM I ship; note the partially open helicopter hangar and the Mark 32 torpedo tubes forward, in place of the No. 2 twin 5-inch/38 gun mount. (U.S. Navy)

A covey of FRAM destroyers: From top, the FRAM II ships *Lyman K. Swenson* (DD 729), *Collett* (DD 730), and *Blue* (DD 744); and the 14-foot longer FRAM I *Shelton* (DD 790) are moored alongside the destroyer tender *Bryce Canyon* (AD 36) in 1961. The "short-hull" ships have both 21-inch torpedo tubes and Mark 32 tubes between their stacks; the "long-hull" *Shelton* has an ASROC "pepper box" launcher as well as Mark 32 tubes. All are fitted to carry two DASH helicopters. FRAM I ships lost one twin 5-inch/38 gun mount to help compensate for the weight of ASROC. (U.S. Navy)

A Gyrodyne DSN-3 DASH carrying a Mark 46 torpedo. The DASH helicopters had no sensors, relying on commands from the surface ship to release its torpedo near the target submarine. The high accident loss rate led the Navy to discard DASH helicopters by the beginning of 1971. (U.S. Navy)

The DASH—Drone Anti-Submarine Helicopter—was an innovative effort to provide surface combatants with a means of delivering lightweight torpedoes out to the first sonar convergence zone, about 30 nautical miles. The Navy acquired several hundred of these unmanned aircraft, which could carry two Mark 44 or one Mark 46 torpedoes, or a Mark 57 nuclear depth bomb. The last were not embarked in ships for DASH delivery. This DSN-3 (later QH-50C) DASH is being serviced aboard the USS *Wallace L. Lind* (DD 703) in 1963. (U.S. Navy)

Airborne Multipurpose System—was applied and, with the designation SH-2, the helicopter entered service on board U.S. warships in 1971.

The SH-2 LAMPS I could carry two Mark 46 lightweight torpedoes. Unlike DASH, the SH-2 carried sonobuoys and had radar to detect submarines. The LAMPS II was a study, and the definitive LAMPS III was a navalized adaptation of the Army's widely flown Sikorsky UH-60 Black Hawk helicopter. Known in the Navy as the SH-60 Seahawk, it entered shipboard service in 1983, serving cruisers, destroyers, and frigates that had provisions for two of the helicopters, and in 1989, on board aircraft carriers. The SH-60s and later MH-60R variants could each carry two Mark 46 or Mark 50 lightweight torpedoes (or two Penguin anti-ship missiles), and were fitted with an array of submarine-detection sensors.

Significantly, with the end of the Cold War and the adoption of the multirole SH-60R variant of the Seahawk, the ASW proficiency of ship-based helicopters declined precipitously. Whereas the earlier SH-60B LAMPS and SH-60F carrier-based helicopters concentrated on the anti-submarine mission, the later variant has been employed in surface surveillance, VIP transport, vertical replenishment, and supporting special operations as well as ASW training and exercises. Further, the employment of cruisers, destroyers, and frigates in coastal—littoral—operations has often deprived their helicopters of being in areas where U.S. or allied submarines were available for ASW training.

The last U.S. Navy ships to have tubes for heavy torpedoes were the nine large guided missile frigates of the *Belknap* (DLG 26) class, completed in 1964–1967; the similar, nuclear-propelled *Truxtun* (DLGN 35) of 1967; and a few of the escort ships of the *Knox* (DE 1052) class, completed in the early 1970s. The guided missile frigates' two Mark 25 tubes fitted in the after deckhouse and were to carry ten heavy ASW torpedoes; in the escort ships two Mark 25 tubes were fitted in the stern transom with 12 torpedoes to be carried. But these ships did not carry heavy torpedoes and their tubes were subsequently removed. Thus, from the early 1960s onward U.S. surface combatants carried only lightweight ASW torpedoes that could be launched from shipboard Mark 32 tubes and from ship-based helicopters.

ATTACK FROM THE AIR

In the immediate postwar years U.S. aircraft carriers continued to embark Mark 13 anti-ship torpedoes for their attack planes.[16] Because of the large number of Mark 13 torpedoes produced during the war—approximately 17,000—a successor weapon was never placed in service; that was the improved-performance Mark 25 (see page 86).

The Mark 13s were employed once against an enemy during the Cold War. While torpedoes are looked to as a naval weapon for use at sea, the U.S. Navy used aerial torpedoes against a land target during the Korean War (1950–1953) at the Hwachon Reservoir. Located just north of the 38th parallel—the division between North and South Korea—the reservoir was an artificial body created by damming the upper Pukham River. At the western end of the reservoir was the Hwachon Dam, a huge concrete structure some 250 feet high. Atop the dam's curving upper surface were 18 sluice gates, each about 40 feet wide and 20-feet high, and 2-feet thick. These gates could be raised or lowered to control the flow of water downstream.

In early 1951 communist troops controlled the Hwachon Reservoir dam, and, with it, could control the water level on several rivers in Korea to facilitate their own troop movements; if the allies attacked, the communists could open the gates, flood the rivers, and impede the Allied advance. U.S. Army Rangers had failed in efforts to capture the dam, and B-29 Superfortress heavy bombers, using six-ton guided bombs, had failed to destroy the sluice gates.

At 2:40 PM on April 30 the Army asked the Navy's carrier force to have a go at the dam. Eighty minutes after the request was received the carrier *Princeton* (CV 37) flew off six AD-4 Skyraiders, each carrying two 2,000-pound bombs, with an escort of five F4U-4 Corsair fighters. The Skyraiders made a dive-bombing attack that holed the dam but left the sluice gates untouched. The Army asked the Navy to try again.

The operational losses of DASH "birds" led to development of the LAMPS—Light Airborne Multi-Purpose System—program to provide manned aircraft for ASW ships. The LAMPS I helicopters, designated SH-2, were initially converted from Kaman HU2K-1 utility/rescue helicopters; they carried radar, a magnetic sensor, and sonobuoys for submarine detection, and could drop lightweight torpedoes. This SH-2F LAMPS I is from Light Helicopter Anti-Submarine Squadron (HSL) 31. (U.S. Navy)

The "ultimate" LAMPS III helicopter was based on the U.S. Army's Sikorsky UH-60 Black Hawk series. The relatively large SH-60B Seahawk carried sensors and ASW torpedoes, operating from cruisers, destroyers, and frigates (escort ships). Subsequently, the MH-60R multipurpose variant replaced the LAMPS III as well as the SH-60F ASW helicopters that served aboard aircraft carriers. The SH-60F had dipping sonar, more effective around carrier task groups than sonobuoys. This SH-60B from HSL-51 was dropping a Mark 46 exercise torpedo when photographed in 2009. (U.S. Navy)

The *Princeton* made a second attempt, this time employing aerial torpedoes. A dozen Mark 13 torpedoes had been loaded in the carrier when she departed the Bremerton (Washington) Naval Shipyard. Early on 1 May the carrier launched five AD-4 and three AD-4N Skyraiders, plus 12 F4U-4 Corsairs for flak suppression.[17] They were accompanied by F9F-2P Panther photo planes. The attack planes each carried a single Mark 13 torpedo set for surface run. The torpedo planes also carried napalm tanks, to be delivered on "soft" targets at the direction of ground controllers after the torpedo attack.

The approach for the bombers was difficult because of the hills surrounding the reservoir. One torpedo was a dud and one ran erratic; the six others struck the target and blasted open the dam's floodgates, destroying communist control of the reservoir's waters. This strike remains the world's only aerial torpedo attack since 1945 and the only U.S. Navy use of torpedoes in combat since the end of World War II.

A short time after the Hwachon dam attack, the Mark 13 torpedoes were taken out of service. This occurred although Western intelligence sources were now reporting on the massive Soviet shipbuilding program that had been initiated by Soviet dictator Josef Stalin immediately after World War II. Soviet shipyards were now producing battle cruisers and lesser surface warships as well as large numbers of submarines. War games conducted at the U.S. Naval War College postulated these powerful surface ships being employed to attack Allied convoys in the North Atlantic during a future conflict.[18] The Navy's plans to counter that surface threat called for carrier air strikes—using primarily bombs—and, when weather prevented air attacks, the 16-inch guns of *Iowa* (BB 61)-class battleships and the 8-inch guns of heavy cruisers. Neither aerial nor surface ship torpedoes were accorded a significant role in those operations. (In the event, Stalin's massive shipbuilding program was terminated shortly after his death in March 1953.)

This marked the end of the 50-year-plus doctrine of surface ship torpedoes being primarily anti-ship weapons. Subsequent surface ship and aerial torpedoes developed and deployed by the U.S. Navy were anti-submarine weapons. In 1960 the new guided missile frigate *Dewey* (DLG 14), under Commander Elmo R. Zumwalt, operating in the Baltic Sea, depended on air cover for defense against Soviet surface ships that were shadowing him. Zumwalt related that he felt that his ship—lacking anti-ship missiles or torpedoes—was "castrated."[19]

The first postwar aerial torpedo to enter service was the Mark 34, initially called the Mine Mark 44 for secrecy. This was an improved, and much larger version of the Mark 24 torpedo, the original U.S. acoustic homing weapon. Developed by the Mine Warfare Test Station at Solomons, Maryland, at 1,150 pounds the Mark 34 weighed almost twice as much as its predecessor; the new torpedo was 19 inches in diameter, 10 feet, 5 inches in length, and had electric propulsion that could drive it for 30 minutes (12,000 yards) at 11 knots or up to 8 minutes (3,600 yards) at 17 knots. While the 116-pound HBX warhead would be sufficient to severely damage a submarine, the weapon was too slow to close with an alerted Type XXI-derivative submarine. The Mark 34 Mod 1 was in service from 1948 to 1958, with some 4,050 weapons having been produced.

The next air-launched torpedo was a "universal" ASW weapon, intended for aircraft, surface ship, and submarine launching. The last dictated a 21-inch diameter and at 1,770 pounds the Mark 35 qualified as a heavy torpedo. The General Electric–developed torpedo was 13 feet, 6 inches long and carried a 98-pound PBNX-103 warhead. This was a fast torpedo—27 knots for 15,000 yards.

About 400 Mark 35s were produced. It served as a principal submarine torpedo from 1949 until 1960, when it was replaced by the highly successful Mark 37. While the Mark 37 was a heavy, 1,430-pound, multirole torpedo for submarines, for surface ships and aircraft the Navy procured a series of lightweight torpedoes.

LIGHTWEIGHT TORPEDOES

The line of U.S. lightweight torpedoes began with the Mark 24 acoustic homing torpedo, developed during

After World War II the Navy developed "attack" aircraft in place of specialized dive and torpedo bombers. Two designs went into production—the Martin AM-1 Mauler, shown on a test flight carrying three Mark 13 torpedoes and 12 5-inch rockets in addition to its four 20-mm cannon, and the Douglas AD (later A-1) Skyraider. Only 144 Maulers were produced, while 3,180 of the multirole Skyraiders were built, with the U.S. Navy and Air Force flying the latter aircraft well into the 1960s in the Vietnam War. (U.S. Navy)

An AD-4 Skyraider from Carrier Air Group (CVG) 19 aboard the USS *Princeton* (CV 37) carrying a Mark 13 torpedo—with protective "collar"—en route to strike the Hwachon Dam in North Korea. The Skyraider was a highly capable aircraft that flew in numerous different mission configurations. The number "506" indicated the sixth plane of the fifth squadron aboard the ship—Attack Squadron 195; the letter "B" indicates CVG-19, which had earlier served aboard the carrier *Boxer* (CV 21). (U.S. Navy)

World War II for use from ASW aircraft. At 608 pounds it was the first Navy torpedo to weigh less than half a ton. Beginning in 1951 a series of lightweight ASW torpedoes entered U.S. naval service—the Mark 43, 44, 46, and 50. All were intended for launching from the "short" Mark 32 tubes found on almost all U.S. and many Allied surface warships, and could be carried internally and externally by ASW aircraft, helicopters, and even blimps. Variants of these torpedoes served as warheads for the ASROC anti-submarine weapon and for an anti-submarine mine.

The post–World War II era saw the development of specialized ship-based ASW helicopters and aircraft. In the U.S. Navy the former were developed along two tracks: helicopters to extend the weapon-delivery ranges of surface combatants, first by unmanned DASH helicopters, and then manned helicopters of the LAMPS program. These were preceded by ASW helicopters that could be based aboard aircraft carriers, principally the Piasecki (later Boeing/Vertol) HRP Rescuer, and Sikorsky HO4S, HSS-1/SH-34 Seabat, and HSS-2/SH-3 Sea King series. Each of these helicopter types was

When conventional bombing attacks failed to breach the Hwachon Dam, the Navy employed carrier-based torpedo planes to breach the dam's floodgates. The attack with Mark 13 torpedoes marked the last combat use of torpedoes by the U.S. Navy. Subsequently the British, Indian, Israeli, North Korean, and Pakistani navies have used torpedoes in combat. (U.S. Navy)

more capable with respect to range and sensor/weapons payload than its predecessor. Their weapons, of course, were lightweight torpedoes plus nuclear depth bombs in the Seabat and Sea King.

Fixed-wing, ship-based ASW aircraft were initially derived from attack aircraft, variants of the Grumman TBM Avenger and Douglas AD Skyraider. But purpose-built ASW aircraft followed: the Grumman AF Guardian "twins" (search and attack variants), and the Grumman S2F/S-2 Tracker, the first carrier aircraft that combined a full ASW sensor suite with a torpedo and rocket capability, the latter for attacking surfaced submarines. The aircraft could also carry B57 and Mark 101 Lulu nuclear depth bombs. The twin-engine Tracker was in U.S. Navy service from 1954 to 1975.

The "ultimate" U.S. Navy fixed-wing carrier ASW aircraft—successor to the Tracker—was the Lockheed S-3 Viking. That twin-turbojet aircraft also served as a combined sensor and weapons platform, with torpedo and nuclear depth bomb payloads. That aircraft was in service as an ASW aircraft from 1974 until about 1992. While it flew for several more years as a carrier-based surveillance and tanker aircraft, the appearance of the non-ASW Viking and multirole SH-60R Seahawk helicopter marked the end of specialized shipboard ASW aircraft in the U.S. Navy.

(U.S. Navy land-based ASW/maritime patrol aircraft also carried lightweight torpedoes—the Martin P5M/P-5 Marlin, Lockheed P2V/P-2 Neptune, and P3V/P-3 Orion. All three of these aircraft also served in Allied and, in some instances, Third World navies and air forces.)

The first of the series, the Mark 43, was developed by the Brush Development Company of Cleveland, Ohio, and the Naval Ordnance Test Station (NOTS), Pasadena, California.[20] The 260-pound torpedo was 10 inches in diameter and only 7 feet, 7½ inches long, with acoustic homing guidance. The slow speed of the Mod 1 (15 knots) was somewhat overcome in the subsequent Mod 2 (21 knots), but the small, 54-pound HBX warhead and range of only 4,500 yards gave the torpedo limited effectiveness. It was in the fleet from 1951 to

One of the first widely flown U.S. ASW helicopters was the Sikorsky HO4S-3 (also designated H-19), which operated from U.S. aircraft carriers beginning in 1950. It carried an AN/AQS-4 dipping sonar or one or two Mark 34 lightweight torpedoes. This HO4S-3 is from Navy Helicopter Anti-Submarine Squadron (HS) 3. This helicopter was also built in large numbers as a transport for the Marine Corps. It was flown in the ASW role by the navies of Britain and the Netherlands. (U.S. Navy)

1959. When launched from 12¾-inch-diameter Mark 32 tubes a special liner was fitted.

More effective was the Mark 44, the first service torpedo with a seawater-activated battery for its electric propulsion. The Navy had earlier experimented with seawater batteries for the Mark 26 and Mark 36, the seawater serving as the electrolyte with a silver chloride cathode, providing roughly three times the energy as comparable lead-acid batteries. Superior in every respect to the Mark 43, the Mark 44, also developed by NOTS Pasadena, initially weighed 425 pounds, was 12¾-inches in diameter, and 8 feet, 4 inches long with a 75-pound HBX-3 warhead. With acoustic homing guidance, the Mark 44 was a 30-knot torpedo with a range of 6,000 yards.

Beyond launching from the ubiquitous Mark 32 torpedo tubes in surface ships and being carried by aircraft, the Mark 44 served as the nonnuclear warhead for ASROC and the Australian-developed Ikara ASW rocket (which was fitted in ships of Australia, Brazil, Britain, Chile, and New Zealand). The Mark 44 also became a standard NATO anti-submarine torpedo, being produced in several European countries. South Africa produced an upgraded version in the 1990s designated A44. The Mark 44 was used in the U.S. Navy from 1957 to 1967.

The intuitive follow-on to the Mark 44 was the Mark 46, the Navy's first high-performance, thermal-powered homing torpedo. Developed by NOTS Pasadena and Aerojet General Corporation of Azusa, California, the Mark 46 has been called the definitive lightweight torpedo, having been in U.S. Navy service since 1963—almost half a century.

The Sikorsky SH-3 Sea King was the U.S. Navy's carrier-based ASW helicopter for much of the Cold War. A large helicopter, it was also flown by several Allied navies. It had dipping sonar and—in theory—could carry four lightweight ASW torpedoes, although only two was a normal load. The Royal Navy and Canadian Navy flew the Sea King from ASW ships as well as aircraft carriers. This SH-3D Sea King from squadron HS-2 has a magnetic anomaly detector pulled up under its fuselage as it drops a torpedo. (U.S. Navy)

A Grumman S-2E (formerly S2F-3) Tracker drops a lightweight ASW torpedo during a 1967 exercise. Navy dive and torpedo planes were pressed into the ASW role during World War II. Subsequently, specialized carrier-based aircraft flew in the (separate) anti-submarine search and attack roles; the Tracker was the first carrier aircraft to combine both roles. A magnetic anomaly detection boom extended from the tail and an AN/APS-88A radar was lowered from the fuselage; weapons were carried internally and on wing pylons. (U.S. Navy)

Also 12¾ inches in diameter, the Mark 46 is 8 feet, 6 inches long and, in the Mod 0 variant, weighed 568 pounds. The Mod 0 had a solid-propellant piston engine. The Mod 1 and later variants had Otto fuel—a liquid monopropellant—to provide increased performance, unofficially reported at 45 knots for 8,000 yards. It carried a 96-pound PBNX-103 warhead.

During its long service life there were several upgrades of the Mark 46. The Mod 2 provided a new gyro system and both the Mod 1 and Mod 2 were used as an ASROC payload. In 1978 a Navy official stated that "The existing Mk 46 torpedo does not, however, meet the Soviet submarine threat."[21] The subsequent Mods 4 and 6 were for the Mark 60 CAPTOR (Encapsulated Torpedo) mine; Mods 5 and 5A were to counter slow-moving, shallow-depth targets; and Mod 5A(S) provided new search capabilities. The Mod 5 torpedoes could also engage surface targets, with other ASW torpedoes having a depth "cutoff" to prevent them from attacking surface ships. (Research on Mod 3 was terminated in 1972.) CAPTOR was an ASW sea mine that was also called a "rising mine." When a submarine target passed within the sensor field the mine was launched at the target. It could be launched from 21-inch submarine torpedo tubes and laid by various naval aircraft as well as Air Force B-52 Stratofortress strategic bombers. Entering the fleet in 1979, the submarine-launched version of CAPTOR weighed 2,056 pounds and the aerial version—with a parachute—weighed 2,370 pounds.

The guided missile destroyer *Mustin* (DDG 89) launches a Mark 46 practice torpedo from Mark 32 triple torpedo tubes in a 2009 exercise. Mark 32 tubes for lightweight ASW torpedoes were fitted to most U.S. cruisers, destroyers, escort ships, and frigates—both the large DLG/DLGN and the small FF/FFG types. The potential effectiveness of such tube-launched weapons against Soviet submarines armed with long-range torpedoes and anti-ship cruise missiles was highly questionable. (U.S. Navy)

Addressing Congress in 1978, the Under Secretary of Defense for Research and Engineering declared, "Analyses show that, within the limits in which it can be employed, CAPTOR will kill more submarines per dollar than any other ASW system."[22]

By the late 1970s the Mark 46 was the only conventional ASW weapon in U.S. surface ships and aircraft. It was also manufactured in larger numbers than any other post–World War II torpedo with about 16,800 being produced.

Beginning in 1969 Western intelligence became aware of an advanced submarine being constructed by the Soviets. Over the next few years the submarine emerged as the titanium-hull, nuclear-propelled Project 705, given the NATO code name Alfa. The Alfa was a high-speed, 40-plus knot submarine, at least ten knots faster than the fastest U.S. nuclear submarines. And, it was believed (incorrectly) that the Alfa was also a deep-diving submarine, able to reach perhaps 2,000 feet while the deepest-diving U.S. submarines had a maximum depth of some 1,200 feet.[23] Although the lead submarine encountered massive technical problems, the Alfa was soon placed in limited production with six units completed from 1977 to 1981.

The Alfa's estimated speed and depth, coupled with the extensive use of decoys and jammers, threatened to defeat existing U.S. anti-submarine torpedoes. In response, major torpedo upgrade efforts were initiated. Details remain classified, but in a 1981 congressional colloquy between a U.S. senator and the Deputy Chief of Naval Operations (Surface Warfare), Vice Admiral William H. Rowden, the senator noted that the Alfa could travel at "40-plus knots and could probably outdive most of our anti-submarine torpedoes." He then asked what measures were being taken to redress this particular situation.

Sailors on the USS *Mustin* move a Mark 46 torpedo. The "hard hats" vice battle helmets show that this was a peacetime exercise. The ubiquitous Mark 46 was carried by most U.S. surface combatants and ASW aircraft during the latter stages of the Cold War, and was produced by NATO countries and used by numerous other navies and air forces. It remains in service with the Mark 50, its nominal replacement, having had a limited production run. (U.S. Navy)

Admiral Rowden replied, "We have modified the Mk 48 torpedo . . . to accommodate to the increased speed and to the diving depth of those particular submarines." The admiral was less optimistic about the Mark 46 torpedo used by aircraft, helicopters, and surface ships: "We have recently modified that torpedo to handle what you might call the pre-Alfa."[24] In private, U.S. naval officers were considerably more vocal in their concern over the ineffectiveness of existing Western ASW weapons against the Alfa submarine.[25] While U.S. officials reacted to press inquiries by citing the high noise levels of the Alfa at high speed, at lower

The Mark 46 lightweight torpedo served as "warhead" for the CAPTOR—encapsulated Torpedo—"rising mine." Although considered a highly effective anti-submarine weapon, it was disliked by submariners, who would rather launch torpedoes than mines and see enemy ships that they sink. CAPTOR could also be laid by naval aircraft and by the Air Force's B-52H Stratofortress. These four CAPTOR mines are on the left wing of a B-52H. The weapon has been discarded. (U.S. Air Force)

Although neither the fastest nor the deepest-diving submarine built by the Soviet Union, the Project 705 (NATO Alfa) nuclear-propelled attack submarine had considerable impact on U.S. Navy ASW strategy and torpedo development. Despite the major technical problems suffered by the prototype, the Alfa SSNs demonstrated unequaled advances in material (titanium) and automation (a 29-man crew) in submarine construction. (U.S. Navy)

speeds the submarine would have been difficult to detect. While only six Alfas were completed, and some suffered technical problems, they demonstrated the challenges of modern submarines that U.S. torpedoes would have to counter.

In an effort to enable Mark 46s to deal with advanced Soviet submarines, the Navy initiated the Near Term Torpedo Improvement Program (NEARTIP) to upgrade several Mark 46 variants. Under NEARTIP replacement "kits" were developed to upgrade almost 3,000 existing torpedoes to the Mod 5 configuration.

At the same time, concern was being publicly voiced about the need to provide ASW torpedoes with "significantly improved countermeasures resistance." And, the Navy revealed that the NEARTIP "is designed to improve countermeasure resistance of the Mk 46 torpedo, and to restore the detection range [classified]."[26]

A successor to the Mark 46 was proposed, with some thought being given to rocket propulsion (for speed) and a shaped-charge warhead (to penetrate double-hull submarines). Initiated as the Advanced Lightweight Torpedo (ALWT) program in 1975 with a design competition subsequently being held between Honeywell (Ex-50 design) and McDonnell Douglas (Ex-51). The former firm was selected to develop the torpedo. Limited production began in March 1989 as the Mark 50. The Mark 50 has a Stored Chemical Energy Propulsion System (SCEPS) that sprays sulfur hexafluoride gas over a block of solid lithium that generates heat that, in turn, produces steam from seawater; the steam, in turn, propels the torpedo in a closed Rankine cycle with pump-jet propulsion. The torpedo's speed—in excess of 40 knots—and range remain classified. With a standard 12¾-inch diameter, the Mark 50 is 9 feet, 4 inches long and weighs 750 pounds. Thus it is compatible with Mark 32 torpedo tubes as well as all Navy ASW aircraft. Its warhead is a 100-pound shaped charge. Like the Mark 46 it has active/passive acoustic guidance. The first successful in-water test of a full-scale development prototype torpedo took place on 30 July 1986. However, delays were caused by initial test failures, hardware and software development problems, contractor management problems, and funding reductions.[27] Upon completion of operational trials and evaluation in 1992, full production began, shared equally between Alliant Techsystems Inc. (formerly Honeywell and now Raytheon) and Westinghouse (now Northrop Grumman Corporation).

Although intended to replace the Mark 46, the Mark 50 had a limited production run and instead complemented the Mark 46, which remains in service at this writing. In its place as the nominal Mark 46 successor the Navy initiated the Lightweight Hybrid Torpedo (LHT) in 1996 in an effort to overcome the continuing limitations of both the Mark 46 and Mark 50 torpedoes. The earlier weapons were initially designed for open-ocean ASW, while in the post–Cold War era the U.S. Navy has increasingly had to operate in coastal or littoral areas. Designated Mark 54, the LHT was developed by Alliant/Raytheon and Hughes Aircraft Company. The LHT guidance is based on the Mark 50 torpedo and Commercial Off-The-Shelf (COTS) components; the warhead is from the Mark 46; the fuel tank is from the Mark 46; and the propulsion system from the Mark 46, Mark 48 ADCAP, and Mark 50.

The Mark 54 combines the active/passive acoustic guidance and warhead portions of the Mark 50 and the propulsion unit of the venerable Mark 46, with improvements for littoral operations. After several delays for technical and program problems, the Mark 50 entered service in 2003, with full-rate production beginning in October 2004. With standard lightweight torpedo dimensions and a weight of 608 pounds, it is suitable for both Mark 32 tubes and aircraft use. It is also used as the payload for a vertical-launch ASROC (RUM-139C) projectile that can be launched from modern cruisers and destroyers (the conventional ASROC launchers having been removed from U.S. ships).[28]

However, Mark 54 production has also been limited. Thus, the U.S. Fleet today has the Mark 46 and limited numbers of the Mark 50 and Mark 54 lightweight torpedoes available for surface ships and aircraft. The only other torpedo in the fleet is the Mark 48 for submarines.

CHAPTER FIFTEEN

The Ultimate Torpedo

The Mark 48 (1972–Present)

The Soviet Navy's continuing development of advanced submarines had led the U.S. Navy as early as November 1956 to begin a program to develop technologies for advanced torpedoes. This effort was labeled the RETORC (Research Torpedo Configuration) program and was undertaken largely by the Ordnance Research Laboratory of Penn State University. The most pressing need was for a submarine-launched weapon to replace the Mark 14, 16, and 37 torpedoes. A specific heavy-weight torpedo project emerged by 1960, initially designated Ex-10 and, subsequently, Mark 48.

The development goals for the new torpedo included a range of 35,000 yards at a speed of more than 55 knots, and a 2,500-foot depth capability. After several firms were qualified and competed, a project definition contract was awarded to Westinghouse in 1964 to develop the Mark 48 Mod 0 torpedo.[1] There were early technical problems in the Mark 48 development. A Westinghouse history of the Mark 48 noted,

> By late 1966, program delays and unanticipated difficulties experienced with in-water tests of development prototype torpedoes brought unusual attention from the customer. At the same time, the urgent need for torpedoes by the Navy brought pressure for early production. . . . By late 1967 the situation had worsened with the emergence of new technical problems in the early production prototype torpedoes. These delays, greatly increased program cost estimates, and the magnitude of the technical problems caused the Government to question the ultimate success of the Westinghouse program.[2]

As insurance against a possible delay or failure of the Westinghouse program, the Navy directed a parallel torpedo effort by the Clevite Corp. of Cleveland, Ohio, under the technical direction of the Naval Ordnance Laboratory. The Clevite firm was pioneering development of a "comb filter" acoustics guidance system that showed promise for shallow-water operation and countermeasures resistance. This was followed by a contract modification for Clevite to build "test vehicles" using a piston-type engine that the firm had developed for an earlier competition with Westinghouse. This led to the Clevite-produced Mark 48 Mod 1 torpedo.

Further, while the Mark 48 Mod 0 was originally primarily intended as an ASW torpedo, the Navy stockpile of obsolescent Mark 14 and Mark 16 torpedoes was being deleted and the decision was made to emphasize the anti-ship capabilities of the Mod 1.[3] In 1970 U.S. submarines carried the following torpedoes:

Mark 14	Mod 5
Mark 16	Mod 8
Mark 37	Mods 0, 1, 2, 3
Mark 45	Mods 0, 1

A Mark 48 torpedo being test launched at the Naval Torpedo Station, Keyport, Washington, in 1967. (U.S. Navy)

A Mark 48 torpedo being prepared for loading aboard the strategic missile submarine *Stonewall Jackson* (SSBN 634) at Kings Bay, Georgia, in 1981. Note the cable-reel unit attached to the after end of the torpedo. Strategic missile submarines carry torpedoes for self-defense; during the Cold War both U.S. and Soviet strategic missile submarines also carried nuclear torpedoes, i.e., the Mark 45 ASTOR in U.S. submarines. (U.S. Navy)

A competition followed between the Mark 48 Mod 1 and the Mod 2, a redesign of the Westinghouse Mod 0. The Mod 2 torpedo was propelled by a Sunstrand turbine (as in the Mod 0) using a monopropellant Otto fuel; the Clevite torpedo also used Otto fuel in an external combustion, axial-piston engine. Interestingly, the Navy had specified that the Westinghouse torpedo had to use a turbine engine on the basis of proven feasibility and performance; but it was significantly more expensive to build, especially in the limited quantities required for test and evaluation torpedoes.

One of several Navy selection factors was apparently the better efficiency of the piston engine, especially when running deep, as opposed to the quieter, but less-efficient turbine. The acoustic detection systems were also different. The Mod 2 effort was terminated in favor of the Mod 1 version. The Westinghouse Mod 0 torpedo was continued with a contract for a pilot production lot of 52 Mod 0 torpedoes being awarded to Westinghouse in 1970.

"AN URGENT REQUIREMENT"

That same year, Secretary of Defense Melvin R. Laird, explaining to Congress the need for the Mark 48 torpedo, stated,

> There is an urgent requirement for [Mark 48] capabilities, particularly the ASW capability. The exist-

ing submarine launched ASW torpedo, the Mark 37, does not have the speed, range, acquisition or depth capability required for use against modern, fast, deep-diving submarines. Our present anti-ship torpedoes are old and not very effective against evasive targets. Moreover, our inventory of such torpedoes is quite limited.[4]

In April–May 1971 the submarine *Trigger* (SS 564) fired five live "war-shot" torpedoes at target ships to evaluate the Mark 48 variants. Three of the ships were sunk, two by Westinghouse Mod 0 torpedoes and one by a Clevite Mod 1 torpedo. After careful analysis, a full-scale production contract was awarded for the Mod 1 torpedo to the Gould Corporation, which had taken over the Clevite Corp.[5] The first Mark 48 Mod 1 torpedoes were delivered to the fleet in 1972, some 12 years after the development characteristics had been approved. The Mark 48 Mod 1 was 21 inches in diameter; 19 feet, 2 inches long; and weighed 3,440 pounds. It carried a warhead of 650 pounds of PBXN-103. The Mod 2 warhead was smaller, but the Westinghouse warhead had a scheme for detonating unused fuel to enhance its destructive force.

While performance data for the Mark 48 remains classified, most published sources—some official—attribute a range of 35,000 yards at 55 knots to the basic torpedo, with an operating depth to 2,500 feet. Its acoustic guidance is estimated to have an acquisition range of 4,000 yards, about four times that of the Mark 37. In place of conventional propellers, the later Mark 48s had a more efficient and quieter "pump-jet" propulsor.

Again, although details are lacking, the Mark 48 has an on-board capability to control search, homing, and reattack maneuvers. It is wire guided during its initial run, enabling the launching submarine to take advantage of its more capable sonar to initially guide the torpedo. The Mod 3 torpedo provided the addition of two-way telecommunication to provide transmission from the torpedo's sonar back to the submarine's fire control system. That variant was introduced in 1977.

TABLE 15-1

SUBMARINES WITH FOUR TORPEDO TUBES

CLASS/TYPE	COMM	COMPLETED
SSN 593 *Thresher*	1961–1967	14
SSN 597 *Tullibee*	1960	1
SSBN 608 *Ethan Allen*	1961–1963	5
SSBN 616 *Lafayette*	1963–1967	31
SSN 637 *Sturgeon*	1967–1975	37
SSN 671 *Narwhal*	1969	1
SSN 685 *Glenard P. Lipscomb*	1974	1
SSN 688 *Los Angeles*	1976–1996	62

Like most advanced weapons—and especially heavy-weight torpedoes—the Mark 48 was plagued with "teething" problems. These included problems with components, the interference of guidance wires in two-torpedo salvos, and major reverberation problems with the acoustic guidance when tested under the Arctic ice. The principal criticism—at least in public forums—is that the Mark 48 is a noisy torpedo, with self-generated noise interfering with its acoustic guidance. Related to this issue, in close-in operations Soviet submarine sonars are reportedly able to detect the opening of torpedo tube shutters and the launch of the Mark 48, providing a few seconds or possibly minutes of warning in which to deploy jammers and decoys. As late as 1973 the Navy's Commander, Operational Test and Evaluation Force, reported "several problem areas that have an effect on the weapon system's overall effectiveness."[6] Thirty problems of varying severity with the Mark 48 were identified in that 1973 classified report.

Still, the Mark 48 was a success, with Gould/Westinghouse producing most of the early torpedoes for the Navy; Hughes Aircraft Company was given a second-source production contract. The initial fleet requirement was for 4,089 service torpedoes plus 20 development models and 85 production prototypes—a total of 4,194 weapons.[7] However, the question of whether or not heavy torpedoes should be carried in surface ASW ships became an issue. The frigate *Talbot* (DEG 4) was fitted to test launch the Mark 48, but the decision was made not to provide the weapon to surface ships. Thus, fewer launch platforms and an across-the-

The rear aspect of a Mark 48 torpedo showing the shrouded pump-jet propulsor used in place of conventional propellers. (Gould)

board reduction in nonnuclear weapons in the 1970s led to a cutback in the Mark 48 inventory objective to 2,963 torpedoes:

1,680 for 84 attack submarines
60 for 41 strategic missile submarines
260 in pipeline
637 for peacetime attrition (tests, losses, etc., during the service life of the Mark 48)

But even that objective was soon reduced—to 2,771 torpedoes by early 1978—and even fewer basic Mark 48s were procured.

A decision by the U.S. nuclear submarine community in the mid-1980s could have significantly increased the demand for Mark 48 torpedoes. From the late 1950s a succession of U.S. submarine classes had four 21-inch torpedo tubes and internal stowage for 22 heavy weapons. Normally one torpedo reload "slot" is left empty to permit a weapon to be withdrawn from a torpedo tube for maintenance or replacement. In addition, the 23 later submarines of the *Los Angeles* (SSN 688) class—of 62 submarines in the class—have 12 vertical-launch tubes for Tomahawk cruise missiles, giving those submarines 34 weapons.

In addition to conventional torpedoes, submarine weapon loadouts in various periods of the Cold War included Mark 45 ASTOR nuclear torpedoes, SUBROC rockets, Harpoon anti-ship missiles, Tomahawk cruise missiles, and mines. (U.S. submarines no longer carry Harpoon missiles or mines, and only the conventional, land-attack variant of the Tomahawk.[8])

The torpedo requirements changed in July 1982 when the Navy made the decision to construct a new class of attack submarines designated SSN 21—indicating a submarine for the 21st Century. That program designation was then (incorrectly) used as a hull number for the first submarine of the class, subsequently named *Seawolf*. This is a large, nuclear-propelled attack submarine displacing 9,140 tons submerged, 2,000 tons larger than the previous *Los Angeles*-class nuclear submarine. Further, the *Seawolf* design has eight 26-inch torpedo tubes and can carry a total of 50 weapons. The *Seawolf* is the first U.S. submarine since the "R" class, completed in 1918–1919, not to have 21-inch torpedo tubes. During the gestation of the *Seawolf* no consideration was given to developing a 26-inch tube-launched weapon, but the popular belief was that if the planned 30 submarines of the *Seawolf* class were built, advanced weapons of 26-inch diameter as well as unmanned underwater vehicles would be developed for the submarines.

The Navy's original intention was for an "almost" all-Mark 48 loaded in *Seawolf*-class submarines. Thus armed, they were to prowl Arctic waters seeking Soviet submarines with the expectation that a single, ultra-quiet *Seawolf* could detect and sink a score of enemy submarines. By carrying double the torpedo load of a *Los Angeles*-class submarine, a *Seawolf* could remain in the operating area twice as long and kill at least twice as many targets without having to return to a friendly port to rearm and replenish.

The Mark 48 torpedo was originally intended for use in both submarines and surface ships, in the latter for the anti-submarine role. Twin Mark 25 tubes for Mark 48 torpedoes were fitted in guided missile frigates (DLG/DLGN) in their after deckhouse and in escort ships (DE/DEG) in their stern counters. These photos show the stern tubes of the USS *Brooke* (DEG 1) and a Mark 48 being launched from the USS *Talbot* (DEG 4). The tubes and Mark 48 support gear were removed from these ships. (Lockheed Shipbuilding; U.S. Navy)

Several hundred additional Mark 48s would be needed to arm the *Seawolfs*. The nuclear community initially sought to build 30 *Seawolfs* at the rate of three per year. The lead ship was laid down in 1989. The cost of a *Seawolf* would be more than twice that of a *Los Angeles*, and the cost factor coupled with the end of the Cold War in 1991 led the George H. W. Bush administration to immediately cancel the entire *Seawolf* program except for the lead submarine. However, within hours of the cancellation, members of the nuclear submarine community, led by Admiral Bruce DeMars, then the director of the Navy's nuclear program, were

The Navy's plans for building 30 nuclear-propelled attack submarines of the *Seawolf* (SSN 21) class were cancelled with the end of the Cold War. Each was to have eight 26½-inch torpedo tubes and carry 50 weapons, which would have been primarily Mark 48 torpedoes. This is the *Connecticut* (SSN 22) in late 2009, one of three submarines of the class that were built; one of them, the *Jimmy Carter* (SSN 23), is configured for special missions—deep-ocean search and recovery. (U.S. Navy)

on Capitol Hill, lobbying Congress to continue the program. In response to their pleadings, Congress did fund two additional *Seawolfs*.

The next U.S. attack submarine program—the *Virginia* (SSN 774) class—returned to four 21-inch torpedo tubes with a total payload of 25 weapons, plus 12 vertical-launch cells for Tomahawk missiles.

With the end of the Cold War due to the demise of the Soviet Union in 1991, and the truncating of the *Seawolf* program the issue of Mark 48 numbers became more convoluted. With the massive reduction in U.S. naval forces in the 1990s, by 2010 the U.S. fleet numbered 45 attack submarines, 4 cruise missile/transport submarines, and 14 strategic missile submarines, a total of 63 torpedo-armed submarines.

The performance improvements in foreign submarines, particularly those of the Soviet Union until 1991, led to new variants of the Mark 48 being developed.

THE ADCAP TORPEDO

In 1975 the Chief of Naval Operations issued a new operational requirement for countering advanced Soviet undersea craft. After extensive tests of the existing Mark 48 with "laboratory modifications," the Navy found that the torpedo could meet *most* requirements to counter the Soviet high-speed Alfa attack submarine.

The Mod 4 added speed to the torpedo as well as a fire-and-forget capability, which enabled the launching submarine to immediately take evasive action after launching a torpedo. Mod 4 torpedoes were produced from 1980 and earlier torpedoes were upgraded with kits. The Mod 5 was an interim upgrade of existing torpedoes pending the availability of the ADCAP (Advanced Capability) version of the Mark 48. The modified Mark 48 designated ADCAP was in reality the Mod 6 version.

The ADCAP had an entirely new nose section fitted with digital electronics and inertial guidance (replacing the gyro system), resulting in a major reduction in the volume devoted to electronics and providing a corresponding increase in fuel capacity. It also had a strengthened outer shell and, of course, inclusion of previous Mod 4 performance "expansion" features. The Mark 48 piston engine and propulsor were retained, but with a greater fuel flow rate to provide an (unofficially) estimated speed possibly as high as 63 knots. Much of the change was made possible by the introduction of integrated circuits, including microprocessors. The guidance wire spool was moved to a position behind the enlarged fuel tank and other layout changes were made.

The new "nose" of the torpedo was described in a Westinghouse brochure as the Ex-13 nose—"the quietest, wide bandwidth torpedo transducer ever built. . . . By modifying the Ex-13 nose, Westinghouse has the low self-noise . . . ruggedness, producibility, and self-test capability needed for ADCAP."[9] With the same dimensions as the basic Mark 48, the ADCAP weighs slightly more—3,695 pounds. The ADCAP required an upgrade of submarine fire-control systems. However, the ADCAP was fully compatible with submarine stowage racks and tubes.

A Mark 48 torpedo is pulled from the Arctic ice by a helicopter after an under-ice test launch by the attack submarine *Helena* (SSN 725) in March 2009. The Arctic was a key area of operations for U.S. as well as Soviet submarines during the Cold War. Torpedo performance under ice is affected by marine life, acoustic reverberations, and other factors, contributing to periodic U.S. Navy torpedo exercises in that region. (U.S. Navy photo by PO1 Tiffini Jones)

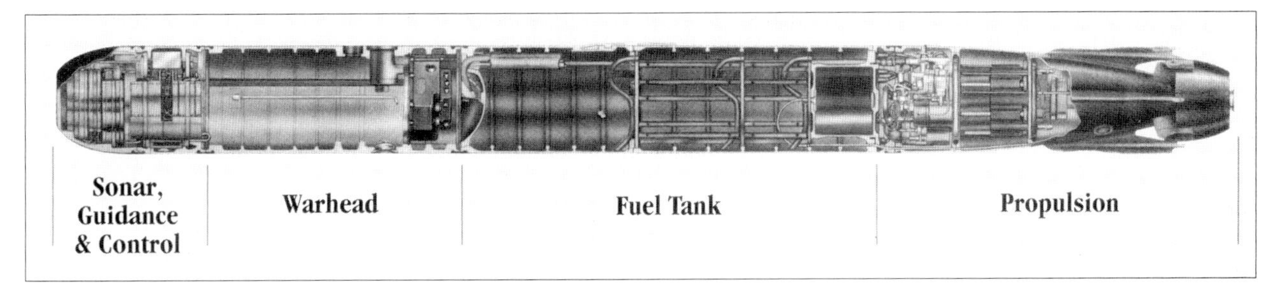

Mark 48 ADCAP torpedo.

The 1985 five-year defense plan called for the acquisition of 1,890 ADCAP torpedoes. Hughes began production of the ADCAP in 1985, followed by a second-source production contract to Westinghouse/Northrop Grumman. And, of course, the ADCAP procurement goal was reduced, to just over 1,000 units. According to Navy data, the total production of Mark 48 torpedoes—not including subsequent modifications—was approximately 5,150.[10]

The first ADCAP torpedoes entered the fleet in 1988, with the USS *Norfolk* (SSN 714) making the first "warshot" launch on 23 July 1988. That torpedo sank the discarded destroyer *Jonas K. Ingram* (DD 938). Full ADCAP production was approved the following year.

Still another upgrade came with the CBASS (Common Broadband Advanced Sonar System), the Mod 7 variant of the ADCAP. This was the outcome of a cooperative U.S.-Australian development program to develop a fully digital, wideband sonar to enable the torpedo to be used in shallow water—the littoral environment (i.e., less than 600 feet). ADCAP production ended in 1993, with kits being provided for the upgrade of earlier torpedoes. The production of the CBASS version was initiated a few years later, but all U.S. Mark 48 production ended about 1996.

Improved versions of the Mark 48 are the standard anti-ship and anti-submarine weapon of U.S. submarines and several allied nations. The only other weapons carried by nuclear powered attack submarines are Tomahawk land-attack missiles and, on rare occasions, submarine-launched mines. Despite its exceedingly long gestation period and numerous problems, the Mark 48 is today well liked and well thought of by the submarine community.

CHAPTER SIXTEEN

An Effective Weapon

The Ship Killers

The torpedo is a very effective weapon. When a torpedo strikes a target and detonates, the damage is severe. Unlike aerial bombs, missiles, and gunfire, which generally damage the upper hull and superstructure of ships, the torpedo—like the mine—opens the lower portion of the ship's hull to the sea. A torpedo that detonates in the proper position under a ship can "break its back"—the keel—with devastating results. Indeed, on occasion a single torpedo has sunk a capital ship—battleship or battle cruiser, as well as aircraft carriers.

Launched from an unseen, undetected submarine, the torpedo often strikes without warning to inflict major damage on a surface ship or submarine. An approaching aircraft or surface ship making a torpedo attack could be more easily detected by the "target" surface ship, and it is possible to destroy or deter the torpedo-carrying attacker. Similarly, a submarine can detect a surface ship more easily than another submarine, and there are means for a submerged submarine to detect an approaching aircraft.[1]

The acoustic homing torpedo has also proved to be a most effective anti-submarine weapon. Beginning in 1943, these weapons invariably outperformed depth charges, hedgehogs, and aerial bombs and rockets. Today the acoustic torpedo—dropped from aircraft, launched by tube, or propelled by rocket—is the principal anti-submarine weapon of all navies.

TORPEDOES IN COMBAT

Tens of thousands of torpedoes have been launched in combat by a score of nations, beginning in 1877. U.S. Navy surface ships, submarines, and aircraft have, since 1918, launched some 17,000 torpedoes against enemy targets.[2]

The first U.S. torpedoes fired in anger were launched by three of the submarines in Submarine Division 5 on anti-submarine patrol in St. George's Channel (between Ireland and Wales) and the western approaches to the English Channel in 1918. The targets for these 11 torpedoes, fired in four attacks, were supposed German U-boats, although the existence of those targets was questionable at best, and no hits were scored.[3]

The next U.S. torpedoes fired in anger were launched by the submarine *Swordfish* (SS 193) against a Japanese cargo ship in the South China Sea on 9 December 1941. No hits were observed. During the subsequent 44 months of combat in World War II U.S. submarines launched 14,748 torpedoes, most against Japanese ships and submarines. (A few torpedoes were launched by U.S. submarines in the Atlantic against supposed U-boats, but, again, without result.) In the Pacific U.S. submarines carried out 1,699 patrols, launching an average of 8.7 torpedoes per patrol. The results of these attacks are shown in Table 16-1. U.S. submarine-launched torpedoes appear to have sunk:

The effectiveness of the modern torpedo is graphically shown in these photos of the 2,750-ton Australian destroyer escort *Torrens* being sunk on 14 June 1999, by an under-the-keel detonation of a Mark 48 Mod 4 torpedo launched by the Australian submarine *Farncomb*. Mark 48 torpedoes arm the submarines of Australia, Brazil, Canada, the Netherlands, and Taiwan as well as all U.S. submarines. (Royal Australian Navy)

1 battleship
4 aircraft carriers
4 escort carriers
13 cruisers
40 destroyers
18 submarines

About 120 lesser naval ships were also sunk by submarine torpedoes. More significant in many respects was the sinking of 1,152 merchant vessels of 4,859,634 gross tons by U.S. submarine torpedoes.[4] (U.S. Mark 14 torpedoes also sank one, or possibly two, U.S. submarines by circular runs.[5] And, one U.S. submarine torpedoed and sank a small U.S. Navy auxiliary ship through misidentification.)

As shown in Table 16-1, U.S. submarine successes increased remarkably in 1943–1944, primarily because of the belated modifications made to submarine torpedoes, the availability of sufficient torpedoes, and the fitting of the Torpedo Data Computer (TDC) and SJ surface search radar in U.S. submarines. Although torpedo effectiveness increased significantly in 1943, U.S. submarines made only limited contributions to the Pacific War in 1942–1943. The Japanese Navy lost the war at sea at the carrier battles of Coral Sea and Midway, and in the surface ship actions in the violent battles for the Solomons. The 1945 submarine scores were low because of the lack of suitable targets. During the war submarines also sank hundreds of small Japanese merchant vessels using their deck guns.

In the Atlantic area, six U.S. fleet submarines formed Submarine Squadron 50 that operated during 1942–1944 from Roseneath, Scotland, carrying out patrols off northwest Africa, in the Bay of Biscay, in the Norwegian Sea, and south of Iceland. They launched 47 torpedoes and were credited with 12 hits that sank a few small merchant ships.[6]

U.S. surface ships—destroyers, destroyer escorts, and motor torpedo boats—launched more than 1,500 torpedoes in all combat theaters. About 40 major Japanese warships were sunk by surface ship torpedoes—mostly by destroyer-launched "fish," and most in the Solomons campaign. Only one destroyer escort is believed to have launched torpedoes against enemy warships, the *Samuel B. Roberts* (DE 413), part of the screening force for U.S. escort carriers off Samar in

TABLE 16-1

U.S. SUBMARINE TORPEDO EFFECTIVENESS

PACIFIC AREAS	FIRED	HITS	EFFECTIVENESS
1941	104	13	13%
1942	1,926	554	28%
1943	3,761	1,298	34.5%
1944	6,108	2,329	38%
1945	2,365	573	24%

ATLANTIC AREAS	FIRED	HITS	EFFECTIVENESS
1942	19	4	21%
1943	24	8	33%
1944	4	0	0

Leyte Gulf on 25 October 1944, that were surprised by a major Japanese surface force.

The hapless DE—armed with two 5-inch guns and a set of triple torpedo tubes—joined larger destroyers in attacking the Japanese cruisers. Launching three torpedoes and opening gunfire at the enemy ships, the *Roberts* was struck by a salvo of 14-inch shells from a Japanese battleship. Ripped open amidships, she was abandoned and soon sank. One torpedo hit on a Japanese cruiser was claimed by the DE sailors.

The PT boats had less success than destroyers in scoring torpedo hits on major enemy warships. Beyond three destroyers (one already damaged by U.S. Army aircraft), PT boats could claim to have sunk only small combat craft and coastal craft, almost all of the latter falling victims to gunfire. But those inshore operations as well as other PT boat missions were of major significance.

U.S. aircraft—land-based and carrier planes, Navy and Army Air Forces—launched more than 1,400 torpedoes during the war. Almost all of these were Mark 13 torpedoes launched against Japanese surface ships. Their success rate was low, especially early in the war. At Midway in June 1942, six Navy TBF Avenger land-based and 41 carrier-based TBD Devastator torpedo planes attacked the Japanese carriers; none of them scored a torpedo hit. The Army Air Forces used torpedoes at Midway, with four B-26 Marauder twin-engine bombers armed with Mark 13 torpedoes flying from Midway atoll against the main Japanese carrier force. They, too, scored no torpedo hits. The single torpedo hit at Midway was scored by a PBY Catalina flying boat in a night attack. The "fish" inflicted minor damage on a Japanese tanker.

Despite this poor beginning, by 1944–1945 the Mark 13 had been improved as had U.S. torpedo plane tactics. Several Japanese warships were sunk by carrier torpedo planes, including the world's two largest battleships, the *Musashi* in October 1944 and the *Yamato* in April 1945.

In addition, U.S. Navy aircraft used 124 Mark 24 acoustic homing torpedoes during the war. They were credited with sinking 26 German and 5 Japanese submarines, plus damaging a few others; Mark 24s released by Canadian and British aircraft had similarly impressive U-boat kill rates. This was a remarkably high success rate for an ASW weapon.

POST-WORLD WAR II COMBAT

Although submarines were a major component of U.S. forces during the 45 years of the Cold War and did have a limited role in the Korean and Vietnam conflicts, no U.S. submarine or surface ship fired torpedoes in anger during this period. Indeed, the only torpedoes that were used in combat by the United States were the eight Mark 13s launched by AD Skyraider aircraft against a North Korean dam on 1 May 1951. Six struck the target, one missed, and one was a dud.

A few other nations did use torpedoes in combat in the post–World War II period. During the Israeli air and torpedo boat attacks on the U.S. spy ship *Liberty* (AGTR 5) off the coast of Sinai in June 1967, three Israeli MTBs launched five torpedoes. One struck the *Liberty*, already damaged by Israeli air attacks. these were 18-inch aerial torpedoes acquired from Italy and modified for use by surface ships.[7]

Four years later the Pakistani submarine *Hangor* torpedoed and sank the Indian frigate *Khukri*. (Also in 1971, an Indian destroyer sank a Pakistani submarine with depth charges.)

The British nuclear-propelled submarine *Conqueror* torpedoed and sank the Argentine cruiser *General Belgrano*, the former USS *Phoenix* (CL 46), in the 1982 Falklands conflict. The cruiser, sunk with heavy loss of life as her two accompanying destroyers fled the scene, was the victim of outdated Mark VIII torpedoes. She was history's first combat victim of a nuclear-propelled submarine attack.

A North Korean submarine sent a torpedo into the South Korean corvette *Cheonan* on 26 March 2010. The 1,200-ton warship broke in two and sank, with 46 of the ship's crew of 104 being killed. The attack took place in South Korean territorial waters off Baengnyeon Island in the Yellow Sea, near the North-South Korean border.

Throughout the Cold War there were confrontations between U.S. and Soviet naval forces, and both sides had torpedoes armed, loaded, and ready for launch. Some of those weapons had nuclear warheads, because throughout the period both nations emphasized nuclear weapons as the final arbitrator should a conflict erupt between the two superpowers.

In the U.S. Navy there was a rapid rise of the nuclear submarine community's role as well as the size of the submarine force in the early 1960s, in large part due to the deployment of the Polaris submarine-launched missile as the West's most survivable deterrent to nuclear conflict. Subsequently, submarines armed with strategic missiles as well as other submarines armed with torpedoes that sought out Soviet missile submarines became a key component of U.S. military strategy. In consequence of Soviet submarine developments, U.S. surface warships and ASW aircraft, land- as well as ships and carrier-based, were armed with anti-submarine torpedoes.

Today the anti-submarine torpedo remains the only torpedo carried by U.S. Navy surface ships and aircraft. For the foreseeable future lightweight torpedoes will remain the principal ASW weapon for surface ships as well as aircraft. The warheads of these weapons are small, about 100 pounds, and may not be able to sink a large submarine, especially a double-hull craft with significant standoff distance between the hulls. Still, the damage inflicted on most submarines by a lightweight torpedo would probably be sufficient to deter the submarine from carrying out its mission—a "mission kill."

The issues related to heavy torpedoes are far more complex. Torpedoes are not used by U.S. surface ships or aircraft in the anti-ship role. Rather, anti-ship missiles and GPS-guided bombs are employed in that role. The arming of ships with advanced anti-aircraft weapons has made it too dangerous for attacking ships or aircraft to close within torpedo range of modern warships.

Thus, in the U.S. Navy only submarines now carry heavy torpedoes, variants of the Mark 48. There is no other weapon on the horizon. Either a very-high-speed torpedo, such as the Russian VA-111 Shkval (squall) or a rocket-boosted weapon that travels through the air to reach a target with a lightweight homing torpedo, such as the Russian RPK-6 Vodopad (waterfall) and RPK-7 Veter (wind) missiles, offers certain advantages over a conventional torpedo travelling through the water at about 50 knots.[8] But the U.S. Navy's submarine community is both committed to the Mark 48 family and complacent about other approaches to solving the problems involved with relying entirely on a conventional heavy-weight torpedo.

The Mark 48 is a powerful weapon. A single Mark 48 torpedo detonating under the keel of a destroyer-size warship will undoubtedly sink the ship. A similar attack against a larger warship or commercial vessel will cripple if not sink the target. But a 50-knot torpedo takes several minutes to travel five or ten miles. An alerted target ship—whose sonar has detected the submarine or the noise of the weapon launch—would have time to deploy decoys or acoustic jammers, or to activate towed countermeasure devices, such as the U.S. Navy's AN/SLQ-25 Nixie.

An alternative is for the submarine to use submerged-launch, anti-ship cruise missiles against surface ships.[9] Modern cruise missiles usually carry a relatively small warhead and damage to the target could be limited, because such missiles usually strike the ship's upper hull or superstructure. Still, there are significant advantages to employing a cruise missile rather than a torpedo against surface ships. First, although countermeasures are readily available to the target, a fast, maneuvering small combatant is more vulnerable to a cruise missile than to a submarine-launched torpedo. This was demonstrated by the Israeli Navy's estimated 13-to-0 score of successes in 1973 against Egyptian and Syrian missile ships and on other occasions using anti-ship missiles.

Second, relatively small cruise missiles can sink large ships. In the 1982 Falklands conflict, an Argentine Exocet missile struck the large British destroyer *Sheffield*. The missile's 360-pound warhead failed to detonate. Still, it ignited fires, fed by the missile's unburned fuel, and the *Sheffield* was burned out. (The ship later sank in rough seas while under tow.) In 1987 the U.S. frigate *Stark* (FFG 31) was struck by two Exocets launched by an Iraqi aircraft; one of the missiles detonated. The ship suffered 37 dead as the crew fought fires for 24 hours, and at one point the accumulation of water on board from fire-fighting efforts caused the ship to list 16 degrees. Although the *Stark* survived her ordeal, she had to be carried back to the United States by a heavy lift ship and was in the yard for nine months. Clearly the Exocets inflicted a "mission kill."

Several navies now have cruise missiles that are launched from submarine torpedo tubes, including the French-built Exocet. But the U.S. Navy has none. Previously two anti-ship cruise missiles were carried in U.S. submarines—the Harpoon and the larger Tomahawk. Neither missile is now used in U.S. submarines. (U.S. attack submarines still carry the long-range Tomahawk in the conventional, land-attack or "strike" version.)

One other submarine-launch weapon could be used against surface ships: the anti-ship ballistic missile. In the 1960s the Soviet Navy developed such a weapon, given the Soviet designation R-27K and the NATO designation SS-NX-13. The missile—with a nuclear warhead of 500 kilotons to one megaton—was estimated by Western intelligence to have a range of 350 to 400 nautical miles and, with external guidance, the ability to strike ship targets within a 27-nautical-mile "footprint." [10] The R-27K missile, which could be launched from standard ballistic missile tubes, became operational in 1975. However, it was not deployed because the Strategic Arms Limitation Talks (SALT) agreements of the 1970s would count every submarine ballistic missile tube as a strategic weapon regardless of whether it held a strategic or tactical (anti-ship) weapon.

The current Chinese development of *land-based*, anti-ship ballistic missiles—for the anti-aircraft carrier role—may provide the opportunity for a future submarine-launched anti-ship ballistic missile. The anti-ship variant of the Dong Feng (East Wind) 21 missile has a conventional warhead with multiple reentry vehicles and a range of up to 1,000 nautical miles. The anti-ship weapon could be adopted for submarine launch against surface ships—if targeting resources were available—as the DF-21 also forms the basis for a submarine-launched ballistic missile.

For the foreseeable future, however, U.S. Navy surface ships, aircraft, and submarines will employ only lightweight and heavy-weight torpedoes against enemy surface ships and submarines. Those complex and deadly weapons had their beginnings in the efforts of Messrs. Whitehead, Bliss, Leavitt, and others. And, those efforts had their conceptual beginnings with Messrs. Bushnell, Fulton, Cushing, and Holland. All were men of foresight, skill, and imagination.

Appendix A

Torpedo Fire Control

No book on torpedoes would be complete without some discussion of the methods used to solve the torpedo fire-control problem. The first of these methods used relatively simple analog devices to provide a graphical solution to the space and velocity triangle of torpedo fire.

In this triangle, the line of sight from own ship to target forms the base of the triangle. The length of this line is proportional to the range of the target. The target track and torpedo track form the other two sides of the triangle. Since the lengths of torpedo track and target track are proportional to the torpedo and target speeds respectively, these lines may be considered as representing torpedo and target speeds. For a given target angle, the torpedo will intercept the target at point of intercept.

Directors for Above-water Torpedo Tubes

In the early days of torpedo warfare the torpedo fire triangle was solved by mechanical torpedo directors. These were simple mechanical devices that automatically provided an aiming point for the proper release of the torpedo. They were made of metal and had sliding arms that could create a miniature model of the space and velocity triangle. A ruled bar, called the speed to torpedo bar, was aligned with the torpedo tube. It had an adjustable slide that could move back and forth on a scale graduated in knots to represent the running speed of the torpedo. Another ruled bar, called the target speed bar, swiveled about a pivot that represented the impact point. It also had an adjustable slide that could be used to set the estimated speed of the target. The last element was the slotted sighting bar that swiveled from a pedestal on top of the torpedo speed bar. It was connected to the target speed bar via a pin running within the slot. When the slide bars on both speed bars were adjusted, the sighting bar was aligned with the sighting line required to produce a hit. Once the director was adjusted the trainer in charge of the torpedo tube looked down the sight line and waited for the target to come into view, at which point the torpedo would be fired.

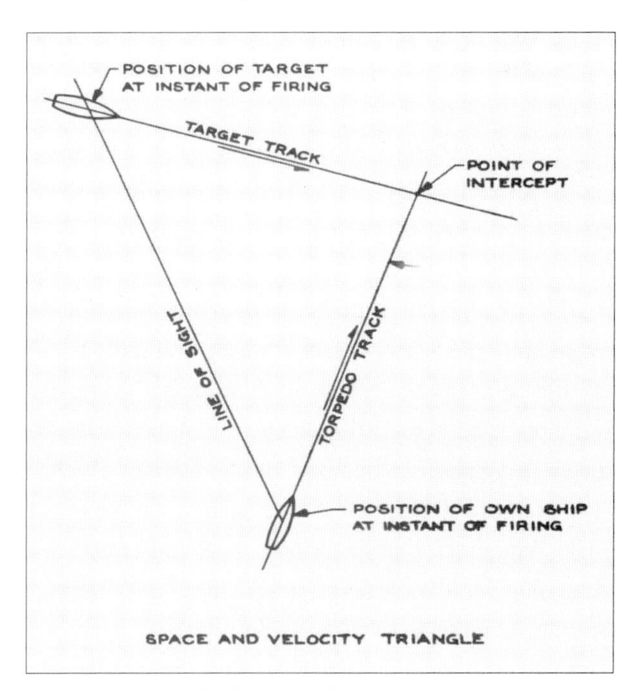

Figure 1. Space and Velocity Triangle

The bow torpedo director shown in figure 2 is a good example of what an early torpedo director looked like. It was used to aim torpedoes mounted in untrainable torpedo tubes mounted in a warship's bow. Directors such as this were used on U.S. warships in the late 19th Century.[1] Torpedo tubes on trainable mounts required a slightly more sophisticated device that provided an adjustment for the training angle of the torpedo tube. An example of a typical director used for this purpose is shown in figure 3. Note that on this device the torpedo speed arm can be rotated about a ruled arc corresponding to the training angle of the torpedo tube.

Although the directors used in World War I were mounted on a more sophisticated pedestal and were equipped with telescopic sights, the basic design for above water torpedo directors remained relatively unchanged until the late 1920s. By then the Navy had fostered the development of automated systems to solve torpedo and anti-aircraft fire control problems. New directors for anti-aircraft guns and torpedo tubes were produced by the Ford Instrument Company and Arma Engineering. Both were "captive contractors," who worked exclusively for the Bureau of Ordnance.

Automated fire control began in the U.S. Navy shortly before World War I with the adoption of "director firing," which controlled all guns on a ship from a centralized location. The Sperry Gyroscope Company connected instruments that collected observed data about a target into a central plotting room. An automatic plotter drew the paths of both the firing ship and the target ship on paper, from which a gunnery officer could read the range and bearing for the guns to fire. He then electrically transmitted these data to gunners in the turrets. In 1915 Sperry's chief designer, Hannibal Ford, left the firm to start the Ford Instrument Company and introduced the Ford Rangekeeper, which both incorporated British technology and added new mechanisms of Ford design. The Rangekeeper, a mechanical analog computer, estimated the course and speed of a target ship based on repeated observations of range and bearing, continually updating the estimate in accord with new observations. The U.S. Navy enthusiastically adopted the Ford Rangekeeper, at first for battleships and then for destroyers and cruisers.

Like Ford, the Arma Engineering Company was formed by an ex-Sperry employee: David H. Mahood. Mahood had worked in the fire control division of the Sperry Company during World War I and then at the Brooklyn Navy Yard. Mahood's position in the Navy Yard brought him into intimate contact with the control systems of the time, as well as their problems. His partner, Arthur P. Davis, was a self-educated engineer who had build switchboards (an integral part of fire-control instrumentation) while working for General

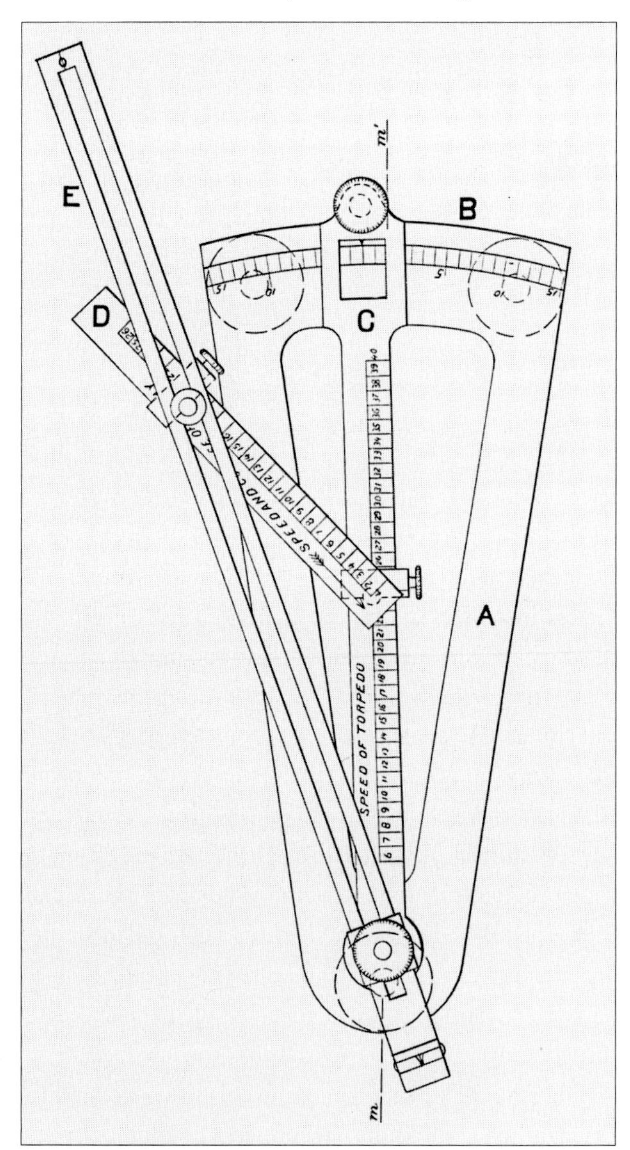

Figure 2. Bow torpedo director was an analog device used to solve the torpedo triangle problem. (U.S. Navy)

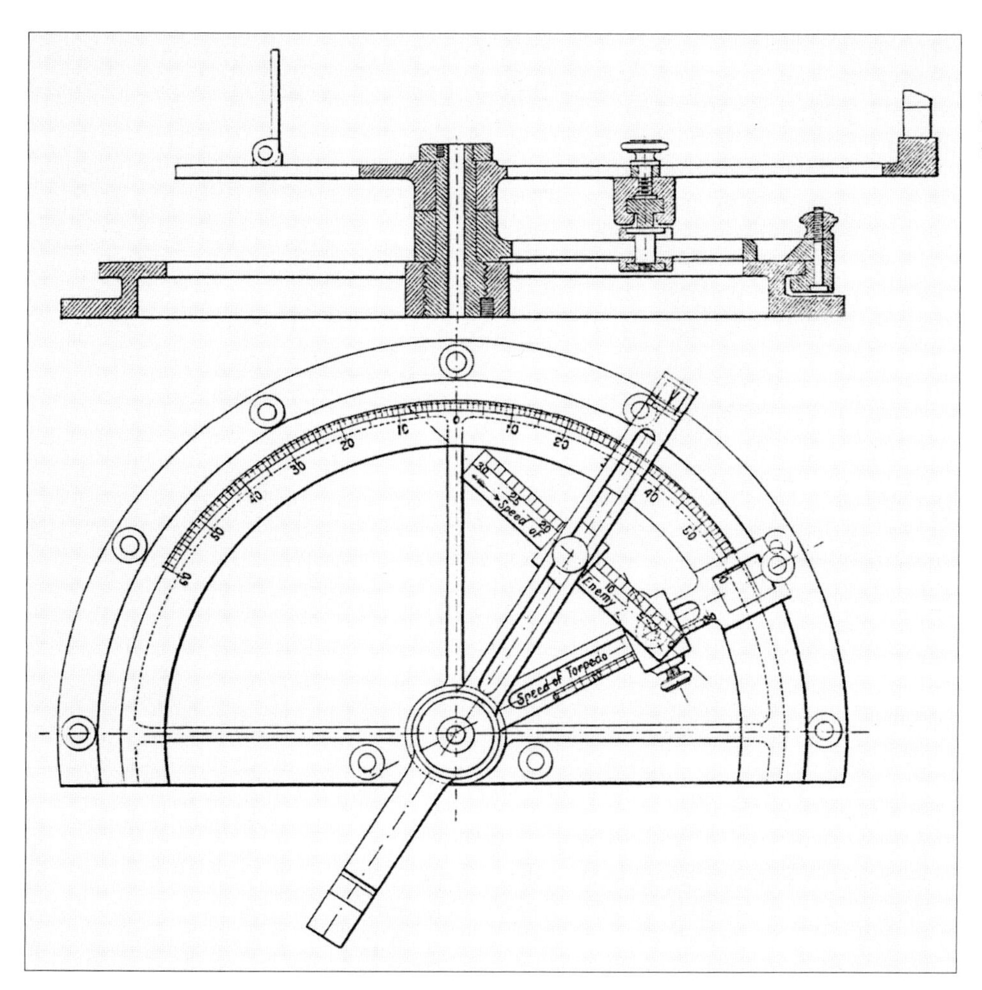

Figure 3. Adding a torpedo bearing indicator and mounting the triangle solver on a pivot point facilitated its use for trainable torpedoes.

Electric. Together, they formed the Arma Engineering Company on 30 January 1918. The name "Arma" was formed by taking the first two letters of *Arthur* and the first two of *Mahood*. While Ford concentrated on gun fire control systems, Arma worked on torpedo control and the development of a "stable element" for main battery fire control.

In 1928 Arma produced a torpedo fire control system for existing light cruisers and destroyers based on a remote director that automatically solved the torpedo triangle problem and transmitted the correct training angles to remote indicators located at the torpedo tubes. The Arma system consisted of the following elements:

- Two Mark 25 bridge mounted torpedo directors, port and starboard
- One Mark 2 tube train indicator on each triple torpedo tube
- One training tube attachment, Mark 2, for each triple torpedo tube
- A self-synchronous electrical transmission system from torpedo directors to torpedo train indicators
- A torpedo firing and ready light indicator

The director was a precision instrument that contained an analog calculator consisting of three logarithmic scales, one curve scale, and one sight-setting scale that was designed to solve any torpedo firing problem to obtain the proper sight angle. Each director was equipped with two transmitters from which indications of tube train were transmitted electrically to zero reading indicators located on the torpedo mounts on the same side of the ship as the director.

The Arma type torpedo fire-control system could be operated in three modes: moving tube, fixed tube—fixed sight, or fixed tube—fixed sight with gyro setting. In the

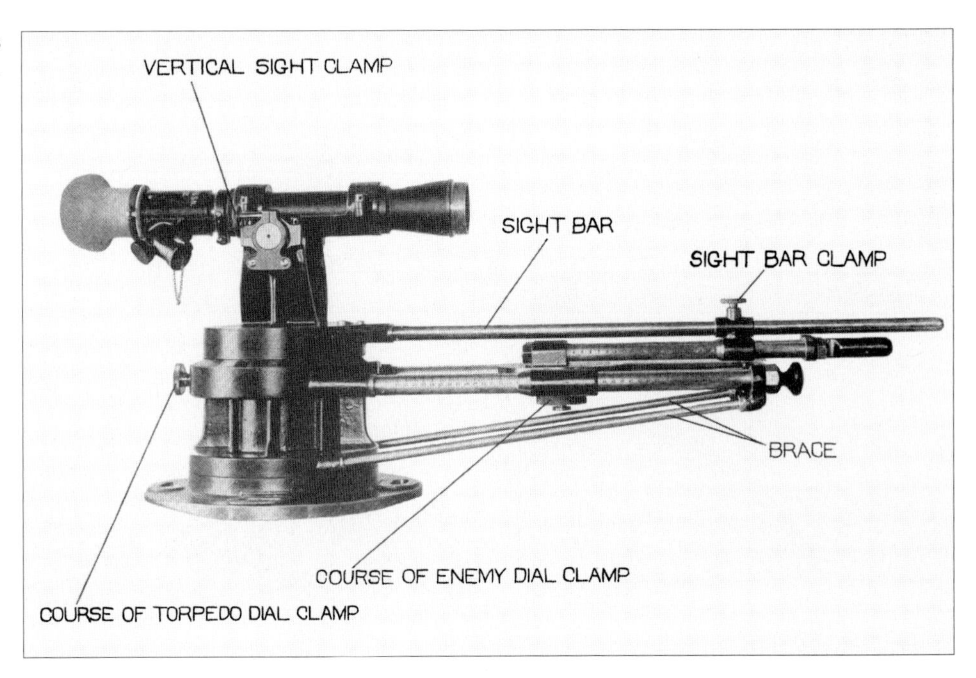

The Mark VI torpedo director was introduced before World War I. (U.S. Navy)

A Mark VI torpedo director directs the twin torpedo tubes of the *Whipple* (DD-15), c. 1917.

moving-tube mode, the telescopic sight on the director was continuously kept on the target. Both the conditions of the problem, i.e., target bearing, target speed, and torpedo speed were set beforehand as was tube offset. Having established these conditions, the director automatically computed the bearing needed for the point of interception. The bearing was then transmitted to the tube train indicator. The tube trainer matched the zero reader pointer on the indicator. When the tube has reached the bearing on which firing can be executed, the tube trainer closed the firing switch. This turned on a ready light at the director, indicating that the tube was ready to fire. The director operator fired the first torpedo by means of the firing key and immediately reset the

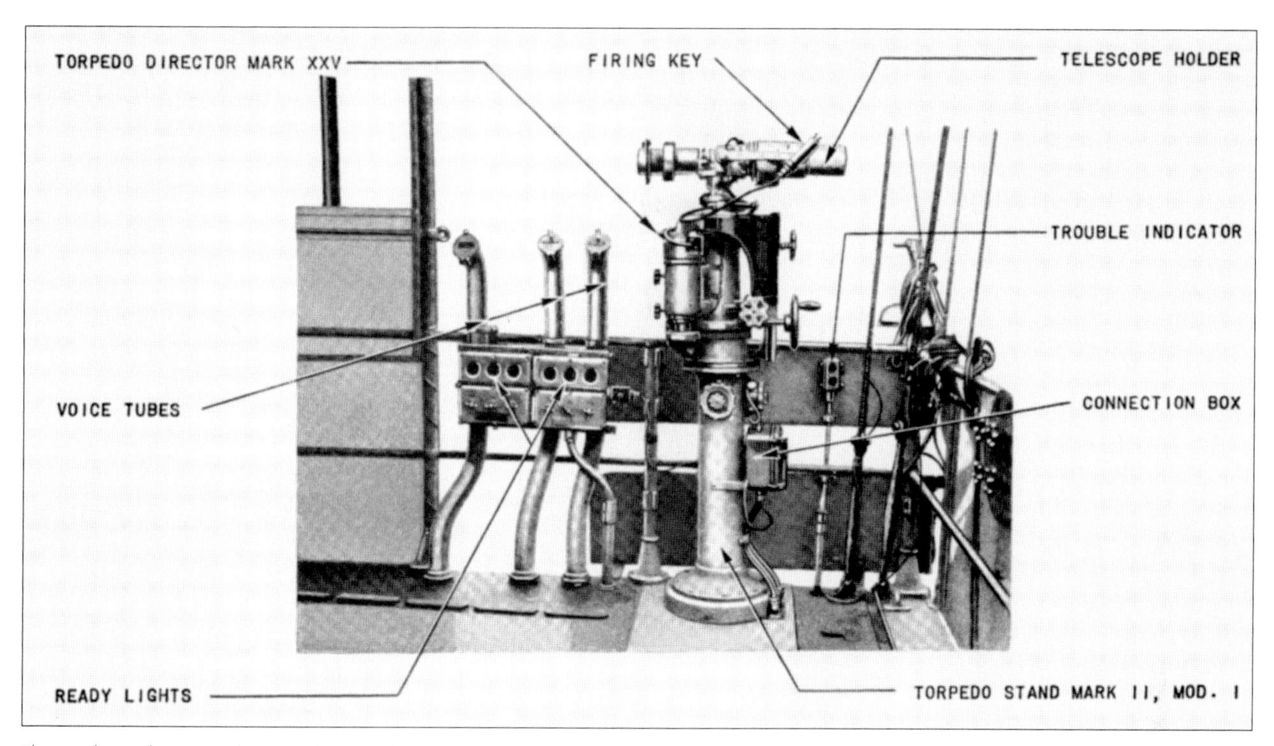

The Mark 25 director in the torpedo control station on the bridge of a destroyer. (U.S. Navy)

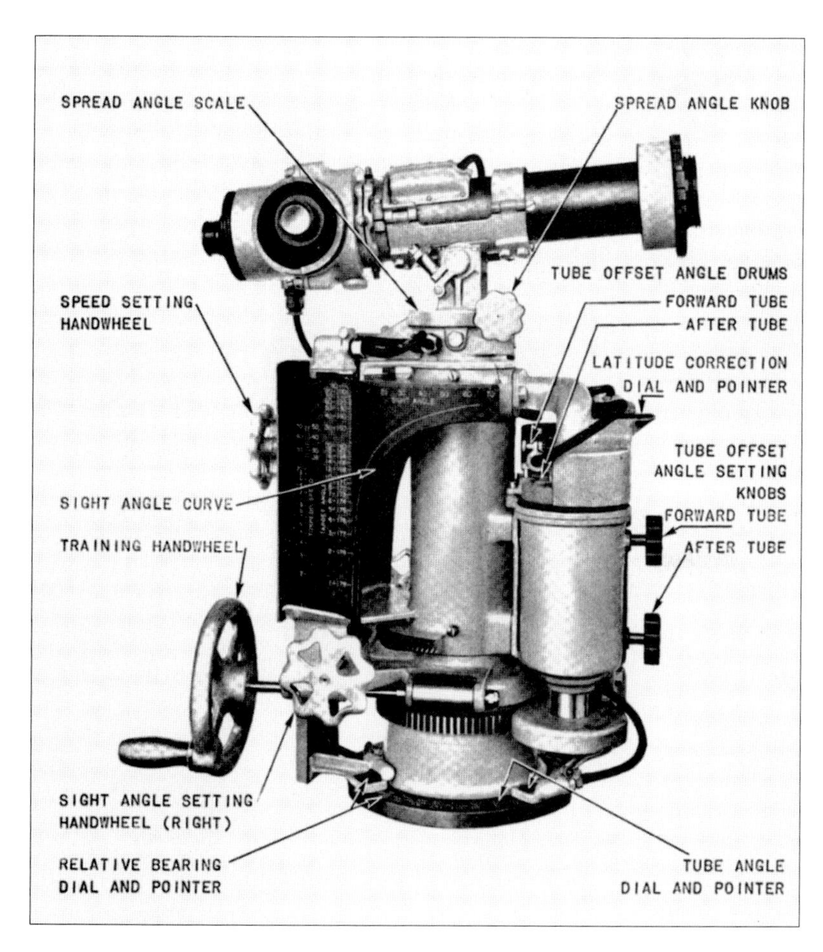

Mark 25 Torpedo Director. (U.S. Navy)

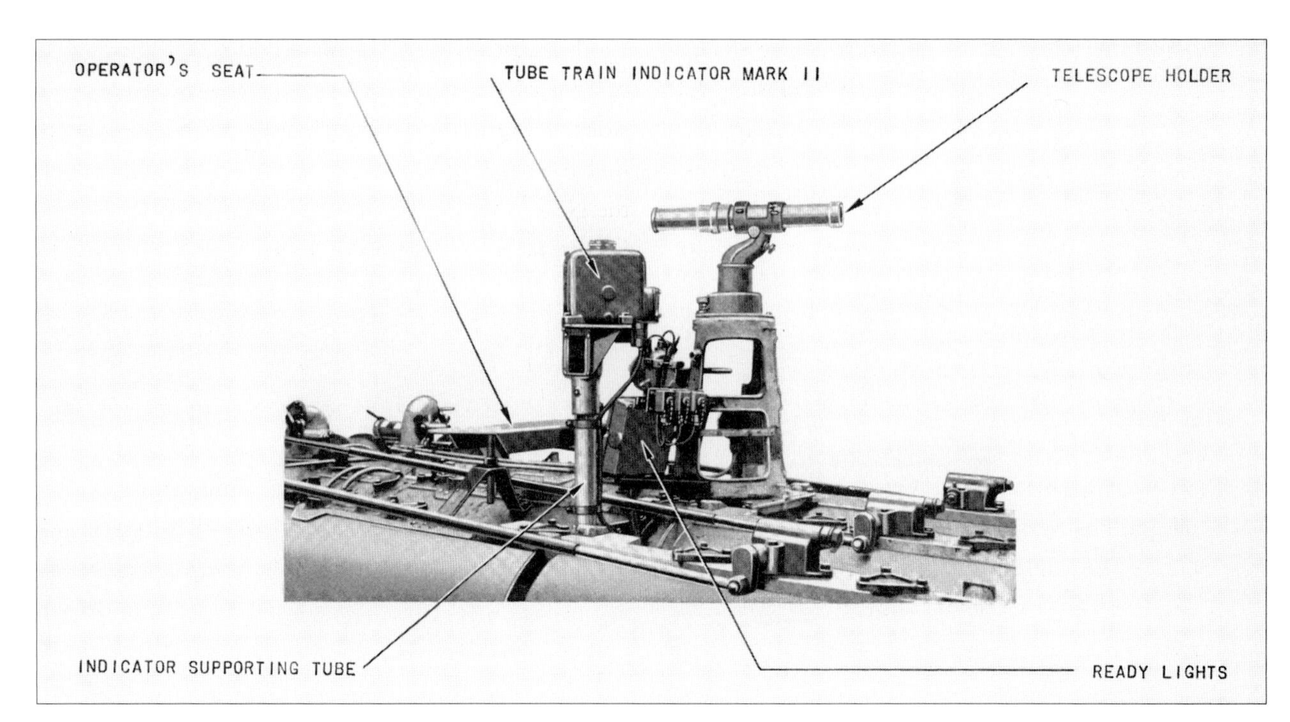

Torpedo tube trainer's station atop a 21-inch triple torpedo tube mount. (U.S. Navy)

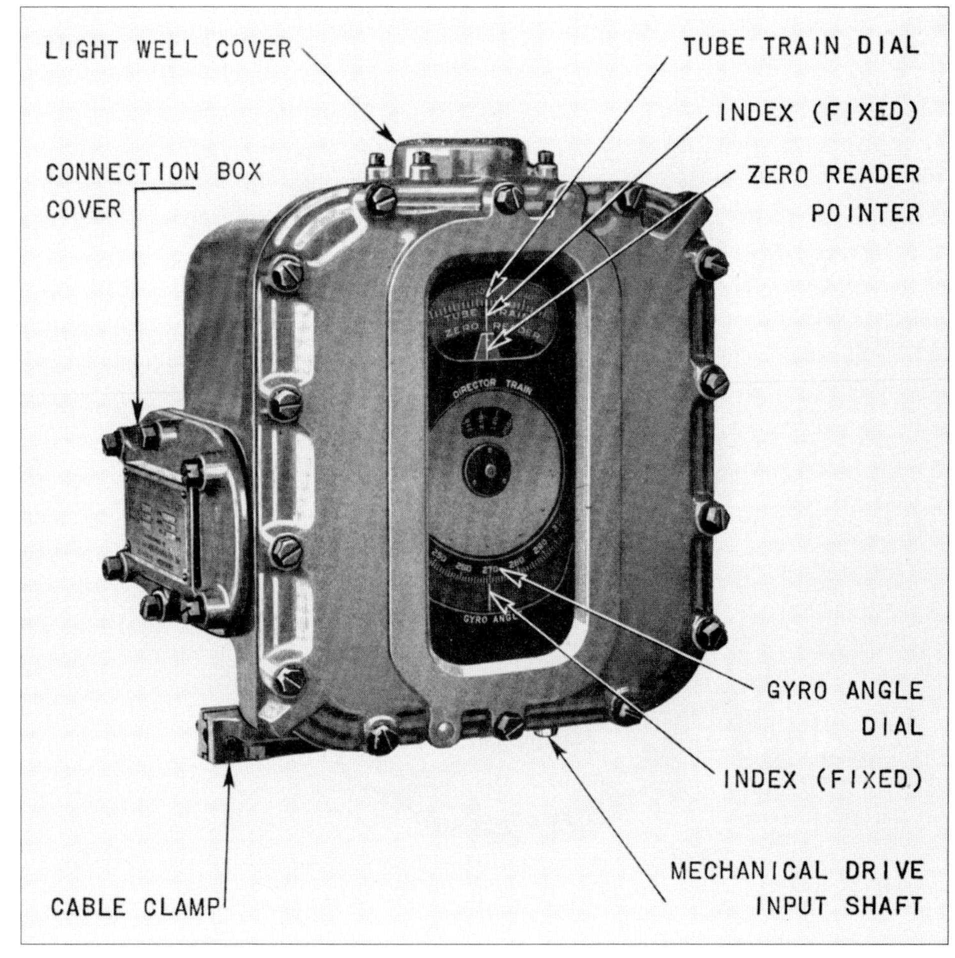

Front view of tube train indicator, Mark II. The tube trainer used this device to adjust the angle of the torpedo tube for firing. When the pointers matched, the torpedo was ready for firing. (U.S. Navy)

offset on the director for the second torpedo. The director firing switch for the second torpedo was closed, and if the tube-ready light was on, the second torpedo was fired. The same procedure was followed for the last torpedo. In the fixed tube—fixed-sight mode, the ship was turned to bring the director telescope on the target. The tube trainer immediately trained and clamped the tube with the zero reader pointer mated. The firing switch on the tube was closed, lighting the ready light at the director. When the ship swung the telescope on target, the firing key at the director was closed and the torpedo was fired. The fixed tube—fixed sight with gyro setting was similar to the fixed tube—fixed sight mode, except that the director sent the gyro angle signal to the gyro angle indicator which was read by the tube trainer who set the gyro mechanism.[2]

It is not known how many Mark 25 torpedo directors where actually installed on U.S. warships. The stock market crash of 1929 and the Great Depression that followed in its wake had catastrophic effects on the Navy's budget, which was slashed to the bone by Congress. There was barely enough money for repairs, let alone the installation of new equipment. Nevertheless, it served as the precursor to the more-advanced Mark 27 torpedo director.

Installations of the Mark 27 director and its associated equipment begin to appear on photographs of U.S. destroyers in the late 1930s.[3] The Mark 27 was a sophisticated analog computing device (see figure 5) that mechanically computed the basic sight angle from the torpedo speed triangle, corrected it for torpedo creep (latitude correction) and then combined the corrected

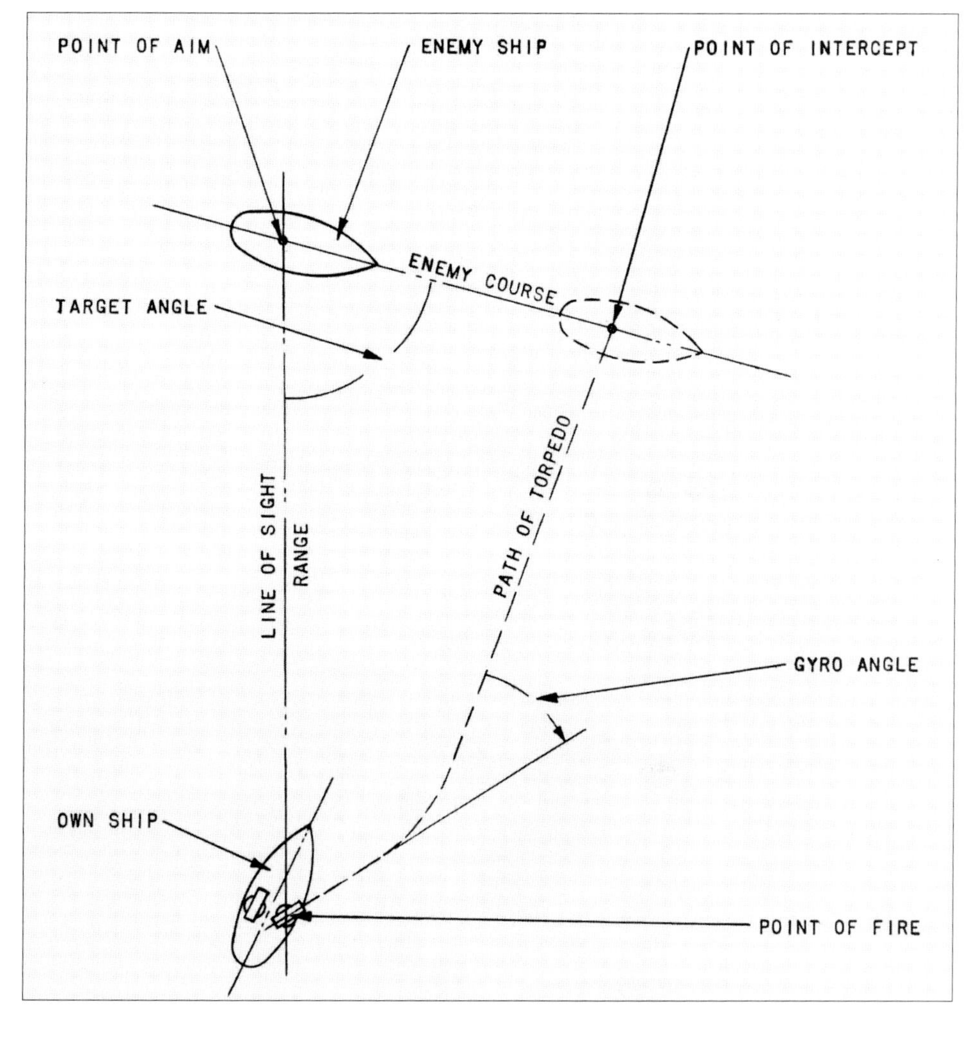

Figure 4. Diagram of fixed tube–fixed sight with gyro setting method of torpedo firing.

sight angle with relative target bearing to produce the basic torpedo course. The Mark 27 had settings for tube offset (to provide a spread angle between the two torpedo tube mounts on the ship) and gyro angle. Synchro generators were used to continuously transmit torpedo course and gyro angle directly to the tube mounts via electric cables.

Synchro generators in the torpedo director continuously transmit (electrically) torpedo course at 1-and 36-speed and gyro angle at 2-speed to the torpedo course indicators at the torpedo tube mounts. Dials at the torpedo director show the values of all the quantities entering in the torpedo fire control problem and the values of torpedo course and gyro angle transmitted to the torpedo course indicators.

Angle Solvers and TDCs

Early submarines had to be maneuvered into position so that their bow tubes aimed directly at the target and the torpedo was fired almost at point-blank range with the submarine itself trained on the target for the firing of a torpedo. Improvement of the gyroscopic steering apparatus permitted the setting of gyro angles, which determined the torpedo's course. According to the gyro angle, which was preset, the torpedo could be made to run dead ahead, or veer off on a tangent. But target

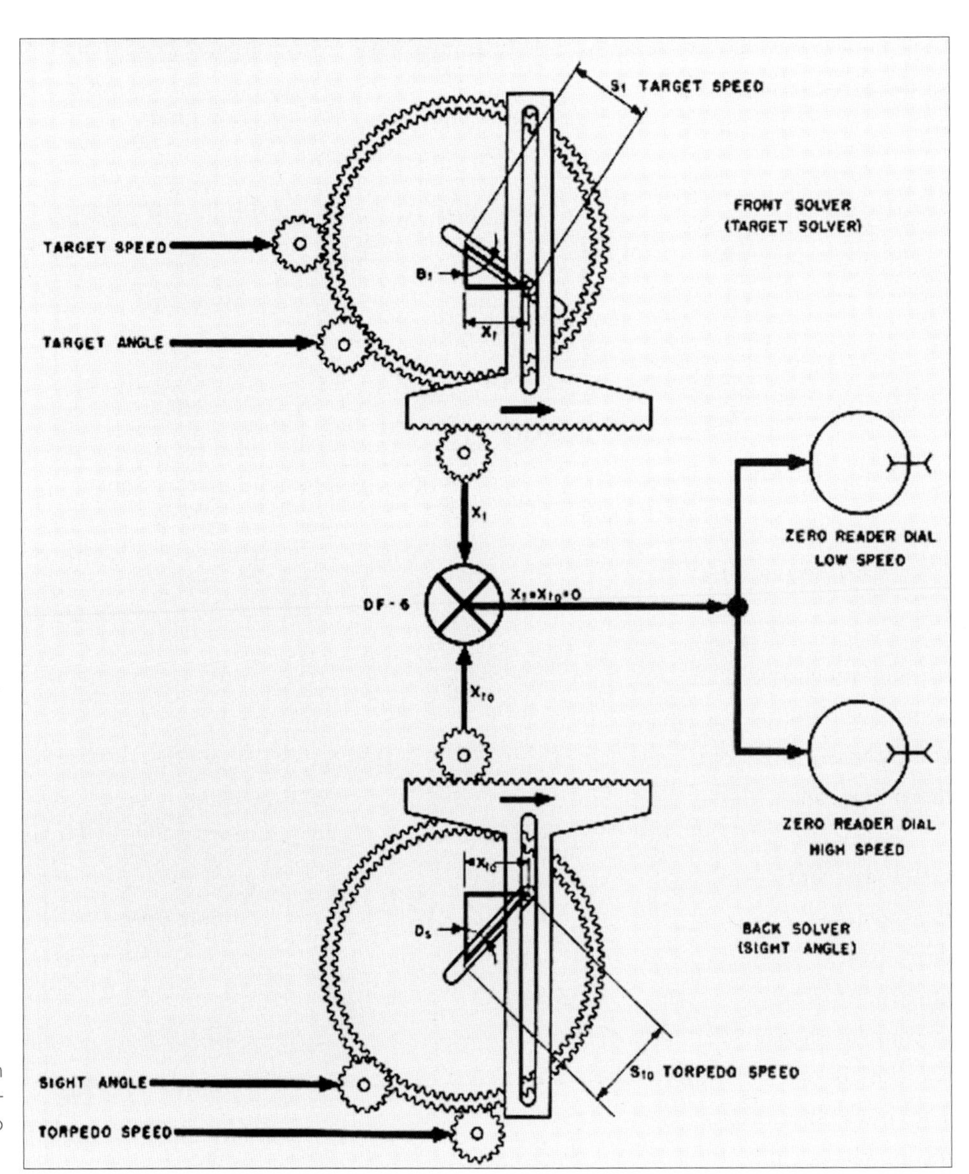

Figure 5. Schematic diagram showing how the Mark 27 torpedo director solved the torpedo speed triangle.

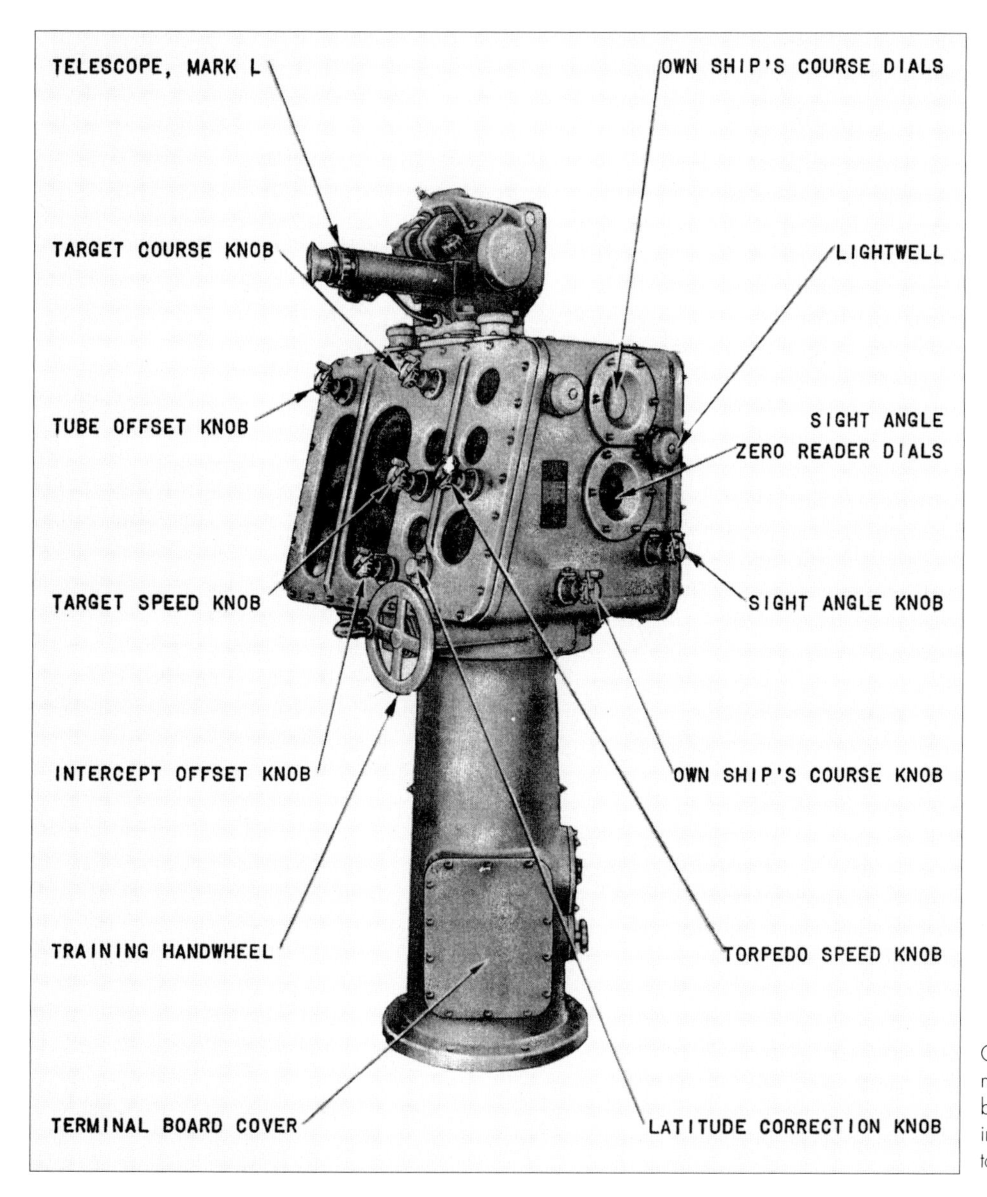

One Mark 27 director was mounted on each side of the bridge. Note the high degree of information shown on the indicator dials. (U.S. Navy)

bearing, range, speed, and accommodating gyro angles were variables that had to be computed by the captain at the periscope and the firing control party that plotted the attack course.

At the end of World War I the general method of attack was for the submarine to get within 500 yards of the target and fire a straight bow shot on about a 90-degree track. Gyro angles could not be set on torpedoes after they were loaded into the torpedo tubes. The only instrument then in use aboard American submarines was the "Is-Was" or Submarine Attack Course Finder.[4] The course finder was a circular slide rule that was used to quickly calculate the bearing for the submarine to take in order to reach a favorable firing position or to ascertain whether or not such a position could be reached. It was based on the principle that a submarine or any vessel desiring to intercept another vessel of probable superior speed would stand the best chance of success if it brought the other vessel abeam and steamed at maximum speed.

By 1925 American submarines were fitted with "outside gyro-setting devices" by means of which the gyro angle could be set on the torpedo after the "fish" was loaded into the tube. The periscope angle (formed by the periscope's line-of-sight to die target, and the track of the torpedo) could only be computed by means

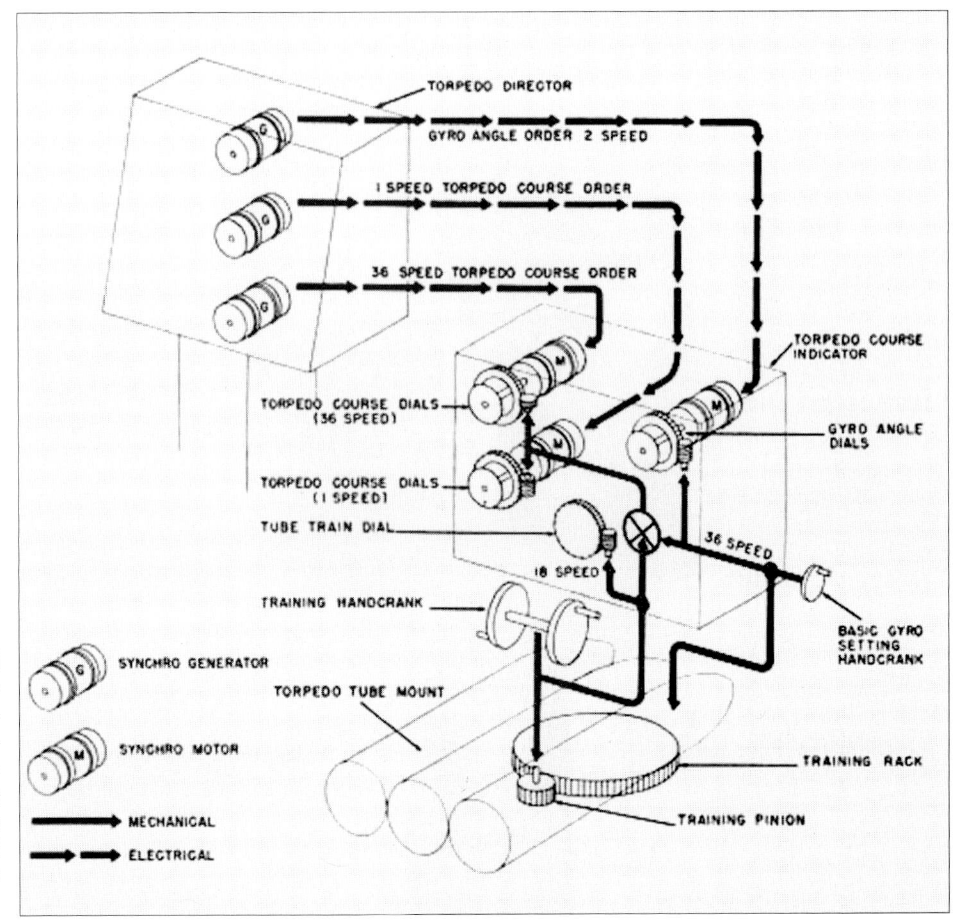

Figure 6. Schematic diagram of a Mark 27 director showing how the torpedo course was transmitted to the torpedo course indicator monitored by the torpedo tube trainer. His job was to set the gyro angle and then turn the hand wheel rotating the mount until the pointers on the tube train indicator matched.

of a series of tables. It was necessary, therefore, to decide what gyro angle was to be used, then compute die required periscope angle, set the gyro angle, set the periscope to the computed angle, and fire when the target passed the cross-wire of the periscope scale.

In practice the gyro angles commonly used were either zero or 90 degrees, with the zero angle generally preferred. To hit with an angled shot it was necessary to know the range, and submarines were particularly weak in range determination. Practice targets of this period were zigzagging, some of them screened, and speeds were low and prescribed within narrow limits for each practice. A sound shot was required at a target on a straight course, and sound gear was used during an approach, enabling the operator to obtain a turn count of the target ship's propellers. Firing ranges were usually about 1,000 yards, and penalties were imposed for firing inside 500 yards.

Practice targets had been speeded up considerably by 1930 and were better screened. Stern tubes had been installed in submarines, adding tremendous firepower to the undersea boats, which had been formerly limited to bow-tube shots. The first torpedo angle solvers were in use, and angled shots other than zero and 90 degrees were frequently made.

The torpedo angle solver, commonly called a "banjo," was a portable instrument designed to compute the required data for firing submarine torpedoes of various marks and powers at gyro angles up to 90 degrees from all classes of submarines. Like the earlier directors from which it was derived, the "banjo" was an analog device that solved the torpedo velocity triangle for a particular torpedo speed (noted as the "Pseudo Torpedo Speed") and gyro angle.[5]

The torpedo angle solver (see figure 7) consisted of three arms adjustable in direction and length but

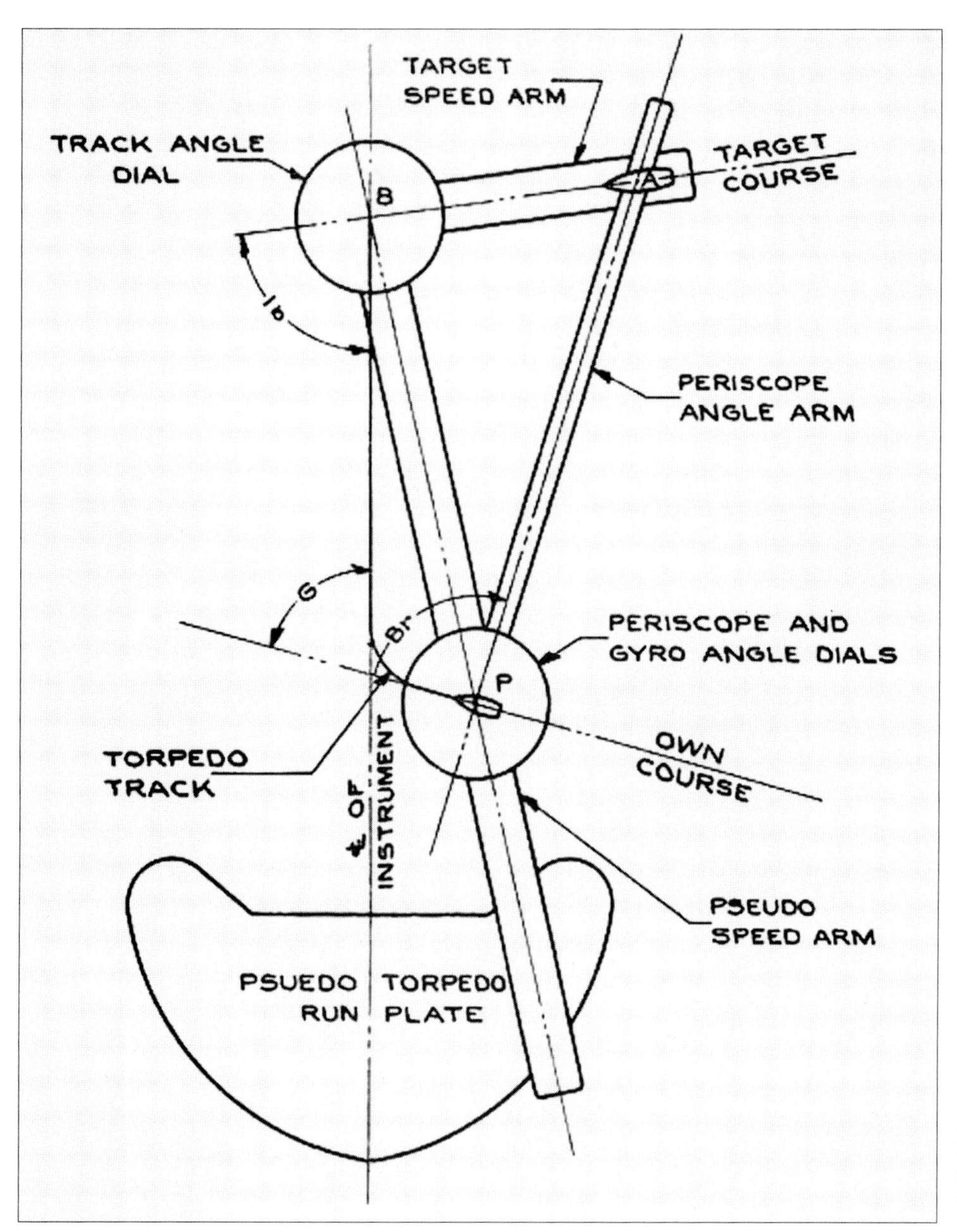

Figure 7. Schematic diagram of the Mark 3 Torpedo Angle Solver, known colloquially as the "banjo."

so joined as to always form the basic graphic-director triangle, together with subsidiary mechanisms indicating periscope and gyro angles. The torpedo angle solver was attached to one of a series of interchangeable pseudo torpedo run plates printed with charts showing the maximum range and gyro angle for a particular torpedo mark number, the power selected, and tube depth. The assembled mechanism looked a little like a banjo, leading to its particular moniker. Torpedo angle solvers continued to be used throughout World War II as a backup to the Torpedo Data Computers (TDCs). The first of these, the Mark 1, was first introduced in the *Cachalot* (SS 170) and *Cuttlefish* (SS 171) submarines completed in 1933–1934. Unlike the "banjo," the TDC provided a continuous solution to the submarine fire control problem, provided a more accurate solution, and did it in a more timely manner.

The TDC Mark 3 was an automated system manufactured by Arma that could track a target and continuously aim torpedoes by setting their gyro angles. Receiving data from the periscope or sonar on the enemy's bearing, range, and angle on the bow (the relative bearing of the submarine to the enemy), the TDC automatically plotted the course of the enemy relative to the course of the submarine and computed and set the proper angle in the torpedo to intercept him. The data were displayed on dials, giving the submarine commander a continuous visual picture of the attack. It was primarily a mechanical computer with some electrical components.

By World War II most submarines in the U.S. Navy had a TDC Mark 3. The Mark 3 gave the U.S. fleet submarine the ability to fire torpedoes without first estimating a future firing position, changing the ship's course, or steering to that position. Instead of hoping that nothing in the setup changed, a fleet submarine with the TDC could fire at the target when the captain judged the probability of making hits to be optimal.

The Mark 3 computer consisted of two sections, the position keeper and the angle solver. The position keeper tracked the target and predicted its current position. To do this, the position keeper automatically received input of the ship's own course from the gyro compass, and own ship's speed from the pit log. The

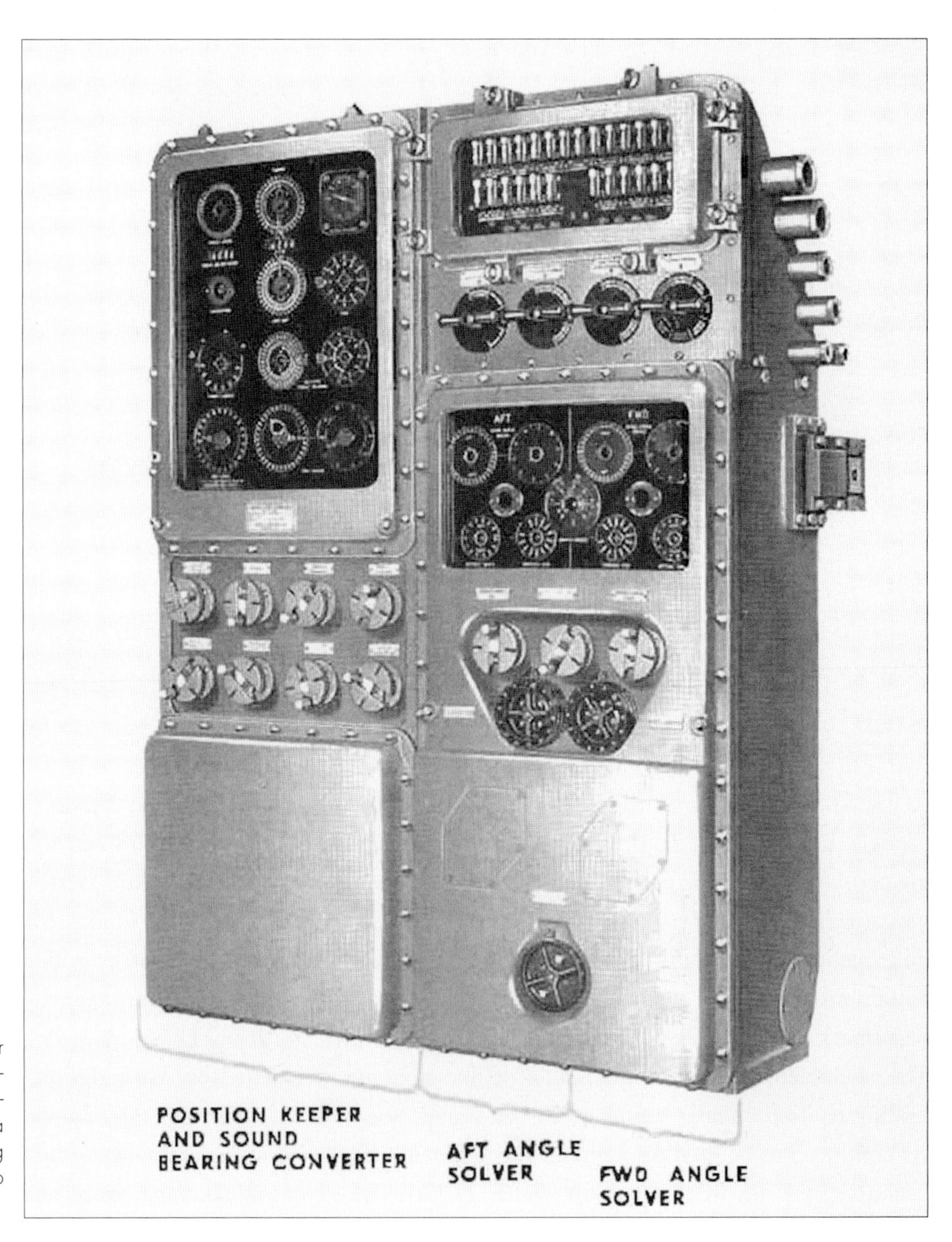

The Torpedo Data Computer Mark 3 gave the U.S. fleet submarine the ability to fire torpedoes without first estimating a future firing position, changing the ship's course, or steering to that position. (U.S. Navy)

position keeper had hand cranks on its face that set the target length, estimated speed, and angle on the bow. It also contained a sound bearing converter that calculated the target's location based on sonar measurements.

The position keeper solved the equations of motion integrated over time. The result was a continuous prediction of where the target was at any instant. Successive measurements of the targets' position were compared to the position keeper predictions and corrections for error were introduced with the hand cranks. The predicted target position became more accurate as more measurements made the corrections smaller. It was typical to get an accurate track on the target after about three or four observations under good conditions.

The angle solver automatically took the target's predicted position from the position keeper, combined it with the tactical properties of the torpedo, and solved for the torpedo gyro angle. Values calculated from this solution were returned to the position keeper in two feedback loops. The gyro angle automatically went to each of the torpedo rooms and was set into the torpedoes continuously. The TDC controlled both torpedo rooms and all ten torpedo tubes at once.

The U.S. Navy thus had a system that would point the torpedoes at a target as the fire-control problem developed. The TDC Mark 3 was the only torpedo targeting system of the time that both solved for the gyro angle and tracked the target in real time. The comparable systems used by both Germany and Japan could compute and set the gyro angle for a fixed time in the future, but did not track the target. Thus the idea of the position keeper, and its iterative reduction of target position error, was unique to the U.S. Navy and represented a distinct advantage.

General Function of the Torpedo Data Computer

In order to direct a torpedo so that when fired it will hit an observed target, it is necessary to have information regarding the position and motion of the target and also the characteristics of the particular torpedo being used.

The target can be initially located by measuring its initial relative target bearing, which is the angle between the line of sight and the forward and after axis of own ship, and initial range. Since own ship course is a known factor, the target's position is then known in terms of initial true target bearing and the angle between north and the line of sight.

The target course can be obtained by estimating the initial target angle and the angle between the forward and after axis of the target and the line of sight and then combining this with the initial true target bearing.

Own ship speed is also a known factor.

If target speed can be ascertained, the location of the target relative to own ship at any time after the initial sighting can be predicted in terms of the present range, the present relative target bearing, and the present target angle, regardless of the motion of own ship and target providing the target does not change course or speed.

When a torpedo is fired it is assumed to travel first in a straight line along own course Co (see figure 8), for a distance called the reach M. A gyro in the torpedo

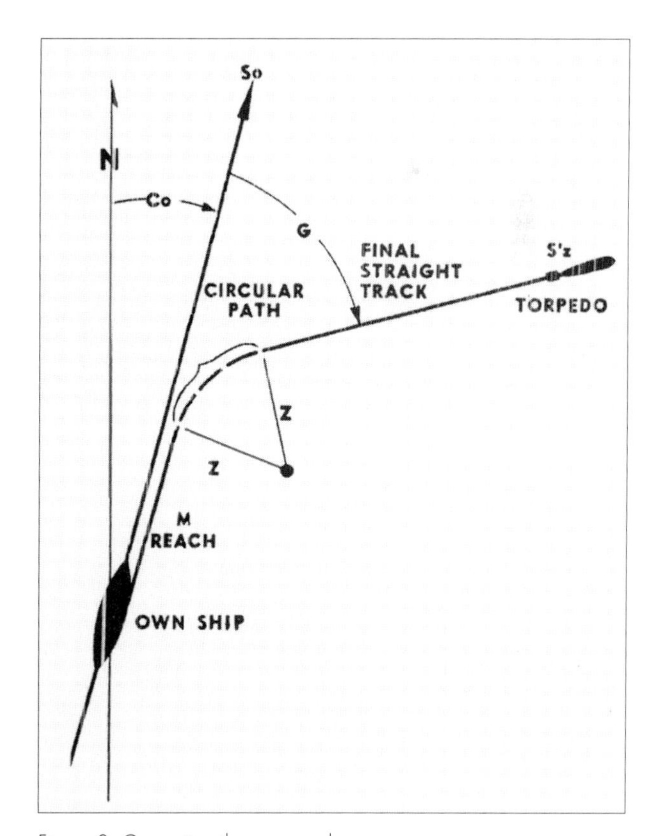

Figure 8. Computing the gyro angle.

then takes control and turns the torpedo on a circular course of radius Z through an angle G, the gyro angle or angle at which the gyro in the torpedo was set at the time of firing. The gyro angle G is the angle between the forward and after axis of own ship and the final track of the torpedo. After turning through the gyro angle G, the torpedo will then continue on a straight course.

From experimental data, the reach M, turning radius Z, and torpedo running speed S'z are known for any particular type of torpedo. The torpedo running speed is different from the speed immediately after firing, and a correction factor is also obtained from experimental data. This factor is known as the torpedo run difference Uy.

The fire control problem then resolves itself into one of finding a gyro angle so that the torpedo will reach a point on the track of the target at the same time that the target reaches that same point. The determination of this gyro angle is the ultimate function of the torpedo data computer.

Appendix B

U.S. Navy Experimental Torpedoes

Station "Fish" Torpedo

A self-propelled torpedo, designed at the Naval Torpedo Station in Newport, R.I., during the winter of 1869–1870, by Lieutenant Commander Edmund O. Matthews and Lieutenant Francis M. Barber.

Length	12 ft, 6 in
Diameter	14 in
Weight	430 lb
Propulsion	gas generated from liquid carbonic acid
Engine	Wheeler rotary engine
Range	unknown
Explosive charge	75 lb

Barber Torpedo

This was a rocket-propelled torpedo constructed in 1873 by Lieutenant Barber. The rocket fuel was stored inside a cast-iron tube wrapped in asbestos with an outer casing of oak.

Length	7 ft
Diameter	12 in
Weight	287 lb
Propulsion	solid-propellant rocket
Engine	unknown
Range	unknown
Explosive Charge	48 lb

Lay Torpedo

John Louis Lay developed spar torpedoes used by the Union during the Civil War. In the 1870s and 1880s he produced a series of "dirigible" torpedoes that were sold to various nations, including the U.S. Navy, which purchased two of his torpedoes for evaluation and testing at the Torpedo Station: No. 1 in 1873 and 1875 and No. 2 in 1875–1876. Both were surface torpedoes powered by carbonic acid gas guided electrically by a trailing cable.

LAY TORPEDO NO. 1 (PURCHASED 1872)

Length	29 ft
Diameter	36 in
Weight	8,000 lb (approx.)
Propulsion	carbonic-acid gas
Engine	unknown
Range	2 miles
Explosive Charge	500 lb

LAY TORPEDO NO. 2 (CONTRACTED IN 1875)

Length	23 ft, 8 in
Diameter	30 in
Weight	4,000 lb (approx.)
Propulsion	carbonic-acid gas
Engine	unknown
Range	1-1/2 miles
Explosive Charge	unknown

Ericsson Torpedo

John Ericsson—designer of the Civil War *Monitor*—developed this rectangular-shaped torpedo propelled by compressed air supplied through a trailing tube. Ericsson presented his proposal for the torpedo to the U.S. Navy in 1870; it was contracted for in 1872, delivered in 1873, and tested in 1875. A redesigned and improved model was also tested in 1877.

Length	15 ft
Diameter	30 in max. (cross section 20 in x 30 in)
Weight	2,000 lb
Propulsion	compressed air, 60 psi
Engine	unknown
Range	800 ft
Explosive Charge	150 lb

Howell Torpedo

In 1870, Captain John A. Howell submitted the design for a self-propelled torpedo powered by the inertia of a high-speed flywheel to the Bureau of Ordnance. The initial design using an iron flywheel weighing 100 pounds rotating at 2,000 rpm was tested in 1873. An improved model having a 115-pound flywheel rotating at 4,000 rpm was tested in 1877. The following data are for the initial Howell design.

Length	4 ft
Diameter	12 in
Weight	245 lb
Propulsion	inertia from rotating mass
Engine	100-lb iron flywheel @ 2,000 rpm
Range	unknown
Explosive Charge	unknown

Lay-Haight Torpedo

This torpedo was designed by George E. Haight, a former employee of John Lay, in 1881 and was similar to the Whitehead torpedo except that it was powered by carbonic acid expanded in external tanks warmed by seawater. A super heater using sulfuric acid and lime was added to the version tested in 1883. The following data are for the 1881 version.

Length	unknown
Diameter	18 in
Weight	2,500 lb
Propulsion	carbonic acid
Engine	unknown
Range	4,000 yd
Explosive Charge	unknown

Sims-Edison Torpedo

The Sims-Edison Torpedo was a joint product of Winfield S. Sims of Newark, N.J., and the famous inventor Thomas Alva Edison. This float-supported torpedo was electrically driven from a shore-based generator through a 4,500 ft. long cable. It was demonstrated to the Navy in 1889.

Length	28 ft
Diameter	21 in
Weight	4,000 lb
Propulsion	electric via cable from the shore
Engine	electric motor
Range	4,500 ft
Explosive Charge	400 lb

Patrick Torpedo

It was similar to the Lay-Haight torpedo and was tested in 1890–1891. It was started, stopped, and guided on its course by means of a two-wire electric cable in connection with a 100-cell bichromate battery. The wire was paid out from the shore station as the torpedo advanced and was charged with 200 pounds of dynamite, which could be fired electrically at will, or on contact.

Length	36 ft
Diameter	22 in
Weight	unknown
Propulsion	carbonic-acid-gas engine
Engine	unknown
Range	5,000 ft
Explosive Charge	200 lb

Hall Torpedo

Lieutenant Martin E. Hall designed this torpedo, which was constructed and tested at the Torpedo Station in 1892. It was similar to the Whitehead in its use of compressed air for motive power, but had a telescoping tube connected to a small tank and righting valve that was used in place of the pendulum for depth control. It did not measure up to service standards, and work on it was discontinued.

Length	unknown
Diameter	unknown
Weight	unknown
Propulsion	compressed air
Range	unknown
Explosive Charge	unknown

Cunningham Torpedo

Invented by Patrick Cunningham, this rocket-propelled torpedo was designed to be fired from submerged tubes and was assembled from four sections each approximately 4 ft. long. It was tested at Beetle's Boat Yard, New Bedford, Mass., on 17 August 1893.

Length	17 ft
Diameter	15.25 in
Weight	1,200 lb
Propulsion	solid rocket composition
Range	1,000 ft (estimated)
Explosive Charge	130 lb

Appendix C

U.S. Navy Torpedoes

Howell Mark I

The first torpedo ordered by the U.S. Navy, the Mark I was the only inertial-powered torpedo acquired by the Navy. A contract was placed in 1899 for 30 torpedoes; an additional 20 were added to order in 1894. It was in service from 1895 to 1903.

Length	11 ft, 1¾ in
Diameter	14.2 in
Weight	518 lb
Warhead weight	99 lb
Explosive charge	82 lb dry guncotton
Propulsion	stored inertia from a 131-lb steel flywheel spun up at 10,000 rpm
Speed	24 knots for first 200 yards
Range	800 yards (max.)
Guidance	pendulum-regulated rudder for anti-roll and direction stability

Whitehead 3.55m x 45cm Mark I, II, and III

The first Whitehead Torpedo was ordered by the U.S. Navy in 1891 as the Whitehead 3.55m x 45cm torpedo. The Mark I entered service in 1895, followed by the Mark II in 1896, and the Mark III in 1898. The Mark III was the first to incorporate a gyroscope. A total of 209 torpedoes were manufactured under license by the E. W. Bliss Company of Brooklyn, New York.

Length	11 ft, 9.6 in
Diameter	17. 7 in
Weight	845 lb
Explosive charge	117 lb wet guncotton
Propulsion	compressed air stored at 1,350 psi
Engine	3-cylinder Brotherhood reciprocating
Speed	26 knots (Mark I); 28 knots (Mark II); 27.5 knots (Mark III)
Range	800 yards
Guidance	preset rudder adjustment (Mark I & II); Mark 1 Mod 1 gyro

Whitehead 5m x 45cm Mark I and II

An order for the first 100 torpedoes was placed in 1896 as the Mark III Whitehead torpedo, with the designation later changed to Mark I. Manufactured under license by the E. W. Bliss Company of Brooklyn, New York. Entered service in 1898.

	MARK I	MARK II
Length	16 ft, 5 in	16 ft, 5 in
Diameter	17.7 in	17.7 in
Weight	1,161 lb	1,230 lb
Explosive charge	212 lb wet guncotton	131 lb
Propulsion	compressed air 1,350 psi	compressed air 1,500 psi
Engine	3-cylinder Brotherhood	3-cylinder Brotherhood
Speed	27.5 knots	28 knots
Range	1,000 yards	1,500 yards
Guidance	Mark 1 Mod 1 gyro	Mark 1 Mod 1 gyro

Bliss-Leavitt 5m x 21-inch Mark I

Experimental torpedo constructed by the Bliss Company to demonstrate the practical use of air heating in 1903. Two ordered by the U.S. Navy, neither issued to the fleet. Note: this torpedo was redesignated the Bliss-Leavitt 5m x 21-inch Mark I Mod 1 in 1908.

Length	16 ft, 5 in
Diameter	21 in
Weight (approx.)	1,900 lb
Explosive charge	no data available
Propulsion	compressed air 2,250 psi; dry heater within air flask
Engine	single vertical turbine
Speed (contract)	28 knots
Range	4,000 yards
Guidance	unknown

Mark 1 Torpedo (Bliss-Leavitt 5m x 21-inch Mark I)

Design based on the experience gained from the experimental Bliss-Leavitt Mark I. Fifty ordered on November 4, 1905. By 1912 all had been converted to the Mark I Mod 2 configuration.

Length	16 ft, 5 in
Diameter	21 in
Weight	1,900 lb (approx.)
Explosive charge	199-lb wet guncotton
Propulsion	compressed air 2,250 psi; dry heater within air flask
Engine	single vertical turbine
Speed (contract)	26 knots
Range (contract)	4,000 yards
Guidance	Mark 2 Mod 2 gyro

Mark 2 Torpedo (Bliss-Leavitt 5m x 21-inch Mark II)

An order for 250 (200 Mark I and 50 Mark I Mod 1) was placed on November 4, 1905, with the E. W. Bliss Company of Brooklyn, New York.

	MOD 0	MOD 1
Length	16 ft, 5 in	16 ft, 5 in
Diameter	21 in	21 in
Weight (approx.)	1,900 lb	1,900 lb
Explosive charge	207-lb wet guncotton	183-lb wet guncotton
Propulsion	compressed air 2,250 psi with dry heater	compressed air 2,250 psi with dry heater
Engine	vertical turbine	vertical turbine
Speed	26 knots at 3,500 yards	26 knots at 3,500 yards
Range (est.)	4,000 yards	4,000 yards
Guidance	Mark 5 gyro	Mark 5 gyro

Mark 3 Torpedo (Bliss-Leavitt Mark 5m x 21-inch Mark III)

Similar to the Bliss-Leavitt 5m x 21-inch Mark II with larger warhead and slightly improved range. Two hundred and eight ordered in 1909–1910.

Length	16 ft, 5 in
Diameter	21 in
Weight	2,000 lb
Explosive charge	218 wet guncotton lb
Propulsion	compressed air 2,250 psi with dry heater
Engine	vertical turbine
Speed	26.5 knots
Range	4,000 yards
Guidance	Mark 5 Mod 2 gyro

Mark 4 Torpedo (Bliss-Leavitt 5m x 45cm Mark III and IV)

Designed and manufactured by the E. W. Bliss Company of Brooklyn, they were the first U.S. Navy torpedo to incorporate a dry heater and a turbine engine to enter service. It was designed to be fired from existing 18-inch torpedo tubes. An order for 50 torpedoes was placed in 1904 as the 5m x 45cm Mark III. The Mark III entered service circa 1907. The Mark IV was similar, but designed specifically for submarines. Fifty Mark IVs were ordered in 1905. All of the Mark IIIs were later converted to the Mark IV Mod 1s.

	MARK 4 MOD 0	MARK 4 MOD 1
Length	16 ft, 5 in (5m)	16 ft, 5 in (5m)
Diameter	17.7 in (45cm)	17.7 in (45cm)
Weight	1,547 lb	1,547 lb

Explosive charge	200-lb wet guncotton	199-lb wet guncotton
Propulsion	compressed air 2,250 psi with dry heater	compressed air 2,250 psi with dry heater
Engine	vertical turbine	vertical turbine
Speed	30 knots	29 knots
Range	2,000 yards	3,000 yards
Guidance	Mark 4-3 gyro	Mark 2-2 gyro

Mark 5 Torpedo (Whitehead 5.2m x 45cm Mark V)

First U.S. Navy torpedo acquired from an overseas manufacturer. Manufactured by the Whitehead Company, Weymouth, England, and at the Torpedo Station under license. Ordered in 1908, entered service circa 1910; 580 produced.

Length	17 ft (5.2m)
Diameter	17.7 in (45cm)
Weight	1,452 lb (1,332 lb Mod 4)
Explosive charge	199-lb wet guncotton
Propulsion	compressed air 2100 psi with dry heating
Engine	4-cylinder engine
Speed	40 knots/36 knots (29 knots Mod 4)
Range	1,000 yards/1,500 yards (4,000 yards Mod 4)
Guidance	Mark 1-3 gyro

Mark 6 Torpedo (Bliss-Leavitt 5.2m x 45cm Mark VI)

This torpedo was designed for above-water tubes and could be launched from destroyers. It featured a horizontal turbine instead of the vertical turbine used on all previous Bliss-Leavitt torpedoes. This would become the standard design configuration for all future turbine-powered U.S. Navy torpedoes. Designed and manufactured by the E. W. Bliss Company of Brooklyn, N.Y., 100 ordered in 1909.

Length	17 ft (5.2m)
Diameter	17.7 in (45cm)
Weight	1,536 lb
Explosive charge	200-lb wet guncotton
Propulsion	compressed air 2,250 psi with dry heater
Engine	horizontal turbine
Speed	35 knots
Range	2000 yards
Guidance	Mark 6 gyro

Mark 7 Torpedo (Bliss-Leavitt 5.2m x 45cm Mark VII)

First steam-powered (compressed air with water spray injected into combustion pot) torpedo in the U.S. Navy. Designed and manufactured by E. W. Bliss Company; 240 ordered in 1912.

	MOD O	MOD 5A
Primary Use	**Submarines**	**Submarines Aircraft**
Length	17 ft	17 ft
Diameter	17.7 in	17.7 in
Weight	1,588 lb	1,628 lb
Explosive charge	205 lb TNT	326 lb TNT or TPX
Propulsion	steam	steam
Engine	turbine	turbine
Speed	32 knots	35 knots
Range	4,000 yards	3,500 yards
Guidance	Mark 7 gyro	Mark 7 gyro

Mark 7 Type D (Short Torpedo)

Shorter version of the Mark 7 developed at the Washington Navy Yard in 1917 to fit smaller submarine torpedo tubes. Never deployed.

Length	12 ft
Diameter	17.7 in
Weight	1,036 lb
Explosive charge	200 lb TNT
Propulsion	steam
Engine	turbine
Speed	35 knots
Range	2,000 yards
Guidance	Mark 7 gyro

Mark 7 Aircraft Torpedo

Mark 7 torpedoes modified in 1920s for use by aircraft by strengthening for shock, installation of exploder safety pin, and attachment of nose drogue.

	MOD A	MOD 2A
Length	17 ft	17 ft, 11 in
Diameter	17.7 in	17.7 in

Weight	1,593 lb	1,736 lb
Explosive charge	205 lb TNT	319 lb TNT
Propulsion	steam	steam
Engine	turbine	turbine
Speed	31 knots	30 knots
Range	3,200 yards	6,000 yards
Guidance	Mark 7 Mod 2 gyro	Mark 7 Mod 2 gyro

Mark 8 Torpedo (Bliss-Leavitt 21-ft x 21-inch Mark IV)

Long-range torpedo initially developed by the E. W. Bliss Company for use on capital ships and destroyers. Development completed circa 1915. First torpedo with warhead containing close to 500-pounds of explosive—introduced in 1923 on Mod 5, 6, and 8.

	MOD 0, 1, 2, 2A, 2B	MOD 3A, 3B	MOD 8
Primary Use	destroyers, MTB	MTB	light cruisers
Length	20 ft, 8 in	20 ft, 10 in	21 ft, 4 in
Diameter	21 in	21 in	21 in
Weight	2,761 lb	3,050 lb	3,176 lb
Explosive charge	321 lb TNT	385 lb TNT	475 lb TNT
Propulsion	steam	steam	steam
Engine	turbine	turbine	turbine
Speed	27 knots	27 knots	29 knots
Range	10,000/12,500 yards	13,500 yards	15,000 yards
Guidance	Mark 8 Mod 1 gyro	Mark 8 Mod 1 gyro	Mark 8 Mod 1 gyro

Mark 9 Torpedo (Bliss-Leavitt 5m x 21-inch Mark III Mod 1)

Long-range torpedo initially developed by the E. W. Bliss Company for use on battleships with 5-meter torpedo tubes, later adopted for submarines. Last torpedo manufactured by the Bliss Company.

	MOD 1	MOD 1B
Primary use	battleship	submarine
Length	16 ft, 3 in	16 ft, 3 in
Diameter	21 in	21 in
Weight	2,059 lb	2,377 lb
Explosive charge	210 lb TNT	395 lb TPX
Propulsion	steam	steam
Engine	turbine	turbine
Speed	27 knots	34.5 knots
Range	9,000 yards	5,500 yards
Guidance	Mark 8 Mod 1 gyro	Mark 8 Mod 1 gyro

Mark 10 Torpedo

High speed, long-range torpedo developed for use on submarines with 21-inch tubes. Last torpedo designed by the E. W. Bliss Company, manufactured at the Naval Torpedo Station.

	MOD 0	MOD 3
Length	16 ft, 5 in	16 ft, 9 in
Diameter	21 in	21 in
Weight	2,050 lb	2,215 lb
Explosive charge	400 lb TNT	497 lb or 485 lb TPX
Propulsion	steam	steam
Engine	turbine	turbine
Speed	30 knots	36 knots
Range	5,000 yards	3,500 yards
Guidance	gyro	Mark 13 Mod 1 gyro

Mark 11 Torpedo

Multi-speed, universal torpedo designed in mid 1920s as an anti-surface ship weapon for use by destroyers, cruisers, and submarines. This was the first torpedo to be entirely designed by the U.S. Navy. It entered production at the Naval Torpedo Station circa 1927.

	MOD 0	MOD 1
Length	22 ft, 7 in	22 ft, 7 in
Diameter	21 in	21 in
Weight	3,511 lb	3,521 lb
Explosive charge	500 lb TNT	500 lb TNT
Propulsion	steam	steam
Engine	turbine	turbine
Speed	27/34/46 knots	27/34/46 knots
Range	15,000/10,000/ 6,000 yards	15,000/10,000/ 6,000 yards
Guidance	Mark 12 Mod 1 gyro	Mark 12 Mod 1 gyro

Mark 12 Torpedo

Improved version of the Mark 11 torpedo designed for improved reliability. Developed by the Naval Torpedo Station and completed in 1928. Approximately 100 torpedoes manufactured by the Torpedo Station for use in destroyers. The following data are for the Mod 0.

Length	22 ft, 7 in
Diameter	21 in
Weight	3,505 lb
Explosive charge	500 lb TNT
Propulsion	steam
Engine	turbine
Speed	27.5/35.5/44 knots

Range	15,000/10,000/7,000 yards
Guidance	gyro

Mark 13 Torpedo

First Navy torpedo designed specifically for aircraft launching. Development was completed in 1936, and it entered service in 1938. More than 17,000 were produced during World War II in several variants.

SUMMARY OF CHANGES AND MODIFICATIONS

Mod 1	Improved tail, strengthened propellers; rudders moved forward
Mod 2	40-knot experimental torpedo; cancelled
Mod 2A	Mod 2 converted to 33.5 knots; water trip delay valve added to delay firing until water entry to prevent turbine runaway when dropped over 300 feet
Mod 3	External gyro angle setting added
Mod 4	Experimental Model; 50 produced with strengthened afterbody
Mod 5	Addition of water trip to 13-1
Mod 6	Addition of shroud ring to 13-2A
Mod 7	Addition of shroud ring to 13-3
Mod 8	Addition of shroud ring to 13-4
Mod 9	Addition of shroud ring to 13-5
Mod 10	Strengthened afterbody, shroud ring, suspension beam, gyro angle eliminated
Mod 11	13-6 Modified to accommodate suspension beam
Mod 12	13-7 Modified to accommodate suspension beam
Mod 13	13-9 Modified to accommodate suspension beam

	MOD 0	MOD 10
Length	13 ft, 5 in	13 ft, 9 in
Diameter	22 in	22 in
Weight	1,949 lb	2,216 lb
Explosive charge	404 lb TPX or 392 TNT	603 lb TNT, or 606 TPX, or 600 HBX
Propulsion	steam	steam
Engine	turbine	turbine
Speed	30 knots	33.5 knots
Range	5,700 yards	4,000 yards
Guidance	Mark 12 Mod 1 gyro	Mark 12 Mod 1 gyro

Mark 14 Torpedo

Developed as a smaller replacement for the Mark X torpedo having a longer range and heavier explosive charge designed specifically for submarines.

	MOD 0	MOD 3
Primary Use	SS	SS
Length	20 ft	20 ft
Diameter	21 in	21 in
Weight	3,000 lb	3,061 lb
Explosive charge	507 lb TNT	668 lb TPX
Propulsion	steam	steam
Engine	turbine	turbine
Speed	32/46 knots	31.5/46 knots
Range	9,000/4,500 yards	9,000/4,500 yards
Guidance	Mark 12 Mod 3 gyro	Mark 12 Mod 3 gyro

Mark 15 Torpedo

An improved Mark 12 having greater horsepower and a lighter air flask. About 9,700 produced between 1940 and 1944.

	MOD 0	MOD 3
Primary Use	DD	DD
Length	22 ft, 7 in	24 ft
Diameter	21 in	21 in
Weight	3,438 lb	3,841 lb
Explosive charge	494 lb TNT	801 lb TNT or 823 lb HBX
Propulsion	steam	steam
Engine	turbine	turbine
Speed	28/34/46 knots	26.5/33.5/45 knots
Range	15,000/10,000/ 6,000 yards	9,000/4,500 yards
Guidance	Mark 12 Mod 3 gyro	Mark 12 Mod 3 gyro

Mark 16 Torpedo

Hydrogen-peroxide (Navol) powered anti-ship submarine torpedo for submarine use. Production began in 1944, but none was used in combat; 1,700 were eventually produced.

	MOD 0	MOD 1
Length	20 ft, 6 in	20 ft, 6 in
Diameter	21 in	21 in
Weight	3,895 lb	3,922 lb
Explosive charge	1260 TPX	960 HBX
Propulsion	Navol	Navol
Engine	turbine	turbine
Speed	46 knots	46 knots
Range	7,000 yards	11,500 yards
Guidance	Mark 12 Mod 3 gyro	Mark 12 Mod 3 gyro

Mark 17 Torpedo

Hydrogen-peroxide (Navol) powered anti-ship destroyer-launched torpedo; 450 were produced before the end of World War II; none was used in combat.

Length	24 ft
Diameter	21 in
Weight	4,800 lb
Explosive charge	880 lb HBX
Propulsion	Navol
Engine	turbine
Speed	46 knots
Range	18,000 yards
Guidance	Mark 12 Mod 3 gyro

Mark 18 Torpedo

Electric powered anti-ship submarine torpedo developed by Westinghouse based on captured German design; 8,500 were produced for submarine use.

	MOD 0	MOD 2
Length	20 ft, 6 in	20 ft, 6 in
Diameter	21 in	21 in
Weight	3,041 lb	3,061 lb
Explosive charge	600 lb TPX	595 lb TPX/HBX
Propulsion	battery	battery
Engine	electric motor	electric motor
Speed	29 knots	29 knots
Range	4,000 yards	4,000 yards
Guidance	Mark 12 Mod 3 gyro	Mark 12 Mod 3 gyro

Mark 19 Torpedo

Improved Mark 18 electric torpedo. Only 10 models built.

Mark 20 Torpedo

Newport designed electric torpedo using a seawater battery. Alternative to Westinghouse Mark 18 and Mark 19. It was completed in 1945, but too late for war service. Performance: 3,500 yards at 33 knots, not considered sufficiently better than Mark 18 to be worth producing in quantity. Only 20 made.

Mark 21 Torpedo

Air dropped anti-ship torpedo for standoff firing using radar guidance in the air using Petrel missile. It was a development program only.

Length	13 ft, 5 in
Diameter	22.5 in
Weight	not determined
Explosive charge	not determined
Propulsion	steam turbine
Engine	turbine
Speed	33.5 knots
Range	6,300 yards
Guidance	acoustic homing (passive)

Mark 22 Torpedo

Acoustically guided, submarine-launched torpedo. An experimental project with only two development models built.

Length	20 ft, 6 in
Diameter	21 in
Weight	3,060 lb
Explosive charge	undetermined
Propulsion	electric battery
Engine	electric motor
Speed	29 knots
Range	4,000 yards
Guidance	acoustic homing (passive)

Mark 23 Torpedo

Simplified single-speed version of the Mark 14 torpedo for submarine use.

Length	20 ft
Diameter	21 in
Weight	3,060 lb
Explosive charge	595 lb TPX/HBX
Propulsion	steam
Engine	turbine
Speed	46 knots
Range	4,500 yards
Guidance	gyro

Mark 24 Torpedo (Fido)

Simplified single-speed version of the Mark 14 torpedo intended for aircraft use against submarines.

Length	7 ft
Diameter	19 in
Weight	680 lb
Explosive charge	92 lb HBX
Propulsion	electric battery
Speed	12 knots
Range	400 yards
Guidance	passive acoustic

Mark 25 Torpedo

Development completed by the end of World War II, but not placed in production. This was an air-launched, anti-ship torpedo.

Length	13 ft, 5 in
Diameter	22 in
Weight	2,306 lb
Explosive charge	725 lb TPX
Propulsion	steam
Engine	turbine
Speed	40 knots
Range	5,700 yards
Guidance	Mark 12 Mod 3 gyro

Mark 26 Torpedo

An improved Mark 18 torpedo with seawater battery and electric "on-off" control for steering and depth control. First use of explosive impulse start gyro. 20 development models built. It was intended for submarine use.

Length	20 ft, 6 in
Diameter	21 in
Weight	3,200 lb
Explosive charge	900- to 1000-lb TPX
Propulsion	battery (seawater)
Engine	Electric
Speed	40 knots
Range	6,000 yards
Guidance	gyro

Mark 27 Torpedo (Cutie)

Acoustic homing torpedo developed by Bell Laboratories. Approximately 1,100 Mod 0 units produced by Western Electric. In service 1943–1946 as a submarine-launched, anti-escort torpedo. The Mod 4, developed by the Ordnance Research Laboratory, Pennsylvania State University, was an anti-submarine torpedo. Approximately 3,000 produced. In service from 1946 to 1960.

	MOD 0	MOD 4
Length	7 ft, 6 in	10 ft, 6 .75 in
Diameter	19 in	19 in
Weight	720 lb	1175 lb
Explosive charge	95 lb HBX	128 lb HBX
Propulsion	Electric battery	Electric battery
Speed	12 knots	15.9 knots
Range	5,000 yards	6,200 yards
Guidance	passive acoustic	passive acoustic

Mark 28 Torpedo

Passive acoustic homing torpedo. Approx. 1750 torpedoes produced. In service from1944–1960 in submarines.

Length	20 ft, 6 in
Diameter	21 in
Weight	2,800 lb
Explosive charge	585 lb HBX
Propulsion	battery (seawater)
Engine	electric
Speed	19.6 knots
Range	4,000 yards
Guidance	passive acoustic

Mark 29 Torpedo

Advanced two-speed passive acoustic homing torpedo for submarine use. Three experimental models constructed before the program was terminated at the end of World War II. This data is for the Mod 1.

Length	20 ft, 6 in
Diameter	21 in

Weight	3,200 lb
Explosive charge	550 lb HBX
Propulsion	battery
Engine	electric
Speed	21 knots/28 knots
Range	12,000 yards/4,000 yards
Guidance	passive acoustic

Mark 30 Torpedo

Aircraft launched acoustic homing torpedo designed as backup to the Mark 24. Development discontinued in 1943. Became a precursor to the active homing Mark 43 Mod 1 and 3.

Mark 31 Torpedo

Passive acoustic version of Mark 18 for submarine use. The project was cancelled in favor of Mark 28 program.

Length	20 ft, 6 in
Diameter	21 in
Weight	2,800 lb
Explosive charge	550 lb HBX
Propulsion	battery
Engine	electric
Speed	29 knots
Range	4,000 yards
Guidance	passive acoustic

Mark 32 Torpedo

Surface ship-launched, anti-submarine torpedo with active acoustic homing. In service after World War II, it was withdrawn after introduction of Mark 43. Ten Mod 1s completed by Leeds and Northrop prior to the end of World War II; 100 Mod 2s produced by General Electric.

Length	6 ft, 11 in
Diameter	19 in
Weight	700 lb
Explosive charge	107 lb HBX
Propulsion	electric battery
Speed	12 knots
Range	24 minutes/9,600 yards
Guidance	active acoustic

Mark 33 Torpedo

Two-speed, passive acoustic homing torpedo (anti-submarine/anti-ship) designed to be launched from standard submarine torpedo tube. Only 30 models were produced and tested between 1943 and 1946. Its features were incorporated in the Mark 35.

Length	13 ft
Diameter	21 in
Weight	1,795 lb
Explosive charge	550 lb HBX
Propulsion	electric battery
Speed	low 12 knots/high 18.5
Range	1,900/5,000 yards
Guidance	passive acoustic

Mark 34 Mod 1 Torpedo

Aircraft-launched, passive acoustic homing torpedo for ASW. It was an improved version of Mark 24. Approximately 4,050 produced.

Length	10 ft, 5 in
Diameter	19 in
Weight	1150 lb
Explosive charge	170 lb HBX
Propulsion	electric battery
Speed	11 knots search/17 knots attack
Range	30 minutes/12,000 yards at 11 knots, 6–8 minutes/3,600 yards at 17 knots
Guidance	passive acoustic

Mark 35 Torpedo

This was a universal torpedo intended for launch by surface ships, submarines, and aircraft against surface ships. Approximately 400 were produced. It was in service from 1949 to 1960.

Length	13 ft, 5 in

Diameter	21 in
Weight	1,700 lb
Explosive charge	270 lb HBX
Propulsion	electric motor, seawater battery
Speed	27 knots
Range	15,000 yards
Guidance	active, passive acoustic; spiral pattern search

Mark 36 Mod 0 Torpedo

An all-electric, submarine-launched, anti-surface ship torpedo developed between 1946 and 1950. Development discontinued due to development of Mark 42 Torpedo.

Length	20 ft, 6 in
Diameter	21 in
Weight	4,000 lb
Explosive charge	800 lb HBX
Propulsion	electric motor, seawater battery
Speed	47 knots
Range	7,000 yards
Guidance	gyro

Mark 37 Torpedo

A two- speed, electrically driven active/passive acoustic homing torpedo. It first entered service in 1956. This was the U.S. Navy's standard submarine-launched, anti-submarine torpedo for 20 years.

	MOD 0, 3	MOD 1, 2
Length	11 ft, 3 in	13 ft, 5 in
Diameter	19 in	19 in
Weight	1,430 lb	1,690 lb
Explosive charge	330 lb HBX	330 lb HBX
Propulsion	electric motor	electric motor
Speed	classified	classified
Range	9,500 yards	9,500 yards
Guidance	active, passive acoustic	wire guided

Mark 38 Torpedo

This was a submarine-launched, anti-ship acoustic homing torpedo planned for post–World War II use as replacement for the Mark 28. Development was terminated before entering service.

Length	20 ft, 6 in
Diameter	21 in
Weight	3,008 lb
Explosive charge	550 lb HBX
Propulsion	battery
Engine	electric
Speed	35 knots
Range	10,000 yards
Guidance	active, passive acoustic

Mark 39 Torpedo

A modified Mark 27 Mod 4, this was the first U.S. torpedo to employ a trailing wire for mid-course guidance through the submarine's fire control system. Approximately 3,000 were produced.

Length	11 ft, 1 in
Diameter	19 in
Weight	1,275 lb
Explosive charge	130 lb HBX
Propulsion	electric
Speed	15.5 knots
Range	13,000 yards/26 minutes
Guidance	wire/acoustic homing

Mark 40 Torpedo

Aircraft- or guided missile-launched, anti-surface ship torpedo. Discontinued due to technology limitations. It was intended for aircraft and possible as a rocket warhead.

Length	8 ft, 9 in
Diameter	21 in
Weight	1,250 lb
Explosive charge	300 lb HBX
Propulsion	turbine, lithium seawater
Speed	80 knots
Range	2,000 yards
Guidance	gyro

Mark 41 Torpedo

This was a compact version of the Mark 35 torpedo

with those features not needed for air launching eliminated. Produced in limited numbers for ASW, it was discontinued in favor of Mark 43.

Length	10 ft
Diameter	21 in
Weight	1,327 lb
Explosive charge	150 lb HBX
Propulsion	electric
Speed	25 knots
Range	8,000 yards
Guidance	active, passive acoustic; spiral pattern search

Mark 42 Torpedo

Submarine-launched, pattern running anti-surface ships torpedo. Development terminated in 1952.

Length	20 ft, 6 in
Diameter	21 in
Weight	4,000 lb
Explosive charge	800 lb HBX
Propulsion	Navol
Engine	turbine
Speed	40 knots
Range	20,000 yards
Guidance	gyro, pattern running

Mark 43 Torpedo

An air-launched, ASW torpedo. Approximately 500 were produced.

	MOD 0	MOD 1	MOD 3
Length	88.25 in	91.5 in	91.5 in
Diameter	12.75 in	10 in	10 in
Weight	370 lb	280 lb	265 lb
Explosive charge	370 lb HBX	54 lb HBX	54 lb HBX
Propulsion	electric	electric	electric
Speed	20 knots	15 knots	21 knots
Range	4,300 yards	9 min/ 4,500 yards	6 min/ 4,500 yards
Guidance	active	active	active

Mark 44 Torpedo

A lightweight ASW torpedo for air and shipboard use with seawater-activated battery.

	MOD 0	MOD 1
Length	100 in	101 in
Diameter	2.75 in	12.75 in
Weight	425 lb	433 lb
Explosive charge	75 lb HBX-3	73 HBX-3
Propulsion	electric	electric
Speed	classified	classified
Range	classified	classified
Guidance	helix search	helix search

Mark 45 Anti-Submarine Torpedo (ASTOR)

A submarine-launched, wire-guided torpedo fitted with a nuclear warhead. It was command detonated by the launching submarine through the control wire. It was carried by U.S. attack and ballistic missile submarines.

	MOD 0	MOD 1, 3
Length	21 ft, 3 in	21 ft, 5 in
Diameter	19 in (21-in guide rail)	19 in (21-in guide rail)
Weight	2,330 lb	2,213 lb
Explosive charge	Mark 34 nuclear	Mark 102 nuclear
Propulsion	electric	electric
Guidance	gyro, wire	gyro, wire

Mark 46 Torpedo

Lightweight ASW torpedo—air and ship launched—it was first deployed in 1968. More than 16,000 produced.

	MOD 0	MOD 5
Length	102 in	102 in
Diameter	12.75 in	19 in
Weight	568 lb	518 lb
Explosive charge	unknown	98-lb PBXN
Propulsion	solid propellant	solid propellant
Motor	piston	piston
Speed	unknown	>28 knots
Range	unknown	8,000 yards
Guidance	active or passive/active acoustic homing	

Mark 47 Torpedo

Anti-ship torpedo cancelled due to Mark 48's anti-ship capabilities.

Mark 48 Torpedo

Standard U.S. submarine-launched torpedo. It was originally intended for surface ship use. It entered service in 1972. A total of 5,150 were produced in several mods.

Length	19 ft
Diameter	19 in (21-in guide rail)
Weight	3,400 lb
Explosive charge	650-lb shaped charge
Propulsion	stored chemical energy
Engine	piston/pump jet
Speed	55 knots
Range	20 miles
Guidance	wire guided, active and passive acoustic homing

Mark 49 Torpedo

Designation not applied to program.

Mark 50 Torpedo

Lightweight ASW torpedo for aircraft and surface ships designed by Honeywell/Garret. Approximately 1,080 produced.

Length	9 ft, 8 in
Diameter	12.75 in
Weight	750 lb
Explosive charge	100 lb shaped-charge
Propulsion	stored chemical energy
Engine	piston
Speed	40+ knots
Range	8,000 yards
Guidance	active, passive acoustic; spiral pattern search

Mark 51 Torpedo

McDonnell Douglas lightweight torpedo design that lost in competition with the Honeywell/Garrett Mark 50.

Mark 52, Mark 53 Torpedoes

Designations not applied to programs.

Mark 54 Torpedo

Developed under the Lightweight Hybrid Torpedo (LHT) project to combine the Mark 46 propulsion system with the Mark 50 guidance system. It was intended especially for shallow-water ASW operations. The Mark 54 entered limited production in 2004, but costs and other factors have severely reduced program plans.

Length	8 ft 11 in
Diameter	12.75 in
Weight	608 lb
Explosive Charge	97 lb shaped-charge
Propulsion	Monopropellant (OTTO fuel)
Engine	reciprocating external combustion
Speed	40+ knots
Range	
Guidance	active and passive acoustic homing

Appendix D

Maintenance Problems Causing Erratic Runs

(a) In depth:

(1) Loose rudder connection.
(2) Dirty depth engine.
(3) Leaky depth engine, causing loss of strength.
(4) Leaky immersion chamber, ruptured diaphragm.
(5) Pendulum interference.
(6) Horizontal depth line not centered.

(b) In deflection:

(1) Gyro out of balance.
(2) Gyro not fully locked.
(3) Gyro gimbal pivots broken.
(4) Balance nut undamped.
(5) Pallet clamp screw loose.
(6) Limiting screw for steering engine valve missing.
(7) Incorrect vertical rudder throws.
(8) Sticky steering engine.
(9) Leaky steering engine.
(10) Loose solder.
(11) Incorrect gyro angle setting.

(c) Cold shots:

(1) Pipe connections not set up tight.
(2) Squeezed, distorted, or clogged piping.
(3) Check valves in bad condition.
(4) Failure to fill fuel compartment.
(5) Water or oil in igniter firing line.
(6) Watered fuel.
(7) Torpedo fired without igniter.
(8) Igniter failure.

(d) In speed:

(1) Improper reducer setting.
(2) Improper nozzle clearance.
(3) Improper turbine clearance.
(4) Reducing valve spring set.
(5) Speed screw deformed.
(6) Leaky or set sylphon bellows.
(7) Fouled or damaged propellers.
(8) Badly dented head.

Source: Bureau of Navigation, United States Navy Department, *Instructions for Use in Preparation for the Rating of Torpedoman 2c* (Washington, D.C.: Government Printing Office, 1939), pp. 8–9.

Loading a 1,600-pound, Mark 7 torpedo required special equipment and muscle power, as shown in this photograph taken in July 1923. The aircraft was a Douglas DT-2, which, according to its "1-T-3" markings, was the third aircraft assigned to Torpedo Squadron 1. These aircraft could also be fitted with wheeled landing gear. (National Archives)

Notes

Notes: ADM = Admiralty records; GPO = Government Printing Office;
HMSO = Her Majesty's Stationery Office; PRO = Public Record Office (now British National Archives).

Chapter 1: The First Torpedoes:
Bushnell, Fulton, and the Civil War (1775–1865)

1. H. W. Dickinson, *Robert Fulton, Engineer and Artist: His Life and Works* (London: John Lane, 1913).
2. Herman H. Henkle, Introduction to *Torpedo War and Submarine Explosions* by Robert Fulton (Chicago: Swallow Press, 1971[reprint]), pp. xii–xiii.
3. Royal B. Bradford, *History of Torpedo Warfare* (Newport, R.I.: U.S. Torpedo Station, 1882), pp. 43–44.
4. William T. Glassell, "Reminiscences of Torpedo Service in Charleston Harbor," *Southern Historical Society Papers* (4, 1877): p. 228.
5. Ibid.
6. A few "Davids" were named; most were identified only by number.
7. G. T. Beauregard to Quartermasters and Railroad Agents on Lines from Charlestown, S.C., to Mobile, Ala., 7 August 1863: "Please expedite transportation of Whitney's submarine boat from Mobile here. It is much needed." *The War of the Rebellion: A Compilation of the Official Records of the Union and Confederate Armies*, Series 1, Vol. 28 (Part II) (Washington, D.C.: GPO, 1890).
8. *Charleston Daily Republican*, 8 October 1870, as cited by Mark K. Ragan in *Submarine Warfare in the Civil War* (Cambridge, Mass.: Da Capo Press, 2002), p. 179.

Chapter 2: False Starts:
The First Self-Propelled Torpedo (1869–1890)

1. Charles William Sleeman, *Torpedoes and Torpedo Warfare*, 2nd ed. (Portsmouth, England: Griffin, 1889), p. 172.
2. R. N. Gallwey, "The Use of Torpedoes in War," *Journal of Royal United Service Institution* XXIX (1885): pp. 471–85.
3. "Extracts from Report of Austrian Commission on the trials of the Whitehead Torpedo in 1868," in F. M. Barber, *Lecture on Whitehead Torpedo* (Newport, R.I.: U.S. Torpedo Station, 1874), pp. 36–38.
4. Lord Paget as cited by Gray, ADM 1. 6049/N219. PRO; Barber, *Lecture on Whitehead Torpedo*, p. 7. Note: Luppis is not mentioned, and his name disappears from all references to the Whitehead torpedo after this date.
5. *The Times*, 6 September 1869 (as cited by Edwyn Gray, *Devil's Device* [Annapolis, Md.: Naval Institute Press, 1991], p. 85).
6. For the drawing, see E. M. Mathews to Chief of the Bureau of Ordnance, Entry 201, Item 14, Records of the Bureau of Ordnance, National Archives, Washington, D.C.
7. Whitehead to Kirkland, 11 November 1873; Kirkland, "Experiments with Whitehead Torpedo," Entry 201, Item 14, Records of the Bureau of Ordnance, National Archives, Washington, D.C.

8. Whitehead to Jeffers, 12 December 1873, Entry 201, Item 14, Records of the Bureau of Ordnance, National Archives, Washington, D.C.
9. *Annual Report of the Secretary of the Navy, 1881* (Washington, D.C.: GPO, 1881), p. 95.
10. Ibid., p. 3.
11. Ibid., Vol. III, p. 7.
12. By 1881 Whitehead's firm had sold more than 1,500 torpedoes to more than a dozen nations.
13. Public Law No. 48 enacted March 3, 1883, 47th Cong. 2nd Session.
14. W. H. Jacques, "Naval Torpedo Warfare," in *Report of the* [Senate] *Select Committee on Ordnance and War Ships* (Washington, D.C.: GPO, 1886), p. 147.
15. Chief of the Bureau [Ordnance] to President of the Torpedo Board, ibid., pp. 151–53.
16. American Ordnance Company, *Handbook of the Howell Torpedo* (Washington, D.C.: American Torpedo, 1896), p. 2.
17. *Annual Report of the Secretary of the Navy, 1891* (Washington, D.C.: GPO, 1891), p. 217.

Chapter 3: The Torpedo Perfected: Bliss-Leavitt Torpedoes (1890–1913)

1. Robert Whitehead to Commodore William Folger, Chief of the Bureau of Ordnance, 14 August 1890, File 3134, Bureau of Ordnance General Correspondence 1885–1903, RG 74, National Archives, Washington, D.C. (hereafter BuOrd File 3134).
2. T. M. Leavitt to E. W. Bliss Company, 27 November 1890, BuOrd File 3134 (Note: this document is marked "copy," and it appears that the initial "T" in Leavitt's name is a typographical error).
3. Chief of the Bureau of Ordnance to Commander G. A. Converse, 15 August 1891, Records of the Torpedo Board of 1891–1893, Records of the Bureau of Ordnance, RG 74, National Archives, Washington, D.C.
4. This torpedo was originally designated as the Whitehead Mark III, leading to much future confusion.
5. Judge Advocate General to Secretary of the Navy, 2 February 1899, File 9492, Secretary of the Navy General Correspondence 1897–1915, RG 80, National Archives, Washington, D.C.
6. U.S. Patent No. 562,235, dated 16 June 1996.
7. "Whitehead 5m x 45cm Mark I Torpedo," File BM, Subject Files of the U.S.N., 1775–1919, Office of Naval Records and Library, RG 45, National Archives, Washington, D.C.
8. L. H. Chandler, "The Automobile Torpedo and Its Uses," U.S. Naval Institute *Proceedings* (March 1900): pp. 47–71.
9. *Annual Report of the Secretary of the Navy 1903*, p. 680.
10. Torpedoes could be fired and recovered numerous times. They were overhauled after each run and the air flask recharged. At the time it was felt that the Brotherhood engine would not stand up to repeated use at the high temperatures produced by the dry heater.
11. *Annual Report of the Secretary of the Navy 1903*, p. 680.
12. In 1908 the designation for these torpedoes was changed to the Bliss-Leavitt 5m x 21 Mark I Modification I torpedo. See: *Nomenclature of U.S. Navy Torpedoes, Torpedo Tubes, Air Compressors, and Miscellaneous Torpedo Gear*, Ordnance Pamphlet No. 316, third ed., March 1908, p. 2.
13. A number or sources, including Jolie's *Brief History of U.S. Navy Torpedo Development*, incorrectly state that the turbine drove a single propeller. See the drawings of same published in Ordnance Pamphlet No. 320 dated May, 1905.
14. U.S. Navy, Ordnance Pamphlet No. 320, *Torpedoes U.S.N. October, 1915* (Washington, D.C.: GPO, 1916), pp. 16–17; *E. W. Bliss Co. v. United States*, 253 U.S. 187; Peter Bethell, "The Development of the Torpedo—IV, *Engineering* (2 November 1945): p. 342.
15. *Annual Report of the Secretary of the Navy, 1903* (Washington, D.C.: GPO, 1903), p. 680.
16. Commander William Sims, USN, memorandum for President Theodore Roosevelt, 23 December 1907, File No. 24970, Secretary of the Navy General

Correspondence 1897–1915, RG 80, National Archives, Washington, D.C.

17. Acting Chief of the Bureau of Ordnance to Secretary of the Navy, "2nd Endorsement," 15 August 1910, p. 3, File 407, Records of the General Board of the Navy, 1900–1947, RG 80, National Archives Washington, D.C. (Hereafter GB File 407.)
18. Table tilted "Number of Torpedoes," dated January 24, 1912, attached to Memorandum, Third Committee, General Board, Future Torpedo Development, 16 July 1918, GB File 407. For conversion of Mk IVs, see Ordnance Pamphlet No. 316, 6th ed. (1913), p. 1.
19. Acting Chief of the Bureau of Ordnance to Secretary of the Navy, "2nd Endorsement," 15 August 1910, p. 4. GB File 407.
20. *Annual Report of the Secretary of the Navy 1911*, p. 223.
21. U.S. Navy, *Nomenclature of U.S. Torpedoes, Torpedo Tubes, Air Compressors, and Miscellaneous Torpedo Gear.* Ordnance Pamphlet No. 316, 6th ed. (1913), pp. 1–2.
22. Twelve Schwartzkopff torpedoes (similar to the Whitehead pattern, but made out of phosphor bronze) were purchased by the U.S. Navy in 1898. Several others of this type were taken from the Spanish ships captured at Santiago.

Chapter 4: Submarines: The Ultimate Torpedo Boat (1900–1918)

1. The U.S. Navy instituted a comprehensive and unique system of ship hull designation and numbers in 1920. These were retroactive to earlier ships dating to about 1900. This system, which continues in use today, is applied to all U.S. Navy ships mentioned in this volume from this point onward.
2. "Hello, Kearsarge! You're Blown to Atoms. This Is the Submarine Boat the Holland," *New York Journal*, 27 September 1900, p. 1.
3. On 17 November 1911 all U.S. submarine names were replaced with class letters and numbers assigned in the place of names; the *Viper* became the *B-1* and the *Octopus* the *C-1*, etc.
4. The five *Octopus* (C-class boats) appear to have been refitted to take a larger torpedo sometime after 1911.
5. Chester W. Nimitz, Senior Group Commander, Atlantic Submarine Flotilla to Secretary of the Navy via Commanding Officer Torpedo Flotilla Atlantic Fleet, Defensive and Offensive Tactics of Submarines, 17 May 1918, GB File 420-15.
6. These boats also had two stern tubes.
7. Electric Boat Drawing SK 4109, Serial File 2227 00E122, Correspondence re "E" Documents 1896–1912, RG 19, National Archives, Washington, D.C.
8. Bureau of Ordnance to Chief of Naval Operations, 2nd endorsement, 8 December 1918, GB File 420-15.
9. President of General Board to Secretary of the Navy, Torpedoes for Submarines, Building Program, 1917, 28 October 1915, GB File 420-15. Admiral Dewey served as its president from the board's inception in 1900 until his death in 1917.
10. President to Secretary of the Navy, Torpedoes for Submarines, Building Program, 1917, 26 November 1915, GB File 420-15.
11. Bureau of Ordnance to Naval War College, 4 September 1918, GB File 407.
12. Bureau of Ordnance to Chief of Naval Operations, 6 December 1918, GB File 420-15.

Chapter 5: Battleships Made Vulnerable: Torpedo Boats and Destroyers (1893–1919)

1. The *Destructor*, a contemporary of the *Rattlesnake* constructed at the J & G Thomson shipyards in Clydebank, Scotland, was turned over to the Spanish Navy on 1 January 1887.
2. Chief of the Bureau of Ordnance to Navy Department (Material Div.), 7 November 1912, GB File 407.
3. Commander-in-Chief to Secretary of the Navy, 31 May 1912, GB File 407.
4. Holloway Frost, *The Battle of Jutland* (Annapolis, Md.: Naval Institute, 1936), p. 380.
5. Keith Yates. *Flawed Victory: Jutland, 1916* (Annapolis, Md.: Naval Institute, 1936), p. 174.

Chapter 6: Between the Wars (1919–1941)

1. J. L. King, "Torpedo Officer's Memorandum Concerning Proposed Navol Torpedo," 18 March 1936, Bureau of Ordnance Secret Correspondence 1923–1941, RG 74, National Archives, Washington, D.C. (Hereafter BuOrdSC 1923–41.) This was not the only problem noted for the Mark 11. When dropped from tubes at a considerable height above the water from fast moving ships, the safety cut-off governor was likely to malfunction and the gyro could not take high angles; these problems were never satisfactorily corrected.
2. Senior Member Present (Andrew T. Long) to Secretary of the Navy, Subj: Torpedo Design, 3rd Endorsement, 3 December 1928, File S75-1, Bureau of Ordnance Confidential Correspondence, 1926–1939, RG 74, National Archives, Washington, D.C. (Hereafter BuOrdCC 1926–39.) According to the instructions given in the *Tactical Instructions Submarines* issued January 1921, "shots should be fired at the closest possible range, from 300 to 500 yards is preferable."
3. "Treaty cruisers" was the term used for U.S. cruisers designed prior to the expiration of the Washington Naval Treaty on 31 December 1936. The *Farragut* class was the first post–World War I design built by the U.S. Navy, with the lead ship being completed in 1934.
4. Senior Member Present to Secretary of the Navy, 1 May 1931, GB File 420-8, RG 80, National Archives, Washington, D.C.
5. Commander Destroyers, U.S. Fleet to Chief of Naval Operations, 5 August 1931, File DD/S75, BuOrdCC 1926–39.
6. Ibid.
7. Senior Member Present to Secretary of the Navy, 30 March 1932, File S75, BuOrdCC 1926–39.
8. These 11 torpedoes were fired at possible German submarines by the A-boats of Submarine Division 5 in European waters.
9. As quoted by Clay Blair Jr., *Silent Victory: The U.S. Submarine War Against Japan* (New York: Lippincott, 1975), p. 56. Hart, as an admiral, commanded the U.S. Asiatic Fleet at the start of World War II in the Pacific.
10. Ibid.
11. Frederick J. Milford, "U.S. Navy Torpedoes, Part Two: The Great Torpedo Scandal, 1941–1943," *The Submarine Review* (October 1996), p. 7.
12. Blair, *Silent Victory*, p. 61–62.
13. Chairman General Board to Secretary of the Navy, 3 November 1941, Files S75, BuOrdSC 1923–41.

Chapter 7: Attack from the Air: Aerial Torpedoes (1917–1945)

1. "Admiral Fiske Tells of Torpedoplane," U.S. Naval Institute *Proceedings* (April 1917), p. 855.
2. The first Yale Unit was composed of student volunteers from Yale University who established the first naval air reserve unit in 1915.
3. Commanding Officer Torpedo Squadron 9-S to Chief of Naval Operations (Division of Fleet Training), 23 March 1928, File A5-1, Bureau of Aeronautics Confidential Correspondence 1922–1944, RG 72, National Archives, College Park, Md.
4. Inspector of Ordnance in Charge to Bureau of Ordnance, 18 June 1936, File S75 13, BuOrdCC 1926–1939.
5. Commander, Aircraft Battle Force to CNO Division of Fleet Training, Subject: VT-3 IBP Torpedoes, 27 January 1940, File A5-2 (1944), RG 313, National Archives, Washington, D.C.
6. Rear Admiral William H. P. Blandy to Rear Admiral J. S. McCain, 1 March 1941, File S75-1, BuOrdSC 1923–41.
7. Read was commander of the Navy's *NC-4* flying boat that made the first trans-Atlantic aircraft flight in 1919.
8. Chief of Bureau of Aeronautics to Senior Naval Aviator Present, Fleet Air Detachment, Norfolk, Virginia, 7 April 1938, File F41-9, Bureau of Aeronautics Confidential Correspondence, RG 72, National Archives, Washington, D.C.
9. Commander Aircraft, Scouting Force, Pacific Fleet

to Commanders Patrol Wing One, Two, and Four, 26 June 1941, File S75-1, Commander Aircraft Scouting Force Confidential Correspondence, RG 313, National Archives, Washington, D.C.

10. Inspector of Ordnance to Chief of the Bureau of Ordnance, 15 March 1942, File S75 (Mk XIII-1), Bureau of Ordnance Confidential Correspondence 1940–1941, RG 74, National Archives, Washington, D.C.
11. John Lundstrum, *The First Team: Pacific Naval Air Combat from Pearl Harbor to Midway* (Annapolis, Md.: Naval Institute Press, 1984), p. 200–201.
12. Capt. Albert K. Earnest, USN, and Comdr. Harry Ferriert, USN (Ret.) "Avengers at Midway," *Foundation* (Spring 1996): pp. 47–55.
13. Comdr. Allan Rothernberg, USN (Ret.), "That Little Ensign in the Back," *Foundation* (Spring 1988): pp. 47–54.
14. Adrian Stewart, *The Battle of Leyte Gulf* (New York: Charles Scribner's Sons, 1980), p. 77.

Chapter 8: They Were Expendable: PT Boats at War (1941–1945)

1. Quoted in Capt. Robert J. Bulkeley Jr., USNR (Ret.), *At Close Quarters: PT Boats in the United States Navy* (Washington, D.C.: Naval History Division, 1962), p. 11. Also see N. Polmar and Samuel Loring Morison, *PT Boats At War: World War II to Vietnam* (Osceola, Wisc.: MBI Publishing, 1999).
2. Quoted in William Manchester, *American Caesar: Douglas MacArthur 1880–1964* (Boston: Little, Brown, 1978), p. 167.
3. Ibid., pp. 167–68.
4. The saga of PT boats in the Philippines is well told in W. L. White, *They Were Expendable* (New York: Harcourt, Brace, 1942), which describes their 1941–1942 operations; the book was the basis for John Ford's classic film of the same name (1945) starring Robert Montgomery, John Wayne, and Donna Reed.
5. The minelayer *Yaeyama*, refitted as an escort ship in late 1943, was sunk off Mindoro in the Philippines by U.S. carrier aircraft on 24 September 1944.
6. MacArthur was recalled to U.S. active duty in July 1941 as a lieutenant general to serve as commanding general of U.S. Army forces in the Far East.
7. Bulkeley, *At Close Quarters*, p. 259.
8. Ibid. p. 297.
9. Capt. Zenji Orita, IJN (Ret.), and Joseph D. Harrington, *I-Boat Captain* (Canoga Park, Calif.: Major Books, 1976), pp. 148–49.

Chapter 9: Torpedoes That Didn't Work: Submarine Failures (1941–1943)

1. The fleet boat *Sealion* (SS 195) was destroyed by aerial bombs when the Japanese struck Cavite on 9 December.
2. The 233 torpedoes destroyed at Cavite on 10 December—including 48 for the submarines *Sealion* (SS 195) and *Seadragon* (SS 194) that were on a barge that was hit and capsized during the raid—amounted to about one-half of the ready Mark 14 torpedoes in the fleet. The submarine *Sealion* was sunk in that raid. The navy yard, on the eastern side of Manila Bay, was the only U.S. submarine support facility west of Oahu.
3. Wilfred J. Holmes, *Undersea Victory* (Garden City, N.Y.: Doubleday, 1966), p. 67.
4. Ibid., p. 68.
5. Adm. King also held the title of Commander-in-Chief, U.S. Fleet; he was a qualified submariner.
6. The *Trigger* had been directed on 9 June specifically to attack the *Hiyo* based on radio intercepts of the carrier's movement. See John Winton, *Ultra in the Pacific* (Annapolis, Md.: Naval Institute Press, 1993), pp. 140–41.
7. Blair, *Silent Victory*, p. 430; Theodore Roscoe, *United States Submarine Operations in World War II* (Annapolis, Md.: U.S. Naval Institute, 1949), p. 258.
8. Blair, *Silent Victory*, p. 431.
9. Ibid., pp. 436–37
10. William H. P. Blandy to Rear Adm. Charles A. Lockwood Jr., 11 August 1943, Official Corres-

pondence Files, Charles A. Lockwood Jr. Papers, Library of Congress, Washington, D.C. (Hereafter Lockwood Papers.)

11. Rear Adm. Charles A. Lockwood Jr. to Rear Adm. W. H. P. Blandy, 30 August 1943, Official Correspondence Files, Lockwood Papers.
12. Roscoe, *United States Submarine Operations in World War II*, p. 252.
13. Ibid., p. 251.
14. Department of the Navy, Bureau of Navigation, *Instructions for Use in Preparation for the Rating of Torpedoman 2c* (Washington, D.C.: 1939), pp. 8–9.
15. Department of the Navy, Bureau of Navigation, *Instructions for Use in Preparation for the Rating of Torpedoman 1c and Chief Torpedoman* (Washington, D.C.: 1940), p. 29.

Chapter 10: Out Ranged:
The Long Lance vs. the Mark 15 (1942–1944)

1. Roscoe, *United States Destroyer Operations in World War II*, p. 89.
2. Comdr. William P. Mack, USN, "Macassar Merry-Go-Round," U.S. Naval Institute *Proceedings* (May 1943): p. 671.
3. Capt. Walter G. Winslow, USN (Ret.), *The Fleet the Gods Forgot: The U.S. Asiatic Fleet in World War II* (Annapolis, Md.: Naval Institute Press, 1982), p. 155.
4. In addition to destroyers and motor torpedo boats, torpedo tubes in U.S. Navy surface forces were also fitted in the ten scout cruisers of the *Omaha* (CL 4) class (two triple mounts) and the new anti-aircraft cruisers of the *Atlanta* (CLAA 51) class (two quadruple mounts). Torpedo tubes had been removed from all U.S. battleships and all other cruisers before World War II.
5. Capt. Roger Pineau, USNR, who worked closely with Morison on the multivolume *History of United States Naval Operations in World War II*, claims that the name was devised by Morison as a fitting identification for this formidable piece of naval ordnance. See David C. Evans and Mark Peattie, *Kaigun: Strategy, Tactics, and Technology in the Imperial Japanese Navy 1887–1941* (Annapolis, Md.: Naval Institute Press, 1997), p. 577, n. 56.
6. Samuel Eliot Morison, *The Struggle for Guadalcanal August 1942–February 1943*, vol. V in *The History of United States Naval Operations in World War II* (Boston: Little Brown, 1949), p. 38.
7. The first *Cushing* (TB 1) was commissioned in 1890; she was followed by the second *Cushing* (DD 55) in 1915; the third *Cushing* (DD 376) in 1936; the fourth *Cushing* (DD 797) in 1944; and a fifth *Cushing* (DD 985) in 1979.
8. Capt. Tameichi Hara claims that there were no Japanese destroyers near where the torpedoes hit. He believes that the *Sterett* mistakenly sunk an American destroyer—perhaps one that was dead in the water. See Tameichi Hara, Fred Saito, and Roger Pineau, *Japanese Destroyer Captain* (New York: Ballantine, 1961), pp. 151–52.
9. Morison, *The Struggle for Guadalcanal*, pp. 221–22.
10. Commander Destroyer Squadron 2 to Commander South Pacific Force, 10 November 1942, File S75, Cominch Secret Correspondence 1942, RG 38, National Archives, College Park, Md.
11. Commander Destroyers, Pacific Fleet to Commander U.S. Fleet via Commander Pacific Fleet, 15 January 1945, File S75, Cominch Secret Correspondence 1943, RG 38, National Archives, College Park, Md.
12. Ibid.
13. Morison, *The Struggle for Guadalcanal*, p. 313.
14. See John A. Lorelli, *The Battle of the Komandorski Islands, March 1943* (Annapolis, Md.: Naval Institute Press, 1984), and Samuel Eliot Morison, *Aleutians, Gilberts, and Marshalls, June 1942–April 1944*, vol. VII in *The History of United States Naval Operations in World War II* (Boston: Little, Brown, 1951), pp. 22–36.
15. Vice Adm. Gerald E. Miller, USN (Ret.), discussion with N. Polmar, Arlington, Va., 31 July 2008.
16. Lorelli, *The Battle of the Komandorski Islands*, pp.

162–63. Lorelli cites 42 Japanese torpedoes being fired; Morison in *Aleutians, Gilberts, and Marshalls* lists 43 (p. 35). The latter number is probably the more accurate, with Morison listing the torpedoes launched by each ship.

17. Russell Sydnor Crenshaw Jr., *South Pacific Destroyer: The Battle for the Solomons from Savo Island to Vella Gulf* (Annapolis, Md.: Naval Institute Press, 1998), p. 207.
18. Ibid.
19. Crenshaw, *South Pacific Destroyer*, p. 213.
20. Hara, *Japanese Destroyer Captain*, pp. 180–81.
21. See Frederick J. Milford, "U.S. Navy Torpedoes: Part Two," *The Submarine Review* (October 1996), pp. 84–86.
22. Capt. Stephen Roskill, RN, *The Period of Reluctant Rearmament 1930–1939*, vol. II, in *Naval Policy Between the Wars* (Annapolis, Md.: Naval Institute Press, 1976), p. 181.
23. For example, at the crucial Battle of Midway in June 1942 in the Japanese force only two battleships had radar; in the opposing U.S. force all three aircraft carriers and all eight cruisers had radar.
24. U.S. and Japanese radars and their capabilities are discussed in Louis Brown, *A Radar History of World War II: Technical and Military Imperatives* (Bristol, England: Institute of Physics Publishing, 1999); also see U.S. Naval Technical Mission to Japan reports, especially *Japanese Radio, Radar, and Sonar Equipment* (February 1946, No. E-17) and *Characteristics of Japanese Naval Vessels: Shipboard Electrical Equipment* (February 1946, S-01-5). The latter notes: "The greatest deficiency in Japanese combatant vessels is the lack of modern automatic fire control equipment and radar. The lack of these two placed a heavy handicap on Japanese vessels" (p. 1).

Chapter 11: Victory Assured: Torpedoes That Did Work (1943–1945)

1. John Terraine, *Business in Great Waters: The U-boat Wars 1916–1945* (London: Leo Cooper, 1989), pp. 364–66. This was the second submarine to surrender to an aircraft, the first having been HMS *Seal* to a German floatplane in the Skagerrak on 4 May 1940.
2. In the German Navy's torpedo designation system the letter "G" indicated 53.3 centimeters diameter, the "7" a length of about seven meters, and the "e" electric propulsion.
3. The *Volendam* was struck by two torpedoes fired by the *U-60*, one of which exploded. The British were able to tow the ship to shore and an unexploded G7e was found lodged in her hull.
4. Appendix A, "Facts and Discussion of Facts," Naval Inspector General to Cominch, 29 April 1943, Development of the Electric Torpedo, p. 1, File S75, Cominch Secret Correspondence 1943, RG 38, National Archives, College Park, Md. (hereafter Cominch Correspondence Secret 1943).
5. Blair, *Silent Victory*, p. 281.
6. Every torpedo manufactured by the United States was tested on the torpedo range in an activity that was known as "proofing."
7. Appendix A, "Facts and Discussion of Facts," Naval Inspector General to Cominch, Subject: Development of the Electric Torpedo, 29 April 1943, p. 7, Cominch Secret Correspondence 1943.
8. The *Wahoo* was attacked and sunk by Japanese aircraft on 11 October 1943.
9. Torpex, short for torpedo explosive, is a military explosive 50 percent more powerful than TNT. It is composed of 42 percent RDX, 40 percent TNT, and 18 percent powdered aluminum.
10. Blair, *Silent Victory*, p. 554.
11. "Submarine Operational History of World War II [Submarine Force, Pacific Fleet]," extract reprinted by R. K. Young, "Torpedo Expenditure Rates in World War II, *The Submarine Review* (October 1987): p. 67.

Chapter 12: Smart Torpedoes: The Fido and the Cutie (1941–1945)

1. The White House, "Order Establishing the National Defense Research Committee," approved by President Franklin D. Roosevelt on 27 June 1940, p. 1.

2. Plan for Handling the Problem of Comprehensive Investigation of Submarine Detection, Folder: NDRC Meeting 18 April 1941, File: C-4 Liaison Conferences, NDRC Div. 6 Records, RG 227, National Archives, College Park, Md.
3. Harvard Underwater Sound Laboratory, *Completion Report on NO-94 (Fido), An Air-Launched Acoustic Antisubmarine Torpedo Mine* (Cambridge, Mass.: OSRD, 1946), p. v, as cited by Robert Gannon in *Hellions of the Deep: the Development of American Torpedoes in World War II* (University Park: Pennsylvania State University Press, 1996).
4. John T. Tate to E. W. Sylvester, Office of the Secretary of the Navy, 17 December 1941, Folder: Project 61, NDRC Div. 6 Records, RG 227, National Archives, College Park, Md. (Hereafter Project 61 Folder.)
5. See Jerome C. Hunsaker to John T. Tate, 16 December 1941, Project 61 Folder.
6. Mark B. Gardner, "Mine Mark 24: World War II Acoustic Torpedo," *Journal of the Audio Engineering Society* 22:18 (March 1968): pp. 614–15.
7. M. D. Fagen (ed.), *A History of Engineering and Science in the Bell System* (Murray Hill, N.J.: Bell Telephone Laboratories, 1978), p. 190.
8. HBX is a form of high explosive made from TNT, RDX, aluminum, lecithin, and wax. It was developed during WW II to replace the more shock-sensitive TORPEX used in depth bombs and torpedoes.
9. A. F. Bennett, Memorandum for File, 24 November 1943, Project 61 Folder.
10. J. A. Furer to John A. Tate, 7 January 1943, Project 61 Folder.
11. Only 18 days after the initial convoy battle in which the GNAT appeared in September 1943, the British began deploying anti-GNAT devices. The British had learned of GNAT development prior to its use through prisoner of war interrogations and through Ultra, the decoding of German radio communications. Forewarned, the British had quickly developed noise-making devices called "Foxer" to decoy the torpedoes away from warships.

Chapter 13: Cold War Torpedoes: Submarines (1946–1991)

1. Supreme Command of the Navy, 2nd Division Naval Staff B.d.U., *Ueberlegungen zum Einsatz des Typ XXI* (Considerations on the Use of Type XXI), 10 July 1944, p. 1.
2. See Fritz Köhl and Eberhard Rössler, *The Type XXI U-Boat* (Annapolis, Md.: Naval Institute Press, 1991), and N. Polmar and Kenneth J. Moore, *Cold War Submarines: The Design and Construction of U.S. and Soviet Submarines* (Washington, D.C.: Potomac Books, 2004), pp. 1–9, 13.
3. Rear Adm. C. B. Momsen, USN, statement to the General Board of the Navy, 8 November 1948.
4. The 1,200 number cannot be found in available Soviet documents. V. P. Semyonov, Central Design Bureau Rubin, letter to N. Polmar, 25 August 1997.
5. Comdr. W. S. Post, USN, before the General Board of the Navy, 19 November 1946.
6. Statement by Rear Adm. R. F. Good, USN, minutes of the Submarine Conference, September 1946, p. 5. The Submarine Conference or, subsequently, Submarine Officers Conference, was initiated in 1926 as a conclave of submarine officers in the Washington, D.C., area to discuss submarine policy, design, weapons, tactics, and personnel matters. During 1946, for example, the meetings were attended by some 70 to 85 officers (mostly lieutenant commanders to rear admirals), plus a few civilian scientists. The conference was reconstituted in late 1950, becoming smaller with representatives from specific Navy offices; it continued in that (less useful) form into the mid-1970s.
7. Charles M. Sternhell and Alan M. Thorndike, *Antisubmarine Warfare in World War II*. OEG Report No. 51 (Washington, D.C.: Office of the Chief of Naval Operations, 1946), p. 1.
8. Admiralty, Historical Section, *Operations in Home, Northern and Atlantic Waters*, vol. I, in *Submarines* (London: 1953), p. 223.
9. GUPPY symbolized Greater Underwater Propulsion Project (the "y" added for phonetic purposes). The GUPPY conversions had their deck guns removed,

streamlined fairwater (sail) structures installed, bows rounded, and battery capacity increased to provide higher underwater speed and endurance; subsequently, snorkels were also installed to enable the submarines to run diesel engines to charge their batteries while at periscope depth.

10. The concept of specialized ASW submarines dated to the British "R" class of World War I, when ten hunter-killer submarines were built. The ten submarines, all launched in 1918, were excellent sea boats and highly maneuverable. Eight were scrapped soon after the war, with two retained to help train ASW forces, the *R-10* until 1928 and the *R-4* until 1932.
11. Capt. L. R. Daspit, USN, statement to the General Board of the Navy, 8 November 1948.
12. Minutes of Submarine Officers Conference, 20 April 1949 (report of 13 May 1949).
13. The *K-1* design was fitted with the AN/BQR-2 and AN/BQR-4 passive sonars, both being array-type sonars derived from the German GHG sonars fitted in the Type XXI submarine; the small AN/BQR-3 (developed from the U.S. Navy's JT set) was also fitted. In the event, only three *K-1* class submarines were built, completed in 1951–1952. Seven larger, World War II fleet submarines were converted in the early 1950s to an SSK configuration and fitted with the large BQR-4 sonar. The decision to build larger, multipurpose "attack" submarines (SS/SSN) led to termination of the SSK program. See Capt. Frank A. Andrews, USN (Ret.), "Submarine Against Submarine," *Naval Review 1966* (Annapolis, Md.: U.S. Naval Institute, 1965): pp. 42–57.
14. Soviet submarine decoys, acoustic jammers, and noisemakers had their origins in German-developed and deployed devices of World war II; those were intended to evade ASW ships and not torpedoes. See Willem Hackman, *Seek & Strike: Sonar, Anti-submarine Warfare and the Royal Navy 1914–1954* (London: Her Majesty's Stationery Office, 1984), p. 321, and Fregattenkapitän Gunther Hessler, *The U-Boat War in the Atlantic 1939–1945*, vol II (London: HMSO, 1989), p. 47.
15. Frederick J. Milford, "U.S. Navy Torpedoes—Part Five: Post WWII Submarine Launched/Heavy weight Torpedoes," *The Submarine Review* (July 1997): p. 75.
16. The loss of the nuclear attack submarine *Scorpion* (SSN 589) in 1968 was at one point believed to have been caused by the malfunction of a Mark 37 torpedo that exploded within or outside of the submarine, possibly caused by a torpedo battery detonation; it was later determined that the torpedo was not the cause of the submarine's loss.
17. Wire guidance was used in the 19th Century in Nordenfeldt and Sims-Edison torpedoes. The German Navy developed the wire-guided T10 Spinne ("spider") version of the G7e torpedo in World War II. That torpedo carried more than 5,000 yards of control wire; it was in service in 1944–1945 but was not considered successful.
18. K. J. Moore discussion with N. Polmar, Arlington, Va., 9 June 2009.
19. Comdr. George P. Steele, USN, "Killing Nuclear Submarines," U.S. Naval Institute *Proceedings* (November 1960): p. 46.
20. Their minority report as well as an article based on the minority report, "Torpedo Warheads: Technical and Tactical Issues in ASW," *Technology Review* (Los Alamos National Laboratory) (January–February 1984), remain classified. An unclassified version appeared under the Polmar and Kerr byline as "Nuclear Torpedoes," U.S. Naval Institute *Proceedings* (August 1986): pp. 62–68.
21. The U.S. anti-submarine torpedoes at the time were the air and surface launched Mark 46 with a 95-pound conventional warhead, and the new Mark 50 with a 100-pound, shaped-charge warhead, and the submarine-launched Mark 48 with a 600-pound warhead.
22. Polmar and Kerr, "Nuclear Torpedoes," p. 68.
23. Ibid.
24. Statement by an attendee at the meeting; letter from the attendee to Norman Polmar, 28 November 1986. Adm. DeMars subsequently served as Director of

Naval Nuclear Propulsion from October 1988 to October 1996.

25. The Department of Energy had replaced the Atomic Energy Commission on 1 October 1977, as the federal agency responsible for the development of nuclear warheads.
26. John J. Englehardt, "The Implications of Sub-Kiloton Nuclear Torpedoes," U.S. Naval Institute *Proceedings* (August 1987): pp. 102–3.
27. UUM = Underwater [to], U = Underwater, M = Missile.
28. The analogous Soviet weapons were the RPK-2 (NATO SS-N-15 Starfish), a 21-inch-diameter weapon that carried a 20-kiloton nuclear depth bomb to more than 20 nautical miles, and the RPK-6/7 (NATO SS-N-16 Stallion), a 25½-inch diameter weapon, that could carry a conventional ASW homing torpedo or nuclear depth bomb to more than 30 nautical miles.

Chapter 14: Cold War Torpedoes: Surface Ships and Aircraft (1946–2010)

1. These designs are discussed in some detail in Norman Friedman, *U.S. Destroyers: An Illustrated Design History* (Annapolis, Md.: Naval Institute Press, 2004).
2. During 1943–1945 German aircraft-launched guided bombs sank an Italian battleship and heavily damaged several U.S., British, and Italian warships.
3. "Heavy" usually means a torpedo weighing more than 1,000 pounds.
4. It was initially designated CLK 1, but changed to DL 1, indicating a "frigate" in a new Navy designation scheme.
5. The *Mitscher*s were originally designed with an all 3-inch/70-caliber gun armament. Chief of Naval Operations Admiral Chester W. Nimitz directed that they also be armed with 5-inch guns for use against surface targets. Still, a major anti-ship torpedo battery was not provided.
6. RUR = surface-to-Underwater Rocket.
7. Capt. William D. Taylor, USN (Ret.), "Surface Warships Against Submarines," U.S. Naval Institute *Proceedings* (May 1979): p. 177.
8. Bat was used in combat in the Pacific in 1945, being employed against surface ships and ground targets; it remained in service, mainly with naval air reserve units, until 1953.
9. The Army Air Forces pursued the aerial torpedo concept, initiating a program in August 1943. Designated GT-1 (for Glide Torpedo), the weapon was a variation of the GB-1 glide bomb carrying a Mark 13 Mod 2A torpedo. The GT-1, after release from the carrying aircraft, flew a preset glide path; entering the water triggered an explosive charge that disengaged the airframe from the torpedo, which followed a present search. A few ineffectual test drops were made in 1945.
10. AUM-N-2 = Air-to-Underwater Missile, Navy, No. 2.
11. On 30 June 1975 all frigates in the U.S. Navy having hull numbers starting with DL, DLG, and DLGN were redesignated as cruisers, CG for conventionally powered ships and CGN for nuclear-powered ships or destroyers (DD/DDG). All escort ships, which had previously been designated as DEs or DEGs, became frigates and were redesignated as FFs and FFGs.
12. Included in the figures for cruisers and frigates category were CGs, CGNs, DLGs, and DLGNs. Included in the figures for destroyers were DDs, and DDGs. Included in the figures for escorts/frigates were AGDE, DE, DEG, AGFF, FF, and FFG types.
13. Of the 97 *Gearing*-class destroyers available in the 1960s, 79 were converted to FRAM I, 10 to FRAM II with DASH helicopter facilities, and 6 to FRAM II without DASH. Only two available *Gearing*s were not rebuilt to FRAM configurations—the *Gyatt* (DD 712), which was converted to a Terrier missile ship (DDG 1), and the *Witek* (DD 848), which was completed as a sonar and propulsion test ship (EDD 848).
14. These helicopters were designated DSN until 1962, when they were changed to QH-50.

15. A few DASH helicopters continued to serve on U.S. warships during the Vietnam War for gunnery spotting (fitted with television cameras), and they were flown for several years by the Japanese Maritime Self-Defense Force.
16. On 11 March 1946, the Navy dropped the traditional "B" for bomber and "T" for torpedo from aircraft designations in favor of the all-inclusive "A" for attack.
17. Attack Squadron 195 provided the five AD-4 aircraft and Composite Squadron 35 provided the three AD-4N aircraft; the latter were included because only three VC-35 pilots had experience dropping torpedoes among the Skyraider pilots on board the *Princeton*.
18. This series of war games and the impact on the U.S. Navy is described in Muir, *Black Shoes and Blue Water*, pp. 45–49.
19. Malcolm Muir Jr., interview with Zumwalt, 20 April 1988, quoted in Muir, *Black Shoes and Blue Water*, p. 97; Adm. Zumwalt served as Chief of Naval Operations in 1970–1974.
20. NOTS Pasadena was an annex of nearby NOTS China Lake, changed in 1967 to Naval Weapons Center China Lake; the facility was reorganized/renamed the Naval Air Weapons Center China Lake in 1992.
21. Rear Adm. Justin E. Langille III, USN, Assistant Deputy Chief of Naval Operations (Surface Warfare), testimony before the Committee on Appropriations, Senate, 7 March 1978.
22. Dr. William J. Perry, Undersecretary of Defense for Research and Engineering, statement before the Committee on Armed Services, Senate, 1 February 1978.
23. Soviet deep-diving combat submarines did not appear until the 1980s:

PROJECT	NATO	TYPE	FIRST COMM.	TEST DEPTH
685	Mike	SSN	1983	3,280 ft
945	Sierra	SSN	1984	1,970 ft
949	Oscar	SSGN	1980	2,000 ft
971	Akula	SSN	1985	1,970 ft

See Gerhardt Thamm, "The ALFA SSN: Challenging Paradigms, Finding New Truths, 1969–79," *Studies in Intelligence* (September 2008), pp. 17–24; and Polmar and Moore, *Cold War Submarines*, pp. 140–46.
24. Vice Adm. William H. Rowden, USN, Deputy Chief of Naval Operations (Surface Warfare), testimony before the Appropriations Committee, Senate, 8 April 1981.
25. Capt. Dominic A. Paolucci, USN (Ret.), discussions with N. Polmar, Alexandria, Va., 15 January 1980 and 26 June 1980; and Capt. S. D. Landersman, USN (Ret.), discussions with N. Polmar, San Diego, Calif., 31 January 1980. Paolucci was a submarine specialist; Landersman was a surface warfare specialist who founded the Navy's tactical training course for prospective commanding officers and task force commanders.
26. Dr. David E. Mann, Assistant Secretary of the Navy (Research, Engineering and Systems), statement to the Subcommittee on Research and Development, Committee on Armed Services, Senate, 14 March 1978, and Rear Adm. Jeffrey Metzel Jr., USN, Deputy Director, ASW and Ocean Surveillance Programs, comments to same committee, 14 March 1978.
27. General Accounting Office, *Defense Acquisition Programs*, GAO/NSIAD-90-159 (June 1990), p. 25.
28. RUM = ship [to] Underwater Missile.

Chapter 15: The Ultimate Torpedo: The Mark 48 (1972–Present)

1. The Mark 48 torpedo was considered a "system" and included the Mark 27 mobile target.
2. Westinghouse Electric Corp., Ordnance Systems Department, "The MK 48 Story," company document dated 10 March 1971, p. 2–3.
3. Ibid., p. 2–6.
4. Melvin R. Laird, Secretary of Defense, testimony before the Committee on Appropriations, House of Representatives, 25 February 1970.
5. Gould emerged in 1969 following a merger with Clevite; Gould's torpedo business was, in turn, pur-

chased by Westinghouse in 1988. The Westinghouse defense and electronics activities—including torpedo projects—were acquired by Northrop Grumman in 1966.

6. Rep. Glenn R. Davis, hearings on the Department of Defense Appropriations for [Fiscal Year] 1974, before the Committee on Appropriations, House, 11 September 1973.
7. "The Mk 48 Torpedo," *International Defense Review* (4, 1 1979): p. 601.
8. Three types of Tomahawk cruise missiles were available—anti-ship, conventional land attack, and nuclear land attack—with different sub-types being in service at various times.
9. Westinghouse brochure, "Westinghouse ADCAP" [n.d.], p. 2.
10. E-mail from Alan Baribeau, Public Affairs Officer, Naval Sea Systems Command, to Norman Polmar, 2 July 2009.

Chapter 16: An Effective Weapon: The Ship Killers

1. Submarines can detect aircraft by radar as well as by towed (passive) hydrophone arrays.
2. Torpedo expenditure numbers in this chapter are based in part on Frederick J. Milford, "U.S. Navy Torpedoes: Part Seven: Torpedoes Fired in Anger," *The Submarine Review* (January 1998): pp. 36–50, and R. K. Young, "Torpedo Expenditure Rates in WWII," *The Submarine Review* (October 1987): pp. 65–68; these articles were based on official Navy data.
3. SubDiv-5 consisted of seven L-class submarines; they were based at Berehaven in Bantry Bay at the southern end of Ireland.
4. The torpedo shortages of 1942–1943 led to submarines often being employed to lay mines in forward areas, which sank additional Japanese ships.
5. Circular runs by Mark 14 torpedoes continued to plague the U.S. submarine force until at least 1968–1969.
6. Six submarines were initially assigned to SubRon-50; two were replaced, making a total of eight U.S. fleet submarines that served in the Atlantic areas from October 1942 through early 1944. All then transferred to the Pacific.
7. Interestingly, Israeli Navy surface ships had fired about 50 torpedoes up to 1967, a few in combat against Arab ships, but most in training exercises. Forty-eight of these torpedoes missed their targets; of the remaining two, one torpedo circled back to strike the Israeli ship that launched it, and the other one struck the *Liberty*. See A. Jay Cristol, *The* Liberty *Incident: The 1967 Israeli Attack on the U.S. Navy Spy Ship* (Dulles, Va.: Brassey's, 2002), pp. 55–57.
8. RPK-6 and RPK-7 missiles were both given a single NATO designation—SS-N-16 Stallion. The RPK-6 was a 21-inch (533-mm) weapon and the RPK-7 a 25½-inch (650-mm) weapon; they could carry either a 15¾-inch (400-mm) ASW homing torpedo or a nuclear depth bomb. The weapons would be torpedo-tube launched; streak to the surface, where their solid-propellant rocket engines would ignite; and fly toward the target. The 21-inch missile had a range of some 20 nautical miles while the larger missile could reach 55 nautical miles.
9. The submerged-launch, anti-ship cruise missile was introduced by the Soviet Navy in 1968 with the Project 670/Charlie-class cruise missile submarine. See Polmar and Moore, *Cold War Submarines*, pp. 162–63.
10. An excellent exposition of the submarine-launched anti-ship ballistic missile is Raymond A. Robinson, "Incoming Ballistic Missiles at Sea," U.S. Naval Institute *Proceedings* (June 1987): pp. 67–71. The R-27K is described in Polmar and Moore, *Cold War Submarines*, pp. 180–82; the Chinese system is described in Office of Naval Intelligence, *Worldwide Maritime Challenges 2004* (Washington, D.C., 2005), p. 22, and Office of the Secretary of Defense, *Annual Report to Congress: Military Power of the People's Republic of China 2009* (Washington, D.C., 2009), pp. 20–21.

Appendix A: Torpedo Fire Control

1. Ingerosll, R. R., *Text-Book of Ordnance and*

Gunnery (Annapolis, Md., United States Naval Institute, 1899).

2. This procedure could only be used for the outboard torpedoes, because the center torpedo tubes were not provided with the gyro-setting device.
3. The directors themselves are difficult to see and are frequently covered with canvas, but the blast shields protecting the tube trainer that accompanied the installation of the Mark 27 fire-control system are readily apparent.
4. The name came from the uncertain position of the target. Surface vessels in submarine zones had adopted the zigzag as an evasive tactic, and when it came to torpedo fire a ship's position was liable to be "was" a lot more frequently than "is."
5. "Pseudo Torpedo Run" was the distance from the periscope to the point of intercept, and "Pseudo Torpedo Speed" was the "Pseudo Torpedo Run" divided by the time of the actual torpedo run to the point of intercept.

Bibliography

Books and Pamphlets

Alden, John D. *U.S. Submarine Attacks During World War II.* Annapolis, Md.: Naval Institute Press, 1989.

Amato, Ivan. *Pushing the Horizon: Seventy-five Years of High Stakes Science and Technology at the Naval Research Laboratory.* Washington, D.C.: Naval Research Laboratory, 1998.

American Ordnance Company. *Handbook of the Howell Automobile Torpedo.* Washington, D.C.: American Ordnance Company, 1896.

Barber, F. M. *Lecture on the Whitehead Torpedo.* Newport, R.I.: U.S. Torpedo Station, 1874.

Blair, Clay Jr. *Silent Victory: The U.S. Submarine War Against Japan.* New York: Lippincott, 1975.

Bradford, Royal B. *Notes on Movable Torpedoes.* Newport, R.I.: U.S. Torpedo Station, 1882.

Brown, Louis. *A Radar History of World War II: Technical and Military Imperatives.* Bristol, England: Institute of Physics Publishing, 1999.

Bulkeley, Robert J. Jr. *At Close Quarters: PT Boats in the United States Navy.* Washington, D.C.: Naval History Division, 1962.

California Institute of Technology. *Aircraft Torpedo Development and Water Entry Ballistics.* Pasadena, Calif.: California Institute of Technology, 1946.

Campbell, R. Thomas. *Hunters of the Night: Confederate Torpedo Boats in the War Between the States.* Shippingsburg, Pa.: Burd Press, 2000.

Christman, Albert B. *Naval Innovators 1776–1900.* Dahlgren, Va.: Naval Surface Warfare Center, 1989.

Coggeshall, W. J., and J. E. McCarthy. *The Naval Torpedo Station Newport, Rhode Island, 1858 Through 1925.* Newport, R.I.: Training Station Press, 1925.

Crenshaw, Russell Sydnor Jr., *South Pacific Destroyer: The Battle for the Solomons from Savo Island to Vella Gulf.* Annapolis, Md.: Naval Institute Press, 1998.

Cristal, A. Jay. *The* Liberty *Incident: The 1967 Israeli Attack on the U.S. Navy Spy Ship.* Dulles, Va.: Brassey's, 2002.

Davis, George T. *A Navy Second to None: The Development of Modern American Naval Policy.* New York: Harcourt Brace, 1940.

Dickinson, H. W. *Robert Fulton, Engineer and Artist: His Life and Works.* London: John Lane, 1913.

Doenitz, Karl. *Memoirs: Ten Years and Twenty Days.* Annapolis, Md.: Naval Institute Press, 1990.

Evans, David C., and Mark R. Peattie. *Kaigun: Strategy, Tactics, and Technology in the Imperial Japanese Navy 1887–1941.* Annapolis, Md.: Naval Institute Press, 1997.

Fagen, M. O., ed. *A History of Engineering and Science in the Bell System.* Murray Hill, N.J.: Bell Telephone Laboratories, 1978.

Fisher, J. A. *Confidential Treatise on Electricity and the Management of Electro-Mechanical Torpedoes*, Second Edition. Portsmouth, England: Lords & Commissions of Admiralty, 1841.

Friedman, Norman. *U.S. Cruisers: An Illustrated Design History*. Annapolis, Md.: Naval Institute Press, 1984.

———. *U.S. Destroyers: An Illustrated Design History*. Annapolis, Md.: Naval Institute Press, 1982.

———. *U.S. Naval Weapons*. Annapolis, Md.: Naval Institute Press, 1988.

———. *U.S. Submarines Through 1945: An Illustrated Design History*. Annapolis, Md.: Naval Institute Press, 1995.

Frost, Holloway H. *The Battle of Jutland*. Annapolis, Md.: United States Naval Institute, 1936.

Fulton, Robert. *Torpedo War and Submarine Explosions*. Chicago: Swallow Press, 1971 (reprint).

Gannon, Robert. *Hellions of the Deep: The Development of American Torpedoes in World War II*. University Park: Pennsylvania State University Press, 1996.

Gardner, W. J. R. *Anti-Submarine Warfare*. London: Brassey's, 1996.

Gerkin, Louis. *Torpedo Technology*. Chula Vista, Calif.: American Scientific Corporation, 1989.

Gray, Edwyn. *The Devil's Device*. Annapolis, Md.: Naval Institute Press, 1991.

Great Britain. Naval Staff, Historical Section. *Submarines*, v. I (pt. 1–2): *Operations in Home, Northern, and Atlantic Waters, Including the Operations of Allied Submarines*. London: Historical Staff, Admiralty, 1953.

Grove, Eric. *Fleet to Fleet Encounters*. London: Arms and Armour Press, 1991.

Hackmann, Willem. *Seek & Strike: Sonar, Anti-submarine Warfare and the Royal Navy 1914–1954*. London: Her Majesty's Stationery Office, 1984.

Hara, Tameichi, Fred Saito, and Roger Pineau. *Japanese Destroyer Captain*. New York: Ballantine, 1961.

Harms, Norman E. *Hard Lessons Vol. 1: U.S. Naval Campaigns Pacific Theater February 1942–1943*. Fullerton, Calif.: Scale Specialties, 1987.

Hervey, John. *Submarines*. London: Brassey's, 1994.

Hessler, Gunther. *The U-Boat War in the Atlantic 1939–1945*, vol. II. London: HMSQ, 1989.

Hoeling, A. A. *Damn the Torpedoes! Naval Incidents of the Civil War*. Winston-Salem, N.C.: John F. Blair, 1989.

Holmes, Wilfred J. *Undersea Victory*. Garden City, N.Y.: Doubleday, 1966.

Honeywell Corporation. *Mk 44 Torpedo Modernization*. Everett, Wash.: December 1986.

———. *MK 50—A New Undersea Weapon*. Washington, D.C.: Honeywell Corporation, March 1985.

Hotchkiss Ordnance Company. *The Howell Torpedo: General Description and Notes*. Washington, D.C.: Hotchkiss Ordnance Co., 1888.

Hughes Aircraft Company. *The Mk-48 Advanced Capability Torpedo*. Buena Park, Calif.: Hughes Aircraft Company, February 1989.

Ingersoll, R. R. *Text-Book of Ordnance and Gunnery*. Annapolis, Md.: United States Naval Institute, 1899.

Johnson, M. A. *Progress in Defense and Space: A History of the Aerospace Group of the General Electric Company*. United States: M. A. Johnson, 1993.

Jolie, E. W. *A Brief History of U.S. Navy Torpedo Development*. Newport, R.I.: Naval Underwater Systems Center, 1978.

Kemp, P. K. *H. M. Destroyers*. London: Herbert Jenkins, 1956.

Köhl, Fritz, and Eberhard Rössler. *The Type XXI U-Boat*. Annapolis, Md.: Naval Institute Press, 1991.

LaVO, Carl. *Back From the Deep: The Strange Story of the Sister Subs* Squalus *and* Sculpin. Annapolis, Md.: Naval Institute Press, 1994.

Library of Congress. Congressional Research Service. *Anti-Submarine Warfare (ASW): Some of the Issues and Some of the Programs*. CRS-63. Washington, D.C.: Library of Congress, 2 September 1971.

Lorelli, John A. *The Battle of the Komandorski Islands, March 1943*. Annapolis, Md.: Naval Institute Press, 1984.

Lundstrum, John B. *The First Team: Pacific Naval Air Combat from Pearl Harbor to Midway*. Annapolis, Md.: Naval Institute Press, 1984.

Manchester, William. *American Caesar: Douglas MacArthur 1880–1964*. Boston: Little, Brown, 1978.

Meigs, Montgomery C. *Slide Rules and Submarines: American Scientists and Subsurface Warfare in World War II*. Washington, D.C.: National Defense University Press, 1990.

Merril, John, and Lionel D. Wyld. *Meeting the Submarine Challenge: A Short History of the Naval Underwater Systems Command*. Washington, D.C.: Department of the Navy, 1997.

Miller, Edward S. *War Plan Orange: The U.S. Strategy to Defeat Japan, 1897–1945*. Annapolis, Md.: Naval Institute Press, 1991.

Morison, Samuel Eliot. *The History of United States Naval Operations in World War II*. Boston: Little, Brown, 15 vol. 1947–1960.

Muir, Malcolm Jr. *Black Shoes and Blue Water: Surface Warfare in the United States Navy, 1945–1975*. Washington, D.C.: Naval Historical Center, 1966.

Orita, Zenji, and Joseph D. Harrington. *I-Boat Captain*. Canoga Park, Calif.: Major Books, 1976.

Peattie, Mark R. *Sunburst: The Rise of Japanese Naval Air Power, 1909–1941*. Annapolis, Md.: Naval Institute Press, 2001.

Poland, E. N. *The Torpedomen: HMS* Vernon*'s Story 1872–1986*. Great Britain: no imprint, 1993.

Polmar, Norman, and Kenneth J. Moore. *Cold War Submarines: The Design and Construction of U.S. and Soviet Submarines*. Washington, D.C.: Brassey's, 2004.

Polmar, Norman, and Samuel Loring Morison. *PT Boats At War: World War II to Vietnam*. Osceola, Wisc.: MBI, 1999.

Ragan, Mark K. *Submarine Warfare in the Civil War*. Cambridge, Mass.: Da Capo Press, 2002.

Reilly, John C. Jr., and Robert L. Scheina. *American Battleships 1886–1923: Predreadnought Design and Construction*. Annapolis: Naval Institute Press, 1980.

Ropp, Theodore. *The Development of a Modern Navy: French Naval Policy 1871–1904*. Annapolis, Md.: Naval Institute Press, 1987.

Roscoe, Theodore. *United States Destroyer Operations in World War II*. Annapolis, Md.: United States Naval Institute, 1953.

———. *United States Submarine Operations in World War II*. Annapolis, Md.: United Naval Institute, 1949.

Roskill, Stephen. *The Period of Reluctant Rearmament 1930–1939*, vol. II in *Naval Policy Between the Wars*. Annapolis, Md.: Naval Institute Press, 1976.

Rowland, Buford, and William B. Boyd. *U.S. Navy Bureau of Ordnance in World War II*. Washington, D.C.: GPO, 1953.

Schultz, Robert, and James Shell. *We Were Pirates: A Torpedoman's Pacific War*. Annapolis, Md.: Naval Institute Press, 2009.

Sleeman, Charles William. *Torpedoes and Torpedo Warfare*, 2nd Ed. Portsmouth, England: Griffin, 1889.

Sternhell, Charles M., and Alan M. Thorndite. *Antisubmarine Warfare in World War II*. OEG Report No. 51. Washington, D.C.: Office of the Chief of Naval Operations, 1946.

Stewart, Adrian. *The Battle of Leyte Gulf*. New York: Charles Scribner's Sons, 1980.

Sueter, Murray F. *The Evolution of the Submarine Boat, Mine and Torpedo, from the Sixteenth Century to the Present Time*. Portsmouth, England: J. Griffin, 1907.

Terraine, John. *Business in Great Waters: The U-Boat Wars 1916–1945*. London: Leo Cooper, 1989.

U.S. Navy. *Antisubmarine Warfare in World War II*. OEG Report No. 51. Washington, D.C.: Operations Evaluation Group, Office of the Chief of Naval Operations, 1946.

———. *History of Torpedo System Development: A Century of Progress, 1869–1969*. Newport, R.I.: Naval Undersea Warfare Center, 1998.

———. *Information from Abroad: Papers on Squadrons of Evolutions and the Recent Development of Naval Materiel, June 1886*. Office of Naval Intelligence, Department of the Navy. Washington, D.C.: GPO, 1886.

———. *The Mk 14 Submarine-Launched Torpedo: Four Decades of Service*. Newport, R.I.: Naval Undersea Warfare Center Division, 1994.

———. *The Mk 15 Destroyer-Launched Torpedo: End of an Era.* Newport, R.I.: Naval Undersea Warfare Center Division, 1994.

———. *Navy Ordnance Activities. World War, 1917.* Washington, D.C.: GPO, 1920.

———. *Nomenclature of U.S. Navy Torpedo Tubes, Air Compressors, and Miscellaneous Torpedo Gear April, 1905.* Ordnance Pamphlet No. 316.

———. *Nomenclature of U.S. Navy Torpedo Tubes, Air Compressors, and Miscellaneous Torpedo Gear June, 1906.* Ordnance Pamphlet No. 316 (Second Edition).

———. *Nomenclature of U.S. Navy Torpedo Tubes, Air Compressors, and Miscellaneous Torpedo Gear March, 1908.* Ordnance Pamphlet No. 316 (Third Edition).

———. *Nomenclature of U.S. Navy Torpedo Tubes, Air Compressors, and Miscellaneous Torpedo Gear March, 1908.* Ordnance Pamphlet No. 316 (Fifth Edition).

———. *NUSC On Torpedoes: Over a Century of Leadership.* Newport, R.I.: Naval Underwater Systems Center, 1988.

———. *The Ship and Gun Drills U.S. Navy.* Washington, D.C.: Department of the Navy, 1914.

———. *Ships' Data: U.S. Naval Vessels.* Washington, D.C.: Navy Department, 1 January 1914.

_______. *Ships' Data: U.S. Naval Vessels.* Washington, D.C.: Navy Department, 1 July 1929.

_______. *Ships' Data: U.S. Naval Vessels.* vols. I & II, Washington, D.C.: Bureau of Ships, 15 April 1945.

U.S. Office of Scientific Research and Development, National Defense Research Committee. *Summary Technical Report of Division 6, NRDC.* Washington, D.C.: U.S. Office of Scientific Research and Development, National Defense Research Committee, 1946.

Westinghouse Electric Corporation. *The Mk 48 Story.* Annapolis, Md.: Westinghouse Electric Corporation, 1971.

_______. *Westinghouse ADCAP* (Mark 48). Annapolis, Md.: n.d.

Wildenberg, Thomas. *Destined for Glory: Dive Bombing, Midway, and the Evolutions of Carrier Airpower.* Annapolis, Md.: Naval Institute Press, 1998.

Wilson, Herbert Wrigley. *Battleships in Action.* Boston: Little, Brown, 1926.

———. *Ironclads in Action.* London: Sampson Low, Marston, 1896.

Winslow, W. G. *The Fleet the Gods Forgot: The U.S. Asiatic Fleet in World War II.* Annapolis, Md.: Naval Institute Press, 1982.

Winton, John. *Ultra in the Pacific.* Annapolis, Md.: Naval Institute Press, 1993.

Articles

"Admiral Fiske Tells of Torpedo Plane." U.S. Naval Institute *Proceedings* (April 1917): p. 855.

"The American Torpedo Situation." *The Navy* (England) (December 1907): pp. 25–29.

Andrews, Frank H. "Submarine Against Submarine." *Naval Review* (1966), Annapolis, Md., U.S. Naval Institute, 1965: pp. 42–57.

Bethell, Peter. "The Development of the Torpedo." (Parts I–VI) *Engineering* Volumes 159–61 (25 May 1945, to 15 March 1946).

Chandler, L. H. "The Automobile Torpedo and Its Uses." U.S. Naval Institute *Proceedings* (March 1900): pp. 47–71.

Cohen, David E. "The Mk-XIV Torpedo: Lessons for Today." *Naval History* 6:4 (Winter 1992): pp. 34–36.

Coletta, Paolo E. "The Perils of Invention: Bradley A. Fiske and the Torpedo Plane." U.S. Naval Institute *Proceedings* (April 1977): pp. 111–16.

Earnest, Albert K., and Harry Ferriert. "Avengers at Midway." *Foundation* (Spring 1996): pp. 47–55.

Engelhardt, John J. "The Implications of Sub-Kiloton Nuclear Torpedoes." U.S. Naval Institute *Proceedings* (August 1987): pp. 102–4.

Enos, Ralph. "The Trouble With Torpedoes." *The Submarine Review* (October 1997): pp. 51–59.

Friedman, Norman. "Submarine Weapons 2008 Update." *Naval Forces* (III 2008): pp. 48–55.

Gallwey, R. N. "The Use of Torpedoes in War." *Journal*

of Royal United Service Institution XXIX (1885): pp. 471–85 (lecture delivered on 6 March 1985).

Gardner, Mark A. "Mine Mark 24: World War II Acoustic Torpedo." *Journal of the Audio Engineering Society* 22:18 (October 1974): pp. 614–26.

Gillette, R. C. "The Passive Acoustic Cutie Torpedo." *The Submarine Review* (April 1987): pp. 69–73.

Glassel, William T. "Reminiscences of Torpedo Service in Charleston Harbor." *Southern Historical Society Papers* 4: pp. 225–35.

Gleaves, Albert. "The Howell Torpedo." U.S. Naval Institute *Proceedings* (January 1885): pp. 125–27.

"Hello, *Kearsarge*! You're Blown to Atoms. This Is the Submarine Boat the *Holland*." *New York Journal* (27 September 1900): p. 1.

Jackson, R. H. "Torpedo Craft: Types and Employment." U.S. Naval Institute *Proceedings* (March 1900): pp. 1–46.

Jacques, W. H. "Naval Torpedo Warfare." In *Report of the* [Senate] *Select Committee on Ordnance and War Ships*. Washington, D.C.: GPO, 1886, pp. 148–91.

Mack, William P. "Macassar Merry-go-Round." U.S. Naval Institute *Proceedings* (May 1943): pp. 669–73.

McCandless, Bruce. "The Howell Automobile Torpedo." U.S. Naval Institute *Proceedings* 92:10 (October 1966): pp. 174–76.

Milford, Frederick J. "U.S. Navy Torpedoes—Part One: Torpedoes Through the Thirties." *The Submarine Review* (April 1996): pp. 50–69.

———. "U.S. Navy Torpedoes—Part Two: The Great Torpedo Scandal, 1941–1943." *The Submarine Review* (October 1996): pp. 81–93.

———. "U.S. Navy Torpedoes—Part Three: Development of Conventional Torpedoes 1940–1946." *The Submarine Review* (January 1997): pp. 67–80.

———. "U.S. Navy Torpedoes—Part Four: WW II Development of Homing Torpedoes 1940–1946." *The Submarine Review* (April 1997): pp. 67–80.

———. "U.S. Navy Torpedoes—Part Five: Post WW II Submarine Launched/Heavyweight Torpedoes." *The Submarine Review* (July 1997): pp. 70–86.

———. "U.S. Navy Torpedoes—Part Six: Post WW II Submarine Launched/Heavy Weight Torpedoes 1940–1946." *The Submarine Review* (October 1997): pp. 41–50.

———. "U.S. Navy Torpedoes—Part Seven: Torpedoes Fired in Anger." *The Submarine Review* (January 1998): pp. 36–47.

"The Mk 48 Torpedo." *International Defense Review* (4 1979).

Nevskiy, Yu. "The Development of Torpedo Weaponry in the NATO Countries." *Morskoy Sbornik* (9 1975): pp. 101–3.

"The New Turbine Torpedo of the United States Navy." *Scientific American* (6 January 1906): pp. 7–8.

Pelick, Tom. "Fido—The First U.S. Homing Torpedo." *The Submarine Review* (January 1996): pp. 66–70.

———. "A Historical Perspective: U.S. Navy's First Active Acoustic Homing Torpedoes." *The Submarine Review* (April 1997): pp. 46–79.

Polmar, Norman, and Donald M. Kerr. "Nuclear Torpedoes." U.S. Naval Institute *Proceedings* (August 1986): pp. 62–68.

Robinson, Raymond A. "Incoming Ballistic Missiles at Sea." U.S. Naval Institute *Proceedings* (June 1987): pp. 67–71.

Rothernberg, Allan. "That Little Ensign in the Back." *Foundation* (Spring 1988): pp. 47–54.

Sears, W. J. "A General Description of the Whitehead Torpedo." U.S. Naval Institute *Proceedings* (1896): pp. 803–9.

Steele, George P. "Killing Nuclear Submarines." U.S. Naval Institute *Proceedings* (November 1960): pp. 44–51.

Tailyour, Patrick. "Torpedo Development." *The Submarine Review* (October 1999): pp. 93–99.

Taylor, William D. "Surface Warships Against Submarines." U.S. Naval Institute *Proceedings* (May 1979): pp. 168–81.

Thamm, Gerhardt. "The Alfa SSN: Challenging Paradigms, Finding New Truths, 1969–1979." *Studies in Intelligence* (September 2008): pp. 17–24.

Vego, Milan. "Torpedoes: Do They Still Have a Role in the Future?" *Naval Forces* (III 2009): pp. 16–23.

Very, E. W. "The Howell Automobile Torpedo." U.S. Naval Institute *Proceedings* (March 1890): pp. 333–56.

Withington, Thomas. "Modern Sophisticated Torpedoes." *Naval Forces* (III 2008): pp. 40–47.

Young, R. K. "Torpedo Expenditure Rates in WWII." *The Submarine Review* (October 1987): pp. 65–68.

Dissertations and Unpublished Documents

Dienesch, Robert M. "Submarine Against the Rising Sun: The Impact of Radar on the American Submarine War in 1943 the Year of Change." Masters Thesis, University of New Brunswick, 1996.

Glascow, Richard Dwight. "Prelude to a Naval Renaissance: Ordnance Innovation in the United States During the 1870s." Dissertation, University of Delaware, 1978.

Ingram, Luther Gates Jr. "The Deficiencies of the United States Submarine Torpedo in the Pacific Theater: World War II." Master's Thesis, San Diego State, 1978.

Williams, Kathleen Broome, "Fido's Ears: An Acoustic Torpedo for Hunting U-boats," unpublished paper, courtesy of the author.

Technical Manuals

U.S. Navy. *The Bliss Leavitt Torpedo: U.S.N. 5-m x 21-in Mark III*. Washington, D.C.: GPO, 1912.

———. *Description, Adjustment, Care and Operation of U.S. Navy Torpedo, Mark XIII*. Ordnance Pamphlet No. 629. Washington, D.C.: Bureau of Ordnance, 1937.

———. *Description of Bliss Leavitt Torpedo, 5m x 21in., Mark I Torpedo*. Ordnance Pamphlet No. 320, May 1905.

———. *Description of Bliss Leavitt Torpedo, 5m x 21in., Mark I Torpedo*. Ordnance Pamphlet No. 320, August 1907.

———. *Description of the Bliss-Leavitt 5.2m x 45cm Torpedo Mark VII and Gyro Gear, Mark VII*. Washington, D.C.: GPO 1914.

———. *Destroyer Doctrine and Manual of Torpedo Control. Destroyer Tactical Bulletin No. 1-43*. Commander Destroyers Pacific Fleet, February 1943.

———. *Fire Control Equipment: Director Firing Installation Arma Type*. Ordnance Pamphlet No. 108, January 1928.

———. *General Description of Torpedo Directors*. Ordnance Pamphlet No. 421, October 1917.

———. *The Howell Torpedo U.S.N., 14.2 Inches Mark I 1896*. U.S. Naval Torpedo Station, 1896.

———. *Manual of Torpedo Fire for Destroyers, 1921*. U.S. Navy Bureau of Ordnance, 1921.

———. *The Mark V Whitehead Torpedo 5.2 Meters By 45 Centimeters With Modifications 1,2,3,4*. U.S. Navy Bureau of Ordnance Pamphlet, June 1912.

———. *Submarine Attack Course Finder Mark I Model 3 Manual*. O.D. 453, 1922.

———. *Torpedo Angle Solver Mark VIII Operating Instructions*. O.D. No. 3518, 1941.

———. *Torpedo Data Computer Mark 3, Mods. 5–12 Inclusive*. Ordnance Pamphlet No. 1056, June 1944.

———. *Torpedo Director Mark 31: Description and Instructions for Use*. Ordnance Pamphlet No. 983, May 1943.

———. *Torpedo Fire Control Equipment* (Destroyer Type), OP 1586, 27 February 1947.

———. *Torpedoes Mark VIII Mods. 4 and 5, General Description*. Ordnance Pamphlet No. 321, June 1925. Washington, D.C.: GPO, 1926.

———. *Torpedoes, United States Navy: A Digest of Torpedoes, Tubes, Directors, and General Torpedo Management, As At Present Practiced in the United States Navy*. Ordnance Pamphlet No. 320, October 1915. Washington, D.C.: GPO, 1916.

———. *U.S. Navy Aircraft Torpedoes OP 1207* (Second Revision) *25 July 1952*. Washington, D.C.: Bureau of Ordnance, 1952.

———. *United States Navy Torpedoes, A Digest of Torpedoes, Tubes, Directors, and General Torpedo Management, As Present Practiced in the United States Navy*. U.S. Navy Bureau of Ordnance Pamphlet No. 320, October 1915.

———. *U.S. Navy Torpedoes: General Data 15 October 1945* (OP 1634). Washington, D.C.: Bureau of Ordnance, 1945.

———. *The Whitehead Torpedo: U.S.N. 45cm x 3.5m Mark I, Mark II, Mark III, and 45cm x 5m Mark I. Notes On Care And Handling, Disassembling and Assembling, etc.* Newport, R.I.: Naval Torpedo Station, 1898.

———. *The Whitehead Torpedo: U.S.N. 45cm x 3.5m Mark I, Mark II, Mark III, and 45cm x 5m Mark I, Mark II.* Newport, R.I.: Naval Torpedo Station, 1901.

Periodicals

Annual Report of the Secretary of the Navy
Annual Torpedo Record
Morskoy Sbornik
Naval Forces
Naval History
Navy Fact File
Navy Register
The Submarine Review
U.S. Naval Institute *Proceedings*

Index

Ship type abbreviations are found on pages xv–xvi

About the Authors

THOMAS WILDENBERG is an independent historian/scholar specializing in the development of naval aviation and logistics at sea. He has written extensively about the U.S. Navy during the interwar period. His articles have appeared in several scholarly journals, including the *Journal of Military History*, *American Neptune*, and the U.S. Naval Institute *Proceedings*. He is also the author of three books on U.S. naval history that cover such varied topics as replenishment at sea and the development of dive bombing. His *All the Factors of Victory* (2003) is a biography of Admiral Joseph Mason Reeves, "the father of carrier warfare" in the U.S. Navy.

His most recent work, with R. E. G. Davies, curator of air transport at the National Air and Space Museum, is *Howard Hughes: An Airman, His Aircraft, and His Great Flights* (2006).

Mr. Wildenberg served as a Ramsey Fellow at the National Air and Space Museum in 1999–2000. He was awarded the Surface Navy Association Literary Award (2005), received the John Laymen Award for best biography from the North American Society for Oceanic History (2003), was awarded the Edward S. Miller Naval War College Research Fellowship (1998), and received an honorable mention in the Ernest J. Eller Prize in Naval History (1994).

NORMAN POLMAR is an analyst, author, and historian specializing in naval, aviation, and intelligence issues. Since 1980 he has been a consultant to several senior officials in the Navy and Department of Defense, and has directed several studies for U.S. and foreign shipbuilding and aerospace firms. For almost 11 years he was a member of the Secretary of the Navy's Research Advisory Committee (NRAC).

Mr. Polmar has consulted to three U.S. senators and two members of the House of Representatives. He has also served as a consultant or adviser to three Secretaries of the Navy and two Chiefs of Naval Operations. Like Mr. Wildenberg, Mr. Polmar held the Ramsey Chair of aviation history at the National Air and Space Museum (1998–1999).

He has written or coauthored 50 books and numerous articles on naval, aviation, and intelligence subjects. From 1967 to 1977, Mr. Polmar was editor of the U.S. and several other sections of the annual *Jane's Fighting Ships*. The first American to ever hold an editorship with that publication, he was totally responsible for almost one-third of the volume in that period. He subsequently became author of the

Naval Institute reference books *Guide to the Soviet Navy* and *Ships and Aircraft of the U.S. Fleet*. He also writes a regular column for the *Proceedings* and *Naval History* magazines, both published by the Naval Institute.

His awards include Outstanding Journalism Graduate (Sigma Delta Chi, 1965); Alfred Thayer Mahan Award for Literary Achievement (Navy League, 1976); Author Award of Merit (Naval Institute, 1986); Rear Admiral Ernest M. Eller Prize (U.S. Naval History Center, 1996), shared with Thomas B. Allen; and Admiral Arthur W. Radford Award (Naval Aviation Museum Foundation, 2004).

The Naval Institute Press is the book-publishing arm of the U.S. Naval Institute, a private, nonprofit, membership society for sea service professionals and others who share an interest in naval and maritime affairs. Established in 1873 at the U.S. Naval Academy in Annapolis, Maryland, where its offices remain today, the Naval Institute has members worldwide.

Members of the Naval Institute support the education programs of the society and receive the influential monthly magazine *Proceedings* or the colorful bimonthly magazine *Naval History* and discounts on fine nautical prints and on ship and aircraft photos. They also have access to the transcripts of the Institute's Oral History Program and get discounted admission to any of the Institute-sponsored seminars offered around the country.

The Naval Institute's book-publishing program, begun in 1898 with basic guides to naval practices, has broadened its scope to include books of more general interest. Now the Naval Institute Press publishes about seventy titles each year, ranging from how-to books on boating and navigation to battle histories, biographies, ship and aircraft guides, and novels. Institute members receive significant discounts on the more than eight hundred Press books in print.

Full-time students are eligible for special half-price membership rates. Life memberships are also available.

For a free catalog describing Naval Institute Press books currently available, and for further information about joining the U.S. Naval Institute, please write to:

Member Services
U.S. NAVAL INSTITUTE
291 Wood Road
Annapolis, MD 21402-5034
Telephone: (800) 233-8764
Fax: (410) 571-1703
Web address: www.usni.org